The Works and Times of Johan Huizinga (1872–1945)

The Works and Times of Johan Huizinga (1872–1945)

Writing History in the Age of Collapse

Thor Rydin

Routledge
Taylor & Francis Group
LONDON AND NEW YORK

The publication of this book was made possible by grants from the Lindroth Fund of Uppsala University and the Humanities Faculty of Utrecht University.

First published in 2023 by Amsterdam University Press Ltd.

Published 2025 by Routledge
4 Park Square, Milton Park, Abingdon, Oxon OX14 4RN
605 Third Avenue, New York, NY 10158

Routledge is an imprint of the Taylor & Francis Group, an informa business

ISBN: 9789463724593 (hbk)
ISBN: 9781041189428 (pbk)
ISBN: 9781003708155 (ebk)
NUR 694

Cover photo: an elderly woman consoles a young lady in *Oud eigen leed herleefd* (1912). This sculpture was made by the Dutch artist Johanna Schreve-IJzerman (1867–1933).

Cover design: Coördesign, Leiden

DOI: 10.5117/9789463724593

Figure 0.1. Johan Huizinga and his daughter Laura in the summer of 1944.

For Kirsti Järvinen (1956–2023)

We had long heard tell of whole worlds that had vanished, of empires sunk without a trace, gone down with all their men and all their machines into the unexplorable depths of the centuries, with their gods and their laws, their academies and their sciences pure and applied, their grammars and their dictionaries, their Classics, their Romantics, and their Symbolists, their critics and the critics of their critics. We were aware that the visible earth is made of ashes, and that ashes signify something. Through the obscure depths of history we could make out the phantoms of great ships laden with riches and intellect; we could not count them. But the disasters that had sent them down were, after all, none of our affair. [...] And we see now that the abyss of history is deep enough to hold us all.

– Paul Valéry (1871–1945) in the opening passage of his essay 'Crisis of the Mind' ('La crise de l'esprit') from 1919.

Table of Contents

List of Illustrations

Foreword

The lifetime of the Dutch cultural historian Johan Huizinga (1872–1945) was marked by dramatic transformations. Cityscapes, aesthetic codes, social orders, political cultures, international travel and means of warfare developed beyond recognition; entire catalogues of hopes and fears were torn asunder and replaced by new ones during not one but two wars. As the tectonic plate of history seemed to recede from that of the present, Huizinga's generation of historians – mostly men – were confronted, sometimes consciously and sometimes not, with the following question: what can the value of the past be to the present in times of rupture? In Huizinga's case, at any rate, these and related experiences of loss dramatically affected how he conceived of the historical research for which he would later gain worldwide renown. His histories were always present ones: his Middle Ages were those of the 1910s, his version of Erasmus was that of the 1920s, his history of 'play' was that of the 1930s, his seventeenth-century Netherlands was that of the 1940s. As a Christian, Huizinga understood himself to have been created in the image of God. This book creates Huizinga anew – not in the image of a God, but in that of a disruptive and relentless loss of an 'old world'.

More commonly celebrated than cited, more often referred to than actually read, Huizinga remains a well-known yet strangely mysterious figure among the international community of historians today. To a significant extent, I believe that Huizinga's contemporary appeal in Europe rests on a renewed ability to relate to themes of uncertainty, instability and vulnerability. Progressive newspaper readers across the continent and beyond learn about the losses of war, the loss of ecological equilibria, the loss of democratic cultures and the loss of women's autonomy over their bodies. Conservative minds, on the other hand, worry about other kinds of losses: those of traditional values, family structures, local communities of various kinds and, in some cases, geopolitical prowess. Finally, the war crimes commited against the Ukrainians remind both sides every day that no political order, no democratic value, no peace, is ever a mere given. 'Loss' is all over the place, almost regardless of one's political identity.

Against this background, one may wonder whether the case of Huizinga could teach us something about how our experiences of loss form and aggregate into beliefs and 'truths'. Could an excavation of his works and times help us understand how our own experiences of loss change the way past and future may appear to us and, subsequently, how our own historical narratives gain or lose authority? Readers interested in the relation between 'crisis' and 'history' shall, I hope, find material for thought and dispute in this book.

Other than as a historical object, however, Huizinga can also be treated as a present interlocutor for present ends. Not unlike us today, Huizinga witnessed how uncertainty and fear in his own times enticed worried minds to rush to new convictions, prophets and perspectives. A confrontation with the boundaries of one's understanding and categories, he learned, can be truly tormenting and degrading and easily invites newer and more intuitive truths that are more resilient to inconvenient facts and detrimental to those communal cultures – those communal 'games', in Huizinga's words – of which academic, political and other kinds of enterprises consist. Just as Huizinga's academic star began to rise, Max Weber (1864–1920) urged that the 'principal task' (*erste Aufgabe*) of a teacher lay in communicating 'uncomfortable facts' (*unbequeme Tatsachen*). Education, Weber held, carried the promise of becoming a 'moral achievement' (*sittliche Leistung*) by deflating all-encompassing systems of thought and the disinterested complacency and arrogance to which they lead. Huizinga was only faintly familiar with Weber but would have agreed on this point: 'knowing', Huizinga urged, is conditional on an awareness of one's limits and fallibility – of precisely one's 'not-knowing'.

In this vein, the vein of the vices and virtues of 'loss', the extra-academic value of this study could be that of a reminder: a reminder not just of how widely experiences of loss may resonate – not even the late Middle Ages or Erasmus were safe from shelling and shellshock – but also of the many ways in which historical authors such as Huizinga have responded constructively to fear, loss and uncertainty. To Huizinga, loss was a teacher. Fear and sorrow may be overwhelming and unbearable at times, but they are not merely destructive. When viewed as a repository of collapse and crisis, Huizinga held, history served as a reminder of finitude, *memento mori*, and was thus able to cultivate a set of virtues and insights. In this sense, Huizinga treated history not as a collection of facts but as a spiritual exercise towards the cultivation of serenity, curiosity, honesty and love for the world. History was a hard case for 'soft' virtues. By studying authors such as Huizinga, we learn perhaps not how to respond best to our present-day losses (our world is different from theirs, after all), but we may be reminded of the next best thing: loss and uncertainty need not lead to just loud-mouthed idols and prophetic systems but can be the beginning of an open-minded modesty and humanism. This book discusses an example.

T. Rydin
August 2022, Amsterdam

Referring to Huizinga

All references to Huizinga's works in this book draw from his Dutch *Collected Works* ('Verzamelde Werken'), edited by L. Brummel, W. R. Juynboll and Th. J. G. Locher, and published 1948–1953 by H. D. Tjeenk Willink & Zoon N.V. in Haarlem, the Netherlands. The references are placed in footnotes and have been composed according to the following logic:

VW [volume]: [*title*[1]] (year of original publication): [page number].

References to Huizinga's correspondence will draw from his published *Correspondence* ('Briefwisseling'), edited by L. Hanssen, W. E. Krul and A. van der Lem and published 1989–1991 by Veen/Tjeenk Willink in Amsterdam/Utrecht, the Netherlands. These references have been composed according to a similar structure:

BW [volume]: [sender–receiver] (year): [letter number].

References to material from the Huizinga Archive (University of Leiden) have been composed according to the archive's guidelines:

HA: [file code] (year).

All other references, both primary and secondary, are to be found in footnotes and have been put together according to the Chicago Manual of Style, 17th edition. An overview of all referenced works can be found in the bibliography.

1 Subtitles are not included in the footnote references.

1. Writing History in Times of Loss: A New Johan Huizinga

Abstract
This introductory chapter presents the central research question and claims of the present book. By investigating the several ways in which 'experiences of loss' informed Huizinga's understanding of various historical periods and the meaning of their study, this book offers a new understanding of Huizinga's relation to his times.

Keywords Johan Huizinga; interwar culture; cultural history; historical experience; history as a way of life

On 30 September 1919, Johan Huizinga took the stage at Leiden University to welcome his students to a new academic year, the first after the Great War. Huizinga had occupied the chair of General History at Leiden since 1914, and in this capacity he was among the most eminent historians of the country, if not Europe as a whole. However, he did not use the occasion to celebrate historical scholarship. He did not, say, offer an account of the discipline's indispensable value to and purpose in a society suffering from turbulent times or laud the virtues and diligence of its canonical authors. No: on the contrary, Huizinga spoke of a sense of 'despair' concerning the meaning of historical research:

> I do not know whether you feel the need for lectures on history, but I need them in order to hold on to history – for a long time I have felt history escape me. Despair [*vertwijfeling*]: how could I continue to teach history? I no longer understand it and I no longer care for it. [...] The consequence of the war? Initially, I also felt that: all the past has lost its importance. [What] now?'[1]

1 'Ik weet niet of gij 't gevoel hebt, college over geschiedenis te behoeven, maar ik heb ze noodig, om mij vast te houden aan geschiedenis, – want lang gevoeld alsof geschiedenis mij

Rydin, T., *The Works and Times of Johan Huizinga (1872–1945): Writing History in the Age of Collapse*. Taylor & Francis Group 2023
DOI: 10.5117/9789463724593_CH01

The audience's response to this lecture is unknown, but Huizinga's words must have made a tremendous impression. The Professor in General History 'no longer understood [history]'. In fact, he 'no longer [even] cared for it'. Yet by some still undefined feature, 'history' – whatever that word now meant – continued to offer Huizinga a certain solace: 'I need [lectures on history] in order to hold on to [history].' Huizinga, it seemed, had lost interest in a particular kind of history, not in history altogether. The edifice of historical scholarship had been rattled by the war, but its future had not been obliterated. A sense of loss had made urgent its reinvention; and so, Huizinga held, a question presented itself to him and his fellow historians: 'what now?'

Huizinga's alarming confessions and unsettling tone were not mere rhetorical devices to create professional urgency or prophetic charisma. Huizinga disliked anecdotal evidence, and it was unlike him to speak in the staccato of unfinished sentences, let alone in public. This lecture was out of character; *something* had shaken him. The lecture was permeated by genuine experiences of loss – a loss of certainties, expectations and ideals. The Great War, alongside many other aspects of 'this modern world', had revealed a new and unprecedented kind of unpredictability, Huizinga thought, and consequently, he held, the value and even the very meaning of the word 'history' had changed.[2] The challenges posed by this apparent need to renegotiate 'history' and the importance of studying it would grow to become a central theme in Huizinga's later thought. The works that ensued solidified his position among the most famous historians of his generation.

This book sets itself the task of showing exactly how 'experiences of loss' shook and affected Huizinga's historical works: What role did these experiences play in Huizinga's authorship? And what image of Huizinga the historian may we extract from these roles? This study is not the first to offer an image of 'Huizinga', but it is the first to do so by giving experiences of change, and particularly those of loss, the leading role. In so far as it could be called an intellectual biography, this project thus does not quite tell the story of a person or a social role. Rather, it traces and tells the

ontzonk. Vertwijfeling: hoe moet ik verder geschiedenis onderwijzen? Ik begrijp het niet meer en het kan mij niet meer schelen. [...] Het gevolg van den oorlog? In 't begin ook dat gevoel: al dat oude heeft zijn belangrijkheid verloren. [En] nu?' See HA 9.I.2 (1919). See p. 11 for further details on the references to Huizinga's collected works and letters as well as to the Huizinga archives. This lecture was first brought to scholarly attention by Anton van der Lem. See A. van der Lem, *Johan Huizinga: Leven en werk in beelden en documenten* (Amsterdam: Wereldbibliotheek, 1993), 154–55. This lecture is discussed again in greater detail in Chapter 3.

2 VW IV: *Over historische levensidealen* (1915): 415.

story of a certain kind of experience belonging to a period in Dutch and even European history. By answering its questions, this project therefore serves both a historical and historiographic purpose. Historically, it sheds new light on a celebrated, iconic historian and his relation to the cultures of the interwar period. Historiographically, it offers a case in point for an 'experiential' approach to authors writing in times of change. Drawing lessons from both angles, this book's single most central contention is this: Huizinga's historical research was to him a way of life in times of loss and change, a way of accommodating experiences of collapse and rupture, and not just during the Great War.

For Huizinga's entire lifetime was one of change – *rapid* change. Within just a few decades, Dutch cities transformed beyond recognition: new sounds, new smells, new speeds and new shapes filled streets, cafés and museums alike.[3] Democracy replaced centuries-old socio-political orders, and women fought their way into universities, politics and new job markets. Modern scientific insights made everything 'shaky and nothing certain', colonial claims eroded and innumerable art movements competed to challenge any aesthetic code imaginable – and all these developments were alloyed with the metal of a new kind of warfare.[4] Huizinga's nineteenth-century dreams of internationalism came crashing down not once but twice. Huizinga was forty-two when the Great War erupted, sixty-eight when Nazi Germany invaded the Netherlands and sixty-nine when the German occupiers detained him. And whilst such events took place, Huizinga, every now and then laying aside his newspapers and magazines, withdrawing from correspondence and debate, would sit down in his study and write about the past – his present's past: 'Whilst writing [*Autumntide of the Middle Ages*], my gaze has been directed to the depths of a nocturnal sky […]. One easily lets one's attention be drawn to the descending, waning, withering, and lets the shadow of death fall on one's writings.'[5] Huizinga's histories were mediated by autumnal times, but not always in the ways he himself imagined.

3 A recent book on the changing landscapes in the Netherlands around 1900 has opened a window into the sensory dimensions of modernization. A. van der Woud, *Het landschap en de mensen: Nederland 1850–1940* (Amsterdam: Prometheus, 2020).

4 'Wij zien voor oogen, hoe bijna alle dingen, die eenmaal vast en heilig schenen, wankel zijn geworden.' VW VII: *In de schaduwen van morgen* (1935): 315.

5 'De blik is bij het schrijven van dit boek gericht geweest als in de diepten van een avondhemel, – maar van een hemel vol bloedig rood, zwaar en woest van dreigend loodgrijs, vol valse koperen schijn. […] Het kan licht gebeuren, dat men, de opmerkzaamheid steeds gericht op neergaan, uitleven en verwelken, te veel van de schaduw des doods over het werk laat vallen.' VW III: *Herfstij der Middeleeuwen* (1919): 3.

Figure 1.1. Huizinga's study at his home on Van Slingelandtlaan 4, Leiden.

To this day, Huizinga is mentioned in English, French and German historiographic literature not only as one of the most influential cultural historians of the previous century but also as an early exponent of an entirely new historical perspective on the past.[6] As a rule, his *Autumntide of the Middle Ages* (1919) and *Homo ludens* (1938) are adduced without contestation as monumental studies in the field of cultural history. However, in order to flesh out my most potent contribution to the field of Huizinga scholarship, a few words on the existing literature on Huizinga's oeuvre, and not its position within the field of cultural history, are in order.

Huizinga was already a recognized historian in Europe and North America during his own lifetime, and soon after his death, his name was further canonized outside the Netherlands.[7] This renown notwithstanding,

6 To name only a few publications: L. B. Dorléac, 'L'histoire de l'art ou l'éloge de la faiblesse', in *L'histoire culturelle en France et en Espagne*, ed. B. Pellistrandi and J. F. Sirinelli (Madrid: Casa de Velázquez, 2008), 78; P. Burke, *What Is Cultural History?* (Cambridge: Polity Press, 2004); U. Daniel, *Kompendium Kulturgeschichte* (Frankfurt: Suhrkamp Verlag, 2001); D. R. Kelley, *Fortunes of History: Historical Inquiry from Herder to Huizinga* (New Haven/London: Yale University Press, 2003); G. Hübinger, 'Konzepte und Typen der Kulturgeschichte', in *Geschichtsdiskurs Band 4: Krisenbewußtsein, Katastrophenerfahrungen und Innovationen 1880–1945*, ed. W. Küttler, J. Rüsen, and E. Schulin (Frankfurt: Fischer Taschenbuch Verlag, 1997); E. Gombrich, *In Search of Cultural History* (Oxford: Oxford University Press, 1969).

7 In 1937 Huizinga received an honorary doctorate at Oxford University, and in the 1930s, Huizinga was invited dozens of times to give lectures in, amongst other countries, England,

no major study of Huizinga appeared in the first four decades after his death. In 1972 a conference to honour the centenary of Huizinga's birth in Groningen called for a re-evaluation of Huizinga, but no major studies resulted from it.[8] In the Netherlands, a historical interest in Huizinga only took proper flight in the 1990s, when books by Wessel Krul (1990), Léon Hanssen (1996) and Anton van der Lem (1993; 1997) debunked a widespread image of Huizinga as a mere conservative, ascetic cynic, a pessimistic 'accuser of his age', as the Dutch historian Pieter Geyl (1887–1966) put it in 1961.[9] In their wake, two German accounts of Huizinga – by Christoph Strupp (2000) and Christian Krumm (2011) – followed.[10] Both offer intelligent and well-researched observations. Yet, whilst Strupp's account does justice to Huizinga's academic identity, it often only glances over Huizinga's lived relations to his own times, and though Krumm's book devotes considerably more attention to Huizinga's lived context, it limits itself to Huizinga's relation to Germany. Two later edited volumes on Huizinga have helped

France, Belgium, Germany, Italy, Spain and Sweden. BW III: Huizinga–Kaegi (1937): 1252. After his death, his contemporary historians judged him respectfully. 'A true historian, which is not quite common,' Lucien Febvre (1878–1956) said of Huizinga; Ernst Kantorowicz (1895–1963) reflected on Huizinga's oeuvre: 'it brings the dead, who died in Nazi exile, back to life'; even those 'who cannot share Huizinga's philosophy can hear that voice *de profundis*,' Ernst Gombrich (1909–2001) stated. To these authors, Huizinga symbolized not only a style of historical research but an ethical and religious perspective on culture. L. Febvre, 'Un moment avec Huizinga', *Annales. Histoire, Sciences Sociales* 6, no. 4 (1951): 493; E. Gombrich, 'Huizinga's *Homo ludens*', in *Johan Huizinga 1872–1972*, ed. W. R. H. Koops, E. H. Kossmann, and G. van der Plaat (The Hague: Martinus Nijhoff, 1973), 154; E. Kantorowicz, 'Review of Johan Huizinga, "Geschichte und Kultur"', *American Historical Review* 60, no. 4 (1955): 855.

8 The conference proceedings were published the following year. W. R. H. Koops, E. H. Kossmann, and G. van der Plaat, eds., *Johan Huizinga 1872–1972* (The Hague: Martinus Nijhoff, 1973).

9 Krul, *Historicus tegen de tijd: Opstellen over leven en werk van J. Huizinga*; Hanssen, *Huizinga en de Troost van de Geschiedenis: Verbeelding en rede*; Van der Lem, *Johan Huizinga: Leven en werk in beelden en documenten*; Van der Lem, *Het eeuwige verbeeld in een afgehaald bed: Huizinga en de Nederlandse beschaving*. For an impression of Huizinga's posthumous reputation in the Netherlands, see P. Geyl, 'Huizinga als aanklager van zijn tijd', *Mededelingen der Koninklijke Nederlandse Akademie van Wetenschappen, Afdeling Letterkunde* 24, no. 4 (1961): 137–77. An English translation of this piece was published two years later. P. Geyl, 'Huizinga as Accuser of His Age', *History and Theory* 2, no. 3 (1963): 231–62. Huizinga's poor posthumous reputation is further illustrated by the fact that fellow historian Jan Romein (1893–1962) mentioned Huizinga only twice in a piece on the state of the Dutch historical discipline during the interwar period. J. Romein, 'De geschiedschrijving in Nederland tijdens het interbellum', in *Carillon der tijden: Studies en toespraken op cultuurhistorisch terrein* (Amsterdam: Querido, 1953).

10 Krumm, *Johan Huizinga, Deutschland und die Deutschen*; Strupp, *Johan Huizinga: Geschichtswissenschaft als Kulturgeschichte*.

contextualize him, but they have considered only his medievalism and *Autumntide* specifically.[11]

In recent literature on Huizinga, only two proper and exhaustive characterizations of Huizinga have been offered, both in Dutch: one by Willem Otterspeer (2006), which has been translated to English, and another by Carla du Pree (2016).[12] Otterspeer, for one, has understood Huizinga in terms of character traits: an author whose conservatism consisted less of unworldly asceticism than of 'reservation', 'duty' and 'courage'. These character traits were, in turn, related to Huizinga's appreciation of 'contrasts', Otterspeer holds: contrasts between inside and outside, emergence and decline, and setback and dream.[13] Otterspeer's 'Huizinga' narrated both his own and historical worlds in terms of their ability to dutifully persevere in times of adversity.[14] Huizinga's cultural history was one of 'conservative revolutions': of cultural forces dragging symbols, rituals, dress and customs against the grain of instrumental logic and biological needs into an open future.[15] His Huizinga is exemplified by Huizinga's adage that 'if we want to retain culture, we must continue to create it,'[16] and drawing from such images of persistence and order, Otterspeer discerns a 'profound unity' in Huizinga's oeuvre, which he takes to be the product of a character's germinal 'pupations' through time.[17]

Du Pree, on the other hand, has understood Huizinga in terms of his social role as a 'public intellectual'.[18] Huizinga corresponded in no less than seven languages, lectured before a wide audience both at home and abroad, edited non-academic journals and submitted innumerable articles on timely

11 É. Lecuppre-Desjardin, ed., *L'odeur du sang et des roses: relire Johan Huizinga aujourd'hui* (Villeneuve d'Ascq: Presses universitaires du Septentrion, 2019); P. Arnade, M. Howell, and A. van der Lem, eds., *Rereading Huizinga* (Amsterdam: Amsterdam University Press, 2019).

12 W. Otterspeer, *Reading Huizinga*, trans. B. Jackson (Amsterdam: Amsterdam University Press, 2010); W. Otterspeer, *Orde en trouw: Over Johan Huizinga* (Amsterdam: De Bezige Bij, 2006); Du Pree, *Johan Huizinga en de bezeten wereld*. Later, another book by Otterspeer was published on Huizinga's *Autumntide* more in particular. W. Otterspeer, *De Kleine Huizinga* (Amsterdam/Antwerp: Atlas Contact, 2019).

13 Otterspeer, *Reading Huizinga*, 97–112; Otterspeer, *Orde en trouw: Over Johan Huizinga*, 95–110.

14 Otterspeer, *Reading Huizinga*, 58; Otterspeer, *Orde en trouw: Over Johan Huizinga*, 54–55.

15 'conservatieve revolutie'. VW II: *Nederland's beschaving in de zeventiende eeuw* (1941): 430.

16 'Wij weten het ten stelligste: willen wij cultuur behouden, dan moeten wij voortgaan met cultuur te scheppen.' VW VII: *In de schaduwen van morgen* (1935): 327.

17 Otterspeer, *Orde en trouw: Over Johan Huizinga*, 38–39; Otterspeer, *Reading Huizinga*, 41–42.

18 In his collected letters, Huizinga can be found to have corresponded in Dutch, English, German, French, Italian, Spanish and Sanskrit. J. Huizinga, *Johan Huizinga: Briefwisseling I–III*, ed. L. Hanssen, W. E. Krul, and A. van der Lem (Utrecht/Amsterdam: Veen/Tjeenk Willink, 1989–91). Du Pree, *Johan Huizinga en de bezeten wereld*, 13.

issues to newspapers and magazines alike.[19] A budding democratic culture, extended means of transportation and a lively Dutch mediatic landscape, Du Pree shows, helped negotiate Huizinga's conservatism, religiosity and even his identity at large.[20] Du Pree has shown that (1) the cultural authority bestowed upon Huizinga as *the* Professor in General History at Leiden and (2) his own personal ethics of duty both pushed him into a public position that was not natural to his character.[21] Once in this position, she shows, Huizinga voiced perspectives that were born from the debates, language and norms of the 1880s and 1890s, and his 1930s cultural critiques must be seen from this generational perspective. Hence, Du Pree's book spends considerable time fleshing out what could be called a parental dynamic.[22] Though Du Pree's socio-generational angle on Huizinga sets itself apart from Otterspeer's more psychological take by its predisposition, both their analyses are synchronized on one substantial front: both their 'Huizingas' were drafted from a character of duty.

The present book draws from, complements and critically challenges these accounts by offering a different kind of Huizinga; the 'Huizinga' presented here consists neither of character traits nor of social roles. Rather, the 'Huizinga' in this book has been drafted from what I shall call 'experiences of loss', such as the loss of a retrospectively imagined *Zeitalter der Sicherheit*.[23] From the 1910s on, the 1880s, 1890s and 1900s appeared as a remarkably calm period; a time before images of war, roaring industry, flashy commercialism and unpredictable political 'masses' had started dominating the public imagination in the Netherlands and beyond. And on a more personal level, this was a world before Huizinga had lost both the love of his life and his oldest son to deadly diseases. I argue that these and other experiences of loss mediated not only Huizinga's analyses of particular historical periods but even Huizinga's understanding of the word 'history' itself. Huizinga's historical engagement with the past became a way of cultivating a Christian-stoic frame of mind in times of turbulence. This experience-based image of Huizinga is not incompatible with Otterspeer's and Du Pree's, but it is qualitatively different: it opens a window into Huizinga's fears, hopes and temporal orientation without edifying his character. The story told here is not one of germinal pupations but of disruptive ends.

19 Du Pree, *Johan Huizinga en de bezeten wereld*, 157–89.

20 Du Pree, *Johan Huizinga en de bezeten wereld*, 256–57.

21 Du Pree, *Johan Huizinga en de bezeten wereld*, 246–47.

22 Du Pree, *Johan Huizinga en de bezeten wereld*, 248.

23 S. Zweig, *Die Welt von Gestern: Erinnerungen eines Europäers* (Frankfurt: Fischer Taschenbuch Verlag, 2019), 15.

But where does one look for disruptive ends in an author's output? Amidst an author's published works, letters, notes, lectures and prayers, where does one start? The question of how authors relate to their words is as old as it is contested, let alone the question of how words relate to experience. Some aspects of these questions shall be addressed and contextualized below. For now, what matters is this: this book shall trace the role of 'experiences of loss' in Huizinga's works by means of the positions he adopted in his disputes with other authors. By showing how such experiences informed and adjusted Huizinga's position in debates, his lived world shall be tied to his academic output. For this purpose, Huizinga is introduced in relation to his most explicitly recognized antagonists – those authors to which he most consistently returned throughout his career. Huizinga sharpened his thoughts on their armour and used their arguments as a banister for the development of his own intellectual identity. And when push came to shove, when he was forced into a corner either by the authors themselves or by the conditions of his times, Huizinga revealed in these disputes his moral and intellectual convictions – sometimes consciously, sometimes not. The antagonists of whom I speak are Alexis de Tocqueville (1805–1859), Jacob Burckhardt (1818–1897), Karl Lamprecht (1856–1915), Oswald Spengler (1880–1936) and Carl Schmitt (1888–1985), and they shall be introduced in the respective chapters.

By reconstructing Huizinga in the image of an experience, this project serves another, second purpose, too. Since the 1990s, authors from a variety of backgrounds have discussed the possibilities of using 'experience' as an analytical term to subvert traditional historical categories.[24] The term 'experience', they have argued, allows historians to challenge certain political narratives, periodic distinctions and actor categories by which past manifestations of uncertainty are otherwise smothered. Retrospectively edified 'crises', 'systems' and 'spirits' cannot take seriously a historical period's undetermined and unstable nature, nor can they do justice to the inevitably multifaceted nature of a lived world. As shall be discussed below in the method section, this book has been inspired by such experiential methods and mobilizes some of their insights to do justice to the unsettling and decentring effects that disruptive events exerted on Huizinga's categories, words, needs, longings and fears. As such, this piece makes a methodological case in point for a historiography of 'experiences of loss'.

24 These authors shall be discussed at length below. For now, it suffices to give just a select few authoritative examples: J. W. Scott, 'The Evidence of Experience', *Critical Inquiry* 17, no. 4 (1991): 773–97; F. R. Ankersmit, *Sublime Historical Experience* (Palo Alto: Stanford University Press, 2005); M. Jay, *Songs of Experience* (Berkeley/Los Angeles: University of California Press, 2005).

In so doing, the present book follows up on a call made by Frits Boterman's recent but already canonical book *Tussen Utopie en Crisis* (Between utopia and crisis). In this book, Boterman argues that Dutch interwar culture deserves being understood in terms of the *Urkatastrophe* ('oerkatastrofe') that was the Great War, despite Dutch neutrality.[25]

By focussing on these experiences of loss and uncertainty and their role in Huizinga's historical works and perspectives, the present project positions itself within a scholarly field that has become known as the 'history of knowledge'.[26] Huizinga's experiences of loss renegotiated the intellectual meaning and moral purpose of historical scholarship, and their study shows 'how classifications and hierarchies of knowledge as well as cardinal epistemic virtues shift over time'.[27] This project investigates, by means of these experiences, how the borders between 'historical knowledge' and other kinds of knowledge shifted. At times, Huizinga's 'history' claimed territory from other academic fields, such as linguistics and anthropology; at other moments, he drew insights from non-academic vernaculars to do with grief, consolation and hope. More particularly, however, this project aligns itself with a specific development within this field of the 'history of knowledge': the study of 'not-knowing'.[28] Through an experiential approach to Huizinga's works, this project reveals how loss and uncertainty were not antithetical to his historical scholarship. Rather, these experiences of loss and uncertainty – experiences of *not-knowing* – informed and rewrote the past and redefined its value to the present.

For these two purposes – the presentation of (1) a new image of Huizinga, and (2) a revisionist case in point for an experientialist historiographic agenda – this book investigates five different 'experiences of loss' Huizinga endured; they shall be listed later, in the method section. This project examines how these experiences informed his cultural-historical works, perspectives and positions in debates. Before these investigations unfold, however, this introductory chapter shall set the stage by offering a number of contexts: (1) Huizinga's personal world, (2) the generational component of

25 F. Boterman, *Tussen utopie en crisis* (Amsterdam: De Arbeiderspers, 2021), 12.

26 The history of this field has been told in a variety of ways. Amongst the most helpful, I have found J. Östling et al., 'The History of Knowledge and the Circulation of Knowledge: An Introduction', in *Circulation of Knowledge: Explorations in the History of Knowledge*, ed. J. Östling et al. (Lund: Nordic Academic Press, 2018), 9–33.

27 L. Daston, 'The History of Science and the History of Knowledge', *KNOW* 1, no. 1 (2017): 145.

28 See e.g. L. Verburgt, 'The History of Knowledge and the Future History of Ignorance', *KNOW* 4, no. 1 (2020): 1–24; S. Dupré and G. Somsen, 'The History of Knowledge and the Future of Knowledge Societies', *Berichte zur Wissenschaftsgeschichte* 42, no. 2–3 (2019): 186–99.

Huizinga's ethics and (3) his academic training and the epistemic convictions it inspired. These three layers (personal, generational and academic) shall prove indispensable to an understanding of Huizinga's experiences of loss and his response to them. Below, I shall illustrate why and how these backgrounds are essential to an understanding of, say, why Huizinga grew interested in medieval 'playfulness' in his own times of urban modernization and mechanical warfare or why he decided to write a book about *homo ludens* whilst the threats of National Socialism swelled. Once these dimensions have been laid out, this chapter concludes by offering an overview of (4) the methods and strategies applied here and (5) the structure of this book.

Repetitions called Huizinga

As has been stated above, this project's central protagonist is not a character, a social role or a diachronic development of either of them. This project's central protagonist is an author cast in the image of a certain kind of experience, the 'experience of loss', and thus this book lies beyond the genre of a chronological biography in the traditional sense of the term. It follows a kind of experience, not a person. Still, the textual manifestations of such experiences do not and cannot exist independently from the textual manifestations of Huizinga's private, generational and academic identity. In later sections, the generational and academic features of Huizinga's perspective and value system are discussed: features related to his socio-generational aesthetic, religiosity, morality and education. First, however, features particular to Huizinga's biography are discussed. This sketch hinges on four peculiarities that reappeared at decisive moments in Huizinga's life and work: (1) his relation to drawing and costume play, (2) his experience with premature death, (3) his relation to women and, lastly, (4) his relation to friendship. What is offered below is thus not a chronological narration of Huizinga's life – which has already been produced by others better equipped for the task[29] – but an image of a personal identity consisting of retrospectively distinguished patterns and repetitions.

From his childhood and adolescence in his native Groningen to late adulthood in Leiden, Huizinga practised drawing and costume play.[30] Even at those moments when life seemed hardly bearable to him, Huizinga continued to engage in and appreciate such playful activities with the greatest of

29 In my personal opinion, the best biographic account to date is Van der Lem, *Johan Huizinga: Leven en werk in beelden en documenten*.

30 Van der Lem, 43–44, 83–107; W. E. Krul, *Historicus tegen de tijd* (Groningen: Historische Uitgeverij Groningen, 1990), 108–22.

seriousness. In the autumn of 1943, upon his detainment by the German Nazi forces occupying the Netherlands, Huizinga reflected on the late medieval costumes in which he had seen a student fraternity parade in Groningen when he was a young child: '[it] was the most beautiful spectacle I had ever seen.'[31] As a student, Huizinga had performed in such spectacles, too, and in 1897 Huizinga obtained his doctoral degree in philology on Sanskrit theatre.[32] In 1914, upon the death of his wife, with whom he had five children, Huizinga took a step back from research to write plays that he and his children acted out at home and in costume, often without audience, sometimes before friends and family.[33] In *Autumntide of the Middle Ages*, Huizinga captured the entirety of late Burgundian culture in terms of a play (*schouwspel*) and its ability to create dreams, explore the forms of the human spirit, and channel and cultivate the passions of the mind.[34] Huizinga's interest in play and theatre was a repeated and lived interest. Costume play inspired him as a child; it consoled him and became a means to engage with his children as an adult and father, and it fascinated him as an academic. Theatre play was a serious matter of the soul, but it was not the only life jacket aboard his ship.

In 1920 Huizinga's first son, Dirk, fell ill, and soon it became clear that Dirk might not survive his condition.[35] Watching over his son's deathbed, Huizinga spent days drawing the increasingly pale face of his first-born.[36] Huizinga had always drawn. As a student he had been infamous among his peers for his caricatures, and both in his personal correspondence and lecture notes, Huizinga did not shy away from including sketches and doodles.[37] Later,

31 VW I: *Mijn weg tot de historie* (1947): 12.

32 VW I: *De Vidûshaka in het Indisch tooneel* (1897): 45–143. Van der Lem, *Johan Huizinga: Leven en werk in beelden en documenten*, 48–52.

33 'The holiday is almost over, and all I have done is chores and [writing and performing] plays with the children,' Huizinga wrote to his friend and law professor Cornelis van Vollenhoven (1874–1933) just after the Christmas holidays in 1917. 'Nu is de vacantie bijna voorbij, en ik heb niets gedaan dan allerlei kleinigheden afdoen en tooneelstukken voor de kinderen schrijven, instudeeren en opvoeren.' BW I: Huizinga–Van Vollenhoven (1917): 185. See also the memories of these plays in his son Leonhard Huizinga's biography of him. L. Huizinga, *Herinneringen aan mijn vader* (The Hague: H. P. Leopold's Uitgeversmij, 1963), 85–90. Otterspeer, *Orde en trouw: Over Johan Huizinga*, 34.

34 VW III: *Herfsttij der Middeleeuwen* (1919): 57.

35 BW I: Henriette Roland Holst–Huizinga (1919): 265; BW I: Richard Roland Holst–Huizinga (1919): 277.

36 Van der Lem, *Johan Huizinga: Leven en werk in beelden en documenten*, 156–58.

37 'no one is safe when you carry a pencil.' BW I: Richard Roland Holst–Huizinga (1914): 121. For discussions of Huizinga's drawing habits and their significance, see e.g. G. de Vugt, 'De droomtekeningen van Johan Huizinga', *De Witte Raaf*, 2021; Van der Lem, *Johan Huizinga: Leven en werk in beelden en documenten*, 83–107. See Figure 1.2 for a few examples of Huizinga's drawings.

Figure 1.2. (A) One of the innumerable colouring pages Huizinga drew for his daughter Laura. (B) An ex-libris for his wife Mary by Huizinga. (C) A cartoon of the academic world by Huizinga.

Figure 1.3. A drawing by Huizinga of his son Dirk on his deathbed (1920).

Figure 1.4. (A) Huizinga's notes. In this document he describes his first car trip. (B) Huizinga on holiday with his children Leonhard, Jakob and Retha, year unknown. (C) Huizinga in costume for a seventeenth-century-themed student masquerade in Groningen in 1894.

in the capacity of professor and author, Huizinga continued to draw – or rather, he drew attention to drawing. His historiographical writings are laced with references to the historian as the draftsman of images, sensitive to aesthetic totality as much as to detail.[38] Similarly, his historical writings, especially *Autumntide*, turn time and again to the 'images' concocted by the respective historical subject's mind. Here Huizinga's native Dutch is relevant to his interest in drawing and images. The Dutch verb 'to draw'

38 In the preface to the first English translation of Huizinga's biography of Erasmus, the British historian George Clark (1890–1979) reminisced over his meeting with Huizinga in Oxford in the 1930s, presumably when Huizinga visited to receive his honorary doctorate in 1937. BW III: Huizinga–Kaegi (1937): 1251. Clark wrote, 'What surprised and delighted me was his seeing eye. [...] His eye was not merely informed but sensitive. I remembered that I had heard of his talent for drawing, and as we walked and talked I felt the influence of a strong, quiet personality deep down in which an artist's perceptiveness was fused with a determination to search for historical truth.' Of course, one needs not and should not uncritically take Clark's words as truth, but they are indicative of the impressions that Huizinga's sensuous perceptiveness would have had on others. J. Huizinga, *Erasmus of Rotterdam*, trans. F. Hopman (London: Phaidon Press, 1952), vii.

(*tekenen*) is directly related to the noun 'sign' (*teken*), and the Dutch noun for 'portrayal' (*afbeelding*) ties directly to 'imagination' (*verbeelding*). In the act of drawing as a historian, as a friend and as a father, Huizinga was engaging in an activity he deemed not only trans-historical but even extra-historical. Like play, the realm of signs and imagination belonged to a human activity capable of arresting time, delaying death and capturing decline.

But the premature deaths of his wife and son were not Huizinga's first encounters with life being cut unexpectedly short. First, his mother had died two years after his birth, too soon for Huizinga to have active memories of her and her loss, and his father does not appear to have spoken much about his son's mother later in life.[39] When the mother of Huizinga's children passed, Huizinga expressed fear over his children forgetting their mother and consequently engaged in not a single romantic nor sexual relation for almost two decades, until he met Auguste Schölvinck in 1937, whom he married.[40] His own father had remarried within two years after the death of Johan Huizinga's mother. Next, Huizinga suffered the loss of his half-brother Herman, who committed suicide in 1902 at the age of seventeen. Judging from their correspondence, Johan Huizinga does not appear to have been very close to Herman, but Johan Huizinga did name two children after Herman (his second son, born in 1908, was named Jakob Herman, and his second daughter, born in 1912, was named Hermanna Margaretha), and when Huizinga discussed medieval practices of suicide in *Autumntide*, his half-brother would most likely have been in mind.[41] Though premature deaths were far more common than they are today in Europe, these deaths prove important to understanding Huizinga's values and virtues.[42]

The death of his first wife Mary Huizinga (1877–1914) in particular plummeted Huizinga into an unspeakable sorrow. Huizinga grew apathetic, and in 1915 he left Groningen and the chairs in Dutch and General History at the university together with his children to take up the chair in General

39 Van der Lem, *Johan Huizinga: Leven en werk in beelden en documenten*, 5–8; Krul, *Historicus tegen de tijd: Opstellen over leven en werk van J. Huizinga*, 43–48.

40 BW I: Johan Huizinga–Jakob Huizinga (1914): 157. For an impression of Huizinga's first introduction to Auguste Schölvinck, see BW III: Huizinga–Schölvinck (1937): 1253. For an impression of the marriage proposal, see BW III: Huizinga–Schölvinck (1937): 1280; BW III: Schölvinck–Huizinga (1937): 1280. This image has been corroborated in Huizinga, *Herinneringen aan mijn vader*, 68–69.

41 For a sense of their relation, see BW I: Johan Huizinga–Herman Huizinga (1898): 8. For Huizinga's discussion of suicide in the Midde Ages, see e.g VW III: *Herfsttij der Middeleeuwen* (1919): 22.

42 This relation shall be expanded on later, especially in Chapters 3 and 4.

History in Leiden.[43] In terms of professional prestige, Huizinga's move to Leiden would have made perfect sense to the outside world. Yet, from his correspondence letters, it is quite clear that Huizinga was also fleeing from Groningen and the memories it bore of his wife.[44] Of course, his memories travelled with him to Leiden, and at least from her death on, Huizinga appears to have demonstrated a new kind of adoration not only for Mary but for women more generally.[45] For example, Huizinga started using the female pronoun both more widely and more consciously: history, the world, fate, theory and other grand concepts were referred to more frequently in the feminine form than before.[46] Much later, in 1940 after he remarried, Huizinga wrote a piece titled *Historical Greatness*, and herein, after a lengthy discussion of all the positions on the term 'greatness', he concluded by referring to how little the term was applied to women and argued that 'based on [its] failure to apply as a rule to the highest and deepest of womanhood, the entire concept of "historical greatness" collapses.'[47] The term exposed 'male delusion' rather than properly understood history, and thus a new term had to be developed.[48] Huizinga suggested 'historical holiness', which 'is commonly used for both men and women'.[49]

Huizinga was a reserved man; he cared about etiquette, spoke calmly and lectured in a collected, monotone voice, and he was therefore frequently experienced as clear but introverted.[50] In his correspondence, too, he was often composed and commonly wrote with personal concern in emotionally

43 Huizinga remained in Leiden for the rest of his career, and eight of his twenty-seven doctoral students here were women. Before coming to Leiden, Huizinga had supervised three doctoral candidates at the University of Groningen, all of whom were men. A. van der Lem, *Inventaris van het archief Johan Huizinga*, ed. A. Th. Bouwman and J. J. Witkam (Leiden: Leiden University Library, 1998), 387.

44 BW I: Johan Huizinga–Jakob Huizinga (1914): 146, 149; BW I: Huizinga–Colenbrander (1914): 152; BW I: Johan Huizinga–Jakob Huizinga (1915): 163; BW I: Huizinga–Veth (1915): 164; BW I: Huizinga–Van Anrooy (1917): 186.

45 Otterspeer, *Orde en trouw: Over Johan Huizinga*, 24–25; C. Santing, 'Het liefkozen van schone vormen. Johan Huizinga en het vrouwenvraagstuk', *De Gids* 168, no. 2 (2005): 121–30.

46 I base this statement on a comparison between Huizinga's doctoral dissertation and monographs, which all appeared after Mary's death. In his dissertation, Huizinga seldomly if ever referred to a noun with 'she' (*zij*) or 'her' (*haar*). In, say, *Autumntide*, several weighty nouns and even the book itself became female. More, and preferably quantitative, analysis would be interesting in this respect.

47 VW VII: *Historische grootheid* (1940): 217.

48 VW VII: *Historische grootheid* (1940): 216.

49 VW VII: *Historische grootheid* (1940): 217.

50 Th. J. G. Locher, 'Johan Huizinga', *Jaarboek van de Maatschappij der Nederlandse Letterkunde* (1946): 104.

detached prose.[51] Yet, when his sadness became too large to carry alone, Huizinga turned to a carefully selected set of friends, and in this space, another side surfaced. Certain friendships in particular were important to Huizinga in this respect: those with André Jolles (1874–1946), Herman T. Colenbrander (1871–1945), Henriette Roland Holst (1869–1952) and Bronisław Malinowski (1884–1942).[52] Each of them shall reappear in the chapters to come.[53] All four were passionate individuals and more comfortable with emotional and vulnerable prose; each of them sought answers to existential concerns through their academic and literary work. The fact that Huizinga sought their companionship and sustained decades-long correspondence with each of them is telling of his dependency on them to cultivate a less-seen yet pressing side of his mind. Like Huizinga, Roland Holst and Malinowski had experienced the pain of the premature death of their dearest loved ones.[54] Jolles was troubled in different ways, and his attempts to navigate his passions through academic work led him to wholeheartedly embrace National Socialism, upon which Huizinga terminated their correspondence, which had lasted no less than four decades.[55]

At least two general lessons about Huizinga can be obtained from this correspondence. First, Huizinga required friendship to express vulnerabilities and explore his emotions in the private sphere. In correspondence with anyone other than his dearest friends, Huizinga's writing was formal, matter-of-fact and sometimes blunt, at times leading his other correspondents wondering where Huizinga stood emotionally. In his correspondence with Roland Holst and Malinowski, and especially in *their* replies to him, another Huizinga shines through. Following his son Dirk's death, Huizinga wrote a letter that is now missing to Roland Holst. In her response, she admits to being unsettled by his 'contradictions and riddles', his disillusionment with life and human agency.[56] Years later, referring to Mary's death, Malinowski writes about the 'essential loneliness' to which he thought he and Huizinga

51 For reflections on how Huizinga might have experienced his own correspondence style, see W. Krul, 'Huizinga in zijn brieven: Beeld en zelfbeeld', *Tijdschrift voor Geschiedenis* 106, no. 1 (1993): 23–37.

52 Other perhaps more or less equally close friends were Jan Veth (1864–1925) and Richard Roland Holst (1868–1938).

53 Several of Huizinga's friendships are discussed in Van der Lem, *Johan Huizinga: Leven en werk in beelden en documenten*, 171–92, 193–212.

54 Henriette Roland Holst lost her younger sister and father in a single traffic accident in 1892. Malinowski lost his wife in 1935. BW III: Malinowski–Huizinga (1935): 1327.

55 BW II: Richard Roland Holst–Huizinga (1933): 1057. Van der Lem, *Johan Huizinga: Leven en werk in beelden en documenten*, 193–212.

56 BW I: Henriette and Richard Roland Holst–Huizinga (1920): 317.

and were tied by fate: 'often when I was feeling very unhappy and very much alone, did I turn to you – silently, for what can one say? But I knew that you were nearer to me in my moods and my loneliness than many people around me.'[57] In these letters, one finds a seldom-revealed Huizinga. In this particular correspondence, at least with Roland Holst and Malinowski, Huizinga reveals a continued sorrow, one that is hard to disconnect from his academic interest in tragedies.

Huizinga was perhaps not the most dynamic of characters, but he was far removed from being static. Huizinga's outlook, writing, values and research were adjusted time and again as he acquired experiences. Still, the four themes discussed above indicate a number of reoccurring themes important to Huizinga's perspective and response to an ever-changing world. Huizinga cared about assembling images and stories and the communal practices involved. He mourned the dead, and it was not intuitive to him to understand the past as 'gone'. Huizinga was as passionate as he was emotionally reclusive, and the economy of his emotional turmoil required strategic practices to not burst when alone. Correspondence with friends, sustained over many decades, was one such strategy, and as shall be explored in greater depth in upcoming sections, academic work was most likely another. Lastly, the image of the 'woman' loomed large in Huizinga's mind. He looked down on what he perceived to be the overly male qualities of confidence and aggression. The virtues he chose to cultivate and advocate instead were 'holy' and, in his own understanding, 'female'.

If these observations had to be tied together more closely in an ideal type of 'Huizinga', one image in particular seems pertinent to me: that of *katechon*, the biblical term for 'the one who withholds', the one who postpones collapse through conservation. In Huizinga's case, it seems to me that this word captures a significant part of both his personal and academic life. In his relations to play, correspondence and friendship, Huizinga, I believe, sought ways to persevere in his world. Congruently, in his historical works, Huizinga practised the perseverance of virtues and ideals he deemed valuable. Huizinga sought to guard the past and protect it against erosive forces of that indifferent non-history called the present. This interest in delaying the past's expiration tied Huizinga the historian with Huizinga the father, husband and friend; the virtues Huizinga celebrated as a historian reflected the virtues he celebrated in the private sphere. Willem Otterspeer made a similar claim when he presented an ideal type of Huizinga in terms of

57 BW III: Malinowski–Huizinga (1937): 1325.

two values: 'order' and 'loyalty'.[58] I wish to add to this ideal type by opening up a complementary dimension. After all, in what kind of world would such values appear 'valuable' in the first place? In Huizinga's case, I think the answer is this: a world of loss. A loss of mothers, brothers, children as well as of societal dreams and political fabric made urgent the virtues of perseverance and conservation. Huizinga's world was one of *katechon*; Huizinga was a conservator, drawing history into an inhospitable future.

Huizinga's moral sympathies

Huizinga's personal interest in the conservation of the past had a pronounced socio-generational dimension. As a student and young adult, in the period 1890–1914, Huizinga adopted a number of moral beliefs and sympathies that together amounted to what could be best described in contemporary terms as a Christian humanism and Eurocentric, patriotic cosmopolitanism. Though his moral beliefs changed and developed over time, a number of basic 'sympathies' remained, or rather, resurfaced time and again in his adult life. In order to understand the impact of certain experiences on his historical outlook, these repeated sympathies need to be spelled out and understood in their socio-generational signature. In most literature, the role of *fin-de-siècle* culture in Huizinga's oeuvre has been traced back to his emphatic writing style and synesthetic metaphors. Yet, to me, this literary influence seems to have taken a back seat soon after the Great War.[59] By

58 For the title of his book, Otterspeer used Huizinga's 'order and loyalty' (*orde en trouw*) from VW V: *Over de grenzen van spel en ernst in de cultuur* (1933): 7. Otterspeer, *Orde en trouw: Over Johan Huizinga*.

59 R. L. Colie, 'Johan Huizinga and the Task of Cultural History', *The American Historical Review* 69, no. 3 (1964): 607–30; L. Dorsman, 'Periodisering als integrale benadering: Nederlandse historici in het Fin-de-Siècle', *Theoretische geschiedenis* 16, no. 3 (1989): 277–96; H. W. von der Dunk, 'Huizinga als Kultuurpessimist', *Groniek*, no. 26 (1973): 150–54; W. Thys, 'Huizinga en de Beweging van Negentig', in *Johan Huizinga 1872–1972*, ed. W. R. H. Koops, E. H. Kossmann, and G. van der Plaat (The Hague: Martinus Nijhoff, 1973), 29–52; R. Anchor, 'History and Play: Johan Huizinga and his Critics', *History and Theory* 17, no. 1 (1978): 63–93; W. E. Krul, 'Het leven der woorden. Taalkunde en geschiedenis in Huizinga's vroegste wetenschappelijke werk', *BMGN – Low Countries Historical Review* 104, no. 3 (1989): 365; Th. J. G. Locher, 'Johan Huizinga', *Jaarboek van de Maatschappij der Nederlandse Letterkunde* (1946): 88–109; J. Noordegraaf, '"On Light and Sound": Johan Huizinga and Nineteenth-Century Linguistics', in *The Dutch Pendulum. Linguistics in the Netherlands 1740–1900* (Münster: Nodus Publikationen, 1996), 130–58; Hanssen, *Huizinga en de troost der geschiedenis: Verbeelding en rede*, 150–52; Van der Lem, *Het eeuwige verbeeld in een afgehaald bed: Huizinga en de Nederlandse beschaving*, 36; Strupp, *Johan Huizinga: Geschichtswissenschaft als Kulturgeschichte*, 118; Otterspeer, *Orde en trouw: Over Johan Huizinga*, 29.

emphasizing and expanding on the socio-generational component of this relation, I follow and extend beyond the few examples of Krul, Kossmann and Du Pree.[60] In order to understand Huizinga's conservatism, one needs to dispense with contemporary notions of the term, if only for a moment. After all, 'conservatism' is relative to that which one imagines to be at risk of losing. So which losses did Huizinga's generation of scholars fear and experience?

For this purpose, a number of pivotal developments must be cited. The first is the urban modernization of the Netherlands, for present purposes best exemplified by Amsterdam between 1903 and 1905, when Huizinga worked as a *privaat-docent* at the University of Amsterdam.[61] During this period, Huizinga witnessed the city's profound and on-going transitions. Crooked and bent medieval streets were straightened and widened, canals were drained and filled, and below the newly recovered surface, an elaborate system of pipes was installed. Above, electrical wires guided new trams and trains. From the late 1870s until the early 1920s, three major urban planning projects – *Plan Kalff* (1877), *Plan Zuid* (1915), *Plan West* (1922) – were either finalized or announced, and they each negotiated anew the city's aesthetic and demographic profile through both exterior and interior design. On one hand, the architects involved, such as Hendrik P. Berlage (1856–1934), mourned the 'certainty', 'harmony' and 'calm conviction' of the city's waning medieval architecture.[62] Yet, on the other hand, consensus existed on architecture's supposed responsibility to answer and reflect the 'activity, passion, longing, [and] struggle' of its times and to adapt to the 'spiritual-psychic development' of the century's

60 E. H. Kossmann, 'Postscript', in *Johan Huizinga 1872–1972*, ed. W. R. H. Koops, E. H. Kossmann, and G. van der Plaat (The Hague: Martinus Nijhoff, 1973), 223–34; W. E. Krul, 'Ter Braak contra Huizinga. Over de grenzen van de ironie', *Groniek*, no. 100 (1988): 61–79; E. H. Kossmann, 'Romeins "Breukvlak" en de Nederlandse geschiedenis', *BMGN – Low Countries Historical Review* 106, no. 4 (1991): 652–58; C. du Pree, *Johan Huizinga en de bezeten wereld* (Leusden: ISVW Uitgevers, 2016), 108.

61 In this period, Huizinga lived in Haarlem, where he worked as a secondary school teacher. He continued to work at the school even after he became a *privaat-docent* in Amsterdam and had to commute to the city by train on a frequent basis. For a sense of the impressions Amsterdam would have made on him, see e.g. Van der Woud, *Het Landschap en de mensen: Nederland 1850–1940*; A. van Veen, *Amsterdam 1900: Foto's van Olie, Breitner, Eilers en tijdgenoten* (Bussum: Thoth, 2016); V. van Rossum, 'De stad gebouwd: De oude binnenstad vernieuwd', in *Amsterdam in de tweede Gouden Eeuw* (Bussum: Thoth, 2000), 36–58; J. de Vries, ed., *Nederland 1913: Een reconstructie van het culturele leven* (Haarlem: Meulenhoff/Landshoff, 1989).

62 H. P. Berlage et al., 'Het doel van deze uitgave', in *Moderne bouwkunst in Nederland* (Rotterdam: W.L. & J. Brusse's Uitgeversmaatschappij, 1932), 6.

turn.[63] 'It would be unwise [to] long for harmony,' Berlage and his colleagues concluded.[64]

'Unwise to long for harmony': new times required not just new spaces and shapes, but new social relations, too. 'What is [the woman] to society? – Almost everything. – What is she to the law? – Nothing. – What does she wish to become? – Something.' These words come from the first Dutch 'feminist manifesto' from 1889, written by the Dutch politician Wilhelmina Drucker (1847–1925).[65] Both within the social-democratic movements and beyond, female suffrage activists laboured indefatigably for women's rights in the written, spoken and implemented word. Male suffrage was passed by parliament in 1917; female suffrage followed in 1919. In tandem with this development, women obtained new professional positions in the developing industrial economy, universities, streetscapes and nightlife, and Huizinga noticed. Through his friendship and correspondence with Henriette Roland Holst, Huizinga became acquainted with the women's rights movement. An increasing number of his students were female, and after his first wife Mary Huizinga died in 1914, Huizinga was confronted in more than one way by the deep implications of nineteenth-century household labour divisions. The gender norms were changing; both personally and publicly, Huizinga applauded this, though he did not go out of his way to actively work towards female emancipation.[66]

The shattering of codes and norms (aesthetic, social, ideological) from the 1880s onwards was observed, commended and actively supported by a significant part of an entire generation of Dutch authors and painters, eager to record and participate in the transitions taking place before their eyes. Many members of this generation of artists self-identified as a *Tachtiger* – literally, 'of the '80s', after the decade of their professional maturation.[67] They celebrated the modern urban developments: new materials, new colours, new department stores, new public spaces and an uprooting of stiff traditional family structures brought new images of freedom and individual creativity.

63 Berlage et al., 'Het doel van deze uitgave', 6.

64 Berlage et al., 'Het doel van deze uitgave', 6.

65 Cited in M. Bosch, *Strijd! De vrouwenkiesrechtbeweging in Nederland, 1882–1922* (Hilversum: Verloren, 2019), 47–48.

66 Perhaps the most explicit example of his public support in his published works is to be found in his aforementioned essay *Historical Greatness* (*Historische grootheid*) from 1940, see above.

67 This generation is most closely associated with authors, artists and composers such as George Breitner (1857–1923), Willem Kloos (1859–1938), Willem Witsen (1860–1923), Frederik van Eeden (1860–1932), Alphons Diepenbrock (1862–1921), Herman Gorter (1864–1927), Lodewijk van Deyssel (1864–1952) and Albert Verwey (1865–1937).

In the reforms enacted by liberal policies, *De Tachtigers* found a political identity complementing their individualist, hedonist and anti-authoritarian convictions and their experience of fleeting movements of change.[68] The marriage of their moral and political interests was captured by Willem Kloos: at heart, what mattered was only 'the most individual expression of the most individual emotion'.[69] By the mid-1880s, *De Tachtigers* had become sufficiently organized to set up and run their own magazine, *De Nieuwe Gids* (The new guide).[70] In this magazine, *De Tachtigers* transmuted on a national scale the industrious age into deeply passionate language and elusive, impressionistic brushstrokes.

As a student in Groningen, Huizinga had read *De Nieuwe Gids* diligently, and he had learned to admire its self-proclaimed generation of artists 'as if they had been demigods'.[71] However, soon after 1900, during his time in Amsterdam, Huizinga's appreciation of *De Tachtigers* and their liberal and individualist politics changed dramatically.[72] Huizinga grew disillusioned with the consequences of liberal reforms towards industrialization, individualism and urbanization.[73] From Huizinga's personal correspondence, it seems that the demolition of historical architecture in particular unsettled him.[74] In the face of architectural change, Huizinga developed *Heimweh* (*heimwee*) for the medieval, 'pre-Sarphatic Amsterdam'[75] and an appreciation for 'stable truths' in tumultuous times.[76] Against this background, it is telling

68 At any rate, Huizinga himself perceived the relation between the liberal reforms and *De Tachtigers* to be most intimate. In a letter from 1928, Huizinga held that *De Tachtigers* had been culturally enabled by the string of liberal governments that had been in charge since the 1860s, see BW Huizinga–Nijhoff (1928): 744.

69 'De allerindividueelste expressie van de allerindividueelste emotie,' see W. Kloos, *Nieuwere literatuur-geschiedenis. Deel* 2, of 5 vols (Amsterdam: L. J. Veen, 1925), 161.

70 This magazine was published in Amsterdam by W. Versluys and distributed on a national level.

71 VW I: *Mijn weg tot de historie*: 19 (1947).

72 In a letter from 1938 to the author Menno ter Braak (1902–1940), Huizinga identified this period as his transitional phase away from *De Tachtigers*, see BW III: Huizinga–Ter Braak 1374 (1938). In his personal correspondence, this disillusionment appears somewhat later, see e.g. BW I: Huizinga–Jolles (1907): 57; BW I: Huizinga–Colenbrander (1916): 177.

73 VW V: *Mensch en menigte in Amerika* (1918): 290–291.

74 Huizinga took significant interest in historical architecture and urban planning, and he objected passionately to the reforms taking place in both domains. Huizinga's response to the urban modernization projects is discussed at length in Chapter 2.

75 Samuel Sarphati (1813–1863) was a Dutch physician and prolific city planner. His name had become synonymous with the larger project of Amsterdam's nineteenth-century modernization. See BW I: Huizinga–Bijvanck (1913): 107.

76 This observation stems from a later reflection by Huizinga on his frame of mind around 1900, see BW III: Huizinga–Ter Braak (1938): 1374.

that Huizinga's magnum opus, *Autumntide of the Middle Ages*, opens with a reflection on the virtues of medieval architecture and urban planning. In the course of only a few years, liberal reforms appeared to Huizinga as uprooting rather than uplifting, and *De Tachtigers*, once 'demigods', now seemed 'snobbish' and a 'minor literary bunch' absorbed by fleeting emotions rather than moral responsibility.[77] In 1915 Huizinga himself became an editor at *De Gids* (not *De Nieuwe Gids*) and started representing a new generation of authors and convictions.[78]

The authors, artists and academics of Huizinga's generation (the self-proclaimed *Negentigers*, 'of the '90s'), such as Henriette Roland Holst, Piet Mondriaan (1872–1944) and André Jolles, were not interested in the depths of a human individual's passions, nor did they celebrate the indulgences of modern city life. They were interested in images of timelessness, cosmopolitan politics, universal ethics and a revival of metaphysics. In opposition to the urban-industrial transitions and supposed moral relativism of his day, Huizinga became increasingly convinced of a non-confessional Christian moral law (*christelijke zedewet*), a super-individual ethical responsibility.[79] Similarly, Mondriaan conceived of his modernist art as an argument against 'individual despotism' and for 'the universal'; Jolles, in turn, sought universal forms of culture in his *Einfache Formen* (1930); and Roland Holst wrote in her poems about the emergence of a new kind of 'friendship'.[80] Huizinga himself later depicted this 'revolution of spirits' explicitly as 'a reaction against the excessive individualism and impressionism of the first *Tachtigers*', a revolution that 'drew from a need for style and certainty, a more fixed direction and belief'.[81] In opposition to a world of acceleration, fleeting

77 BW I: Huizinga–Colenbrander (1916): 177.

78 VW I: *Mijn weg tot de historie* (1947): 19.

79 Huizinga accepted the offer after initially hesitating BW I: Huizinga–Veth (1915): 164; Du Pree, *Johan Huizinga en de bezeten wereld*, 178–81.

80 See respectively BW III: Huizinga–Ter Braak (1938): 1374; T. van Doesberg and P. Mondriaan, 'Manifesto I of De Stijl (1922)', in *100 Artists' Manifestos: From the Futurists to the Stuckists*, ed. A. Danchev (London: Penguin Books, 2011), 216; A. Jolles, *Einfache Formen* (Berlin: De Gruyter, 2006); H. Roland Holst, *Tusschen twee werelden* (Rotterdam: W. L. & J. Brusse, 1923), 63.

81 'De wending der geesten, die zich omstreeks 1890 in het kunst- en letterkundig leven van Nederland begon te doen gevoelen, berustte voor een deel op een reactie tegen het overmatig individualisme en impressionisme der eerste Tachtigers, en sproot voort uit een behoefte aan meer stijl en stelligheid, meer vaste richting en geloof.' VW VI: *Leven en werk van Jan Veth* (1927): 372. Krul has argued that the generational antagonism described by Huizinga concerned not only an older generation of artists and authors but also his liberal father, see Krul, *Historicus tegen de tijd: Opstellen over leven en werk van J. Huizinga*, 25–61. This point shall be examined in greater detail in Chapter 3.

impressions, elusive moments and erratic passions, a younger generation sought to create a space for serenity and timelessness.

Huizinga's moral, aesthetic and academic perspectives around 1900 are emblematic of how the Dutch *fin-de-siècle* has been depicted since at least the 1990s.[82] Since then, studies on the late nineteenth-century Netherlands have roughly imparted the following image of the Dutch *fin-de-siècle*:[83] (1) temporally, it is the period between 1880 and 1905; (2) spatially, it took place between generations and within four domains – politics, academia, architecture and art; (3) thematically, it centred around the need for new political, philosophical, historical and spiritual ideals in times of perceived change and fleetingness; and, lastly, (4) mentally (and this shall prove of utmost importance), the Dutch *fin-de-siècle*, unlike its French, German and Austrian counterparts, was not pessimistic.[84] It was not dominated by images of 'collapse' and utter 'exhaustion' but rather by tropes of 'transition' and a 'need for meaningfulness'.[85] Images of confidence and hopefulness transpired through the artistic, academic and political attempts to narrate new ideals.[86] The Dutch *fin-de-siècle* 'was more than pessimism'.[87]

To conclude, a number of Huizinga's moral convictions discussed above – those to do with communitarianism, universalism and religiosity – can

82 This new scholarly impetus problematized and dispensed with the canonical standard account of the Dutch *fin-de-siècle* by Jan Romein. Romein had presented the 1880s and '90s as a response to a total and all-encompassing experience of doubt. His image was criticized for its totalitarian grasp and dark diagnosis. J. Romein, *Op het breukvlak van twee eeuwen: De westerse wereld rond 1900* (Amsterdam: Querido, 1976).

83 Dorsman, 'Periodisering als integrale benadering: Nederlandse historici in het Fin-de-Siècle'; Kossmann, 'Romeins "Breukvlak" en de Nederlandse geschiedenis'; Krul, 'Nederland in het *fin-de-siècle*. De stijl van een beschaving'; P. de Rooy, 'Een hevig gewarrel. Humanitair idealisme en socialisme in Nederland rond de eeuwwisseling', *BMGN – Low Countries Historical Review* 106, no. 4 (1991): 625; F. van Vree, 'De stad van het betere leven. Cultuur en samenleving in Nederland rond 1900', *BMGN – Low Countries Historical Review* 106, no. 4 (1991): 641–51; N. C. F. van Sas, '*Fin-de-Siècle* als nieuw begin. Nationalisme in Nederland rond 1900', *BMGN – Low Countries Historical Review* 106, no. 4 (1991): 595–609; R. Aerts, 'Op zoek naar een Nederlands fin de siècle', *De Gids*, no. 156 (1993): 91–101; H. te Velde, 'Fin de siècle in de Nederlandse politiek', *Leidschrift* 1, no. 14 (1998): 45–55; W. Krul, 'Het *fin-de-siècle* als vertoning', *BMGN – Low Countries Historical Review* 117, no. 4 (2002): 519–25.

84 Krul, 'Nederland in het *fin-de-siècle*. De stijl van een beschaving', 581–82.

85 Aerts, 'Op zoek naar een Nederlands fin de siècle', 94; Dorsman, 'Periodisering als integrale benadering: Nederlandse historici in het Fin-de-Siècle', 283.

86 As such, these more recent historians have made relevant a significantly older account of the Dutch *fin-de-siècle*, published first in 1955: B. Spaanstra-Polak, 'De atmosfeer van het fin-de-siècle', in *Het symbolisme in de Nederlandse schilderkunst 1890–1900* (Bussum: Uitgeverij Thoth, 2004).

87 Velde, 'Fin de siècle in de Nederlandse politiek', 45.

Figure 1.5. Modernity brought new shapes to the Netherlands. Most Dutch cities, including Amsterdam, had been constructed according to a medieval urban anatomy: layers of circular streets lay around a city's central square. These circular structures did, however, not answer to the industrial needs of public transport and cost efficient housing schemes. In Amsterdam, new neighbourhoods such as (A) the Rivierenbuurt and (B) the Mercatorpleinbuurt answered to the needs of industrial geometry. (C) Along the Dutch coast shapes changed: plans to cut off the Zuiderzee from the sea had been in the making since 1848. In 1920 the construction of a dam started in order to reclaim agricultural land from the sea and to fend existing lands from recurring flooding.

Figure 1.6. *De Tachtigers* mediated the industrial transformation of Dutch society through an impressionist style. This style was meant to capture the fleeting nature of time amidst accelerated change. (A) Richard N. Roland Holst's *Construction Site in Amsterdam* (1891). (B) Willem Witsen's *Warehouses at a Canal on the Uilenburg Island in Amsterdam* (1911). (C) George H. Breitner's *The Singel Bridge at the Paleisstraat in Amsterdam* (ca. 1897).

be said to fit into the ethical fashions of those Dutch academics who had grown up in the *fin-de-siècle* period. At pivotal instances, these convictions were mobilized by certain experiences of loss. For instance, as is explored in Chapter 2, Huizinga's aesthetic conservatism can be seen to have influenced his reading of Van Eyck's *The Arnolfini Wedding* (1434) after the grand Dutch urban modernization projects gained momentum; or, as is explored in Chapter 5, Huizinga's cosmopolitanism fed straight into his analysis of Rembrandt *Syndics of the Drapers' Guild* (1662) after

Figure 1.7. ***De Negentigers*** **launched their criticism against liberal individualism, amoralism and industrialization by rejecting impressionism and turning either to symbolism or socialist realism. The symbolist attempt to 'slow down' a history supposedly lost in acceleration was heavily gendered. Female figures were commonly depicted to convey a sense of resignation and timelessness. For this purpose, images of sorrow and an inward gaze were frequently employed. (A) Piet Mondriaan's** ***Passionflower*** **from 1907. (B) Jan Toorop's** ***Prayer*** **from 1924. (C) Vincent van Gogh's** ***Memory of the Garden in Etten*** **from 1888.**

the Great War had erupted. On a more general level, these convictions mobilized Huizinga to almost completely revise his understanding of not only Dutch history but also the very meaning of 'culture' after Nazi Germany occupied the Netherlands. In these cases, his analyses followed from an exchange between experiences and the moral beliefs of a generation of authors.

Huizinga's academic training and intellectual perspective

Both Huizinga's personal habits and perspectives and his socio-generational ethic were laced with another, third fabric: his philological education in the 1890s. At the time, most Dutch historians had been trained in philology, and only in the 1920s did history become an independent course of study at Dutch universities.[88] Unlike most of his fellow historians in Dutch and general history, however, Huizinga had received the bulk of his postgraduate training not in general Germanic philology but in Sanskrit. Certain parts of this education returned consistently throughout his mature output, and an awareness of these more persistent academic inclinations shall prove to be of great importance to understanding Huizinga's written response to experiences of rupture and loss later on in life. The influence of his philological training on his historical works has been widely acknowledged in the literature but it has been common to frame it solely in terms of Huizinga's interest in words as historical objects.[89] To my knowledge, only Wessel Krul and Jan Noordegraaf have stressed the role of Huizinga's specific philological training in his later historical method.[90] In the present section, I look to contribute to their arguments not by exploring in even higher resolution Huizinga's historical-philological methods but rather by considering the role these methods played in his historical narratives.[91]

88 J. Romein, *Geschiedenis*, ed. K. F. Proost and J. Romein, *Geestelijk Nederland 1920–1940 II* (Amsterdam: Kosmos, 1948), 25.

89 E.g. W. Kaegi, *Das historische Werk Johan Huizingas* (Leiden: Universitaire Pers Leiden, 1947), 9–13; Colie, 'Johan Huizinga and the Task of Cultural History'; Krul, 'Het leven der woorden. Taalkunde en geschiedenis in Huizinga's vroegste wetenschappelijke werk'; Noordegraaf, '"On Light and Sound": Johan Huizinga and Nineteenth-Century Linguistics'; Otterspeer, *Orde en trouw: Over Johan Huizinga*, 161; J. Edwards, 'Play and Democracy: Huizinga and the Limits of Agonism', *Political Theory* 41, no. 1 (2013): 90–115.

90 Krul, 'Het leven der woorden. Taalkunde en geschiedenis in Huizinga's vroegste wetenschappelijke werk'; Noordegraaf, '"On Light and Sound": Johan Huizinga and Nineteenth-Century Linguistics'.

91 Most accounts of Huizinga's narratology, in turn, have not taken his philological perspective into consideration. See for example B. Vanwesenbeeck, 'Huizinga, Theorist of Lateness?', in *Rereading Huizinga: Autumn of the Middle Ages, a Century Later*, ed. P. Arnade, M. Howell, and A. van der Lem (Amsterdam: Amsterdam University Press, 2019), 245–58; H. White, *Metahistory* (Baltimore: Johns Hopkins University Press, 2014), 16; J. den Hollander, H. Paul, and R. Peters, 'Introduction: The Metaphor of Historical Distance', *History and Theory* 50, no. 4 (2011): 1–10.

For this purpose, a brief outline of Huizinga's education, both at a secondary and university level, is helpful.[92] During his secondary school education in Groningen, Huizinga received training in Greek, Latin, Hebrew, Arabic, medieval Dutch and historical Germanic consonant shifts. Upon completing secondary school, Huizinga initially wanted to study Semitic languages and cultures at Leiden University, but, due to professional prospects, his father convinced (if not pressured) him to opt for Dutch literature and history (*Nederlandsche letteren*) instead. Huizinga complied, and as an undergraduate student, he took courses in general post-classical history, Dutch national history, Dutch and German linguistics, historical grammar and comparative linguistics at the University of Groningen. Later, in postgraduate studies, Huizinga took courses in Arabic and Sanskrit in Groningen and studied the *junggrammatische Methode* under the leading *Junggrammatiker* Karl Brugmann (1849–1919) during a term abroad in Leipzig.[93] Huizinga concluded his formal training by obtaining a PhD in the arts (*Letteren*) from the University of Groningen with an investigation into the comedic role of the *Vidûshaka* figure in Sanskrit theatre, completed under the supervision of the Dutch linguist Barend Sijmons (1853–1935).

In retrospect, the time Huizinga spent in Leipzig studying *junggrammatische Forschung* appears to have been formative for his later intellectual convictions. By the end of the nineteenth century, the *junggrammatische Methode* had become the leading tradition in Dutch philology, also among the professors Huizinga most respected.[94] In practice, this meant that most linguists attempted to reconstruct the historical development of, say, Indo-Germanic languages 'from the times of the first human community [*Urgemeinschaft*] until now' by mapping the development of 'common starting points [*gemeinsamen Ausgangspunkte*]' – that is, the historical migrations and developments of the roots and suffixes of the spoken vowels.[95] These vowels, the *Junggrammatiker* commonly held, travelled and developed within and between languages along lawful patterns, such as those already delineated by Grimm's and Verner's laws. These 'sound

92 VW I: *Mijn weg tot de historie* (1947) and Krul, 'Het leven der woorden. Taalkunde en geschiedenis in Huizinga's vroegste wetenschappelijke werk'.

93 BW I: Huizinga–Brugmann (1898): 7; VW I: *Mijn weg tot de historie* (1947): 21–22.

94 J. Noordegraaf, 'Uit het verleden van een historicus: De taalkundige ambities van de jonge Huizinga', *Voortgang*, no. 13 (2011): 199; P. J. Meertens, 'Nederlandse filologie', in *Geestelijk Nederland 1920–1940*, ed. K. F. Proost and J. Romein (Amsterdam/Antwerp: Kosmos, 1948), 1.

95 '[...] der Zeit der [Urgemeinschaft] bis heute'. K. Brugmann, *Grundriss der vergleichenden Grammatik der indogermanischen Sprachen* (Strasbourg: Karl J. Trübner, 1886), 1.

laws [*die Lautgesetze*]' were the 'cornerstones [*Grundpfeiler*] of [the] entire science [*Wissenschaft*]'.[96] Yet, whilst the *Junggrammatiker* experienced their academic heyday, Huizinga grew disillusioned with their formal comparative methods. In a draft article from 1897 titled 'On the Neglect of Semantics in Comparative Linguistics' ('Über die Vernachlässigung der Wortbedeutung in der vergleichenden Sprachwissenschaft'), Huizinga wrote dismissively about any supposed 'mathematical certainty' in comparative philology:[97]

> [The philologist's] knowledge would come to life for him only if he were also self-conscious of the spirit [*Geist*] of past times, which, once felt, leaves behind a homesickness [*Heimweh*] for other times, one which is not satisfied by mere philological occupations [*bloß philologischen Thättigkeit*]. History is not an exact science [*exacte Wissenschaft*], nor should it ever become one. This is even more strongly the case with linguistics [*Sprachwissenschaft*].[98]

In this draft, Huizinga suggested exploring new methods in order to investigate the dynamics by which words acquire meaning (*Bedeutung*) in the first place rather than 'endlessly' mapping their formal etymological origins. Huizinga sent this draft for review to Brugmann, who unsurprisingly responded dismissively, and consequently, the manuscript was never published.[99] The draft reveals not only an early and unmistakable instance of Huizinga's dislike of formal historical methods but also, in concealed terms, an intellectual ally. Consider another passage from the draft:

> [As long as linguistics] steers clear of poetic feeling [*dichterischen Gefühl*] – for language is, after all, in every case a poetic creation [*Schöpfung*] – its satisfaction cannot be deeper than that of the mere archaeological [*bloß*

96 H. Osthoff and K. Brugmann, *Morphologische Untersuchungen auf dem Gebiete der indogermanischen Sprachen* (Cambridge: Cambridge University Press, 2013), xiv. This book was first published in 1878.

97 'mathematische Sicherheit'. HA 33.III.1 (1897). For a transcript and discussion of this draft article, see Noordegraaf, 'Uit het verleden van een historicus: De taalkundige ambities van de jonge Huizinga'.

98 'Aber leben würde ihm sein Wissen erst, wenn ihm ausserdem der Geist früherer Zeiten von selbst bewußt wäre, das, einmal empfunden, ein Heimweh nach anderen Zeiten zurücklässt, von keiner bloß philologischen Thättigkeit befriedigt. Die Geschichte ist keine exacte Wissenschaft, und soll es auch nicht werden. Schon mehr ist dies der Fall mit Sprachwissenschaft.' HA 33.III.1 (1897).

99 See BW I: Huizinga–Brugmann (1898): 7; BW I: Huizinga–Brugmann (1899): 13.

> *archäologischer*] sentiment of historians. Linguistics should stand close to the art of poetry [*Dichtkunst*], but instead it stands miles away from it.[100]

Linguistics cannot advance as long as it does not take heed of language's poetic dimension, Huizinga argued. The confidence he displayed by sending the manuscript to its stated and celebrated adversary – Brugmann – might have been buttressed by Huizinga's awareness that he did not stand entirely alone in his poetic predisposition. Though Huizinga did not elaborate on the poetic methods he suggested, his tentative remarks in this direction allude to an author who had become terribly dear to him since he was a teenager: the German-British philologist F. Max Müller (1823–1900). Huizinga had studied Müller's *The Biography of Words* (1887) carefully,[101] and in this book's introduction, Huizinga encountered an emphatic argument against any 'archaeological' understanding of language:[102]

> Our words are not rough, unhewn stones, left at our door by a glacial moraine; they are blocks that have been brought to light by immense labour, that have been carved, shaped, measured and weighed again and again, before they became what we find them to be. Our poets make poems out of words, but every word, if carefully examined, will turn out to be itself a petrified [poem].[103]

When we open our mouths, history speaks not through trans-historical etymological lineages but through the historical dynamic of a given word's spiritual maturation; this was Müller's baseline conviction. By studying the historical roots of words (their 'biographies') in this latter sense, Müller imagined to have uncovered two general historical patterns. First, words have a natural tendency to extend beyond their original particular meaning and improperly stretch themselves towards the denotation of a general,

100 '[Solange die Sprachwissenschaft] sich dem dichterischen Gefühl – denn die Sprache ist doch jedenfalls eine dichterische Schöpfung –, ferne hält, kann ihre Befriedigung keine tiefere sein als die bloß archäologischen des Historikers. Die Sprachwissenschaft sollte der Dichtkunst am nächsten stehen, in der That steht sie meilenweit davon ab.' HA 33.III.2 (1897).

101 Huizinga mentioned the importance Müller had for him in his autobiographical essay, see VW I: *Mijn Weg tot de Historie* (1947): 16. In the secondary literature, this influence has been established, too; see especially. Krul, 'Het leven der woorden. Taalkunde en geschiedenis in Huizinga's vroegste wetenschappelijke werk'; Noordegraaf, '"On Light and Sound": Johan Huizinga and Nineteenth-Century Linguistics'.

102 HA 33.III.2 (1897).

103 F. M. Müller, *Biographies of Words and The Home of the Aryans* (London: Longmans, Green & Co., 1912), x.

abstract concept. Second, this process of natural abstraction is one of loss: as they 'petrify', words lose their meaning and function, becoming formulaic artefacts and platitudes. Müller made these observations in the context of the vivid late nineteenth-century anthropological debates on myth and religion formation and fitted these observations into an explicitly pathological narrative:

> Mythology, which was the bane of the ancient world, is in truth a disease of language. A myth means a word, but a word which, from being a name or an attribute, has been allowed to assume a more substantial existence.[104]

Language's natural tendency to transgress the boundaries of functionality and become itself an object of platitude was a 'disease' inherent to words, Müller argued. The history of human culture could be reconstructed according to the digressive pathological history of language. The images of language's tendency to become 'petrified', universal and hollow – and eventually in need of revival to regain potency – stuck with Huizinga.[105] Müller spoke of words becoming 'petrified poems', while Huizinga described them as 'petrified flowers' (*versteende bloemen*); Müller diagnosed this process as 'the disease of language', while Huizinga spoke of 'a disease of thought' (*ziekte der gedachte*).[106] In *Autumntide of the Middle Ages*, Huizinga later put these perspectives into practice when he described fifteenth-century cultural collapse as 'a spirit lethally exhausted by allegory and flamboyance', unable to speak and understand suffering meaningfully.[107] In fact, as a whole, the late Middle Ages could be approximated as 'the transition from symbolism to realism and allegory', a process wherein 'a passionate cry' became 'a grammatically correct sentence'.[108] This historical-semantic transition was one of 'decay' (*verval*), 'degeneration' (*ontaarding*), 'decadence'

104 This comes from a lecture Müller held in 1860, which was later published in F. M. Müller, *Lectures on the Science of Language* (London: Longmans, Green & Co., 1864), 11.
105 Huizinga later came to reject Müller's general historical anthropology for several reasons, mostly based on Edward Burnett Tylor's (1832–1917) criticism of Müller. Tylor's theory of animism had explicitly and famously opposed Müller: the abstraction from which myth and religion commenced was not a result of cultural development but rather its very start, and the collapse of mythological systems was one of 'disenchantment' rather than the culmination of disease. Despite his later sympathy for Tylor, Huizinga continued to use Müller's vocabulary discussed here. Krul, *Historicus tegen de tijd: Opstellen over leven en werk van J. Huizinga*, 136–37.
106 VW III: *Herfsttij der Middeleeuwen* (1919): 252.
107 VW III: *Herfsttij der Middeleeuwen* (1919): 389.
108 VW III: *Herfsttij der Middeleeuwen* (1919): 249.

(*decadentie*), and 'disease' (*ziekte*), with a cultural 'necropolis' awaiting at the other end.[109] Müller would have agreed.

Huizinga's pathological metaphors repeatedly informed his historical narratology and could not have been further removed from the cool and collected *junggrammatische* language trees: an encroaching disease looms from within any given culture's bowels, awaiting its historical maturation. These tropes and narratives reappeared time and again in Huizinga's later cultural-historical works to depict the 'tragic person' that is humanity.[110] In terms of Hayden White's historiographic vocabulary, good reason indeed exists to analytically define Huizinga's philological perspective on a culture's history trajectory as 'tragic.' The 'fall of the protagonist' (culture) is inevitable and threatening (an ever-looming disease of decadence). Yet amidst its sorrow, it offers 'the epiphany of the law governing human existence' (an inherently insatiable longing for meaning), and through this lesson, it offers its readers 'security and sanity in the world' (a sense of acceptance through self-awareness).[111] Huizinga's tragedy of the eternal defeat of culture's pursuit of meaningfulness was meant to offer consoling 'therapy' to the reader, a soothing recognition of the human condition.[112] In 1938 Huizinga again emphasized the importance of the tragic form to his work when he wrote that 'a history that resists its condensation into tragedy, has lost its form', the form of Huizinga's 'human'.[113]

To conclude, three points on Huizinga's academic sympathies should be stressed. First, Huizinga's philological training accommodated his resolute rejection of formal analytical approaches to cultural artefacts. Second, his particular antipathy towards 'mechanical' explanations as well as his sympathy towards an existentialist understanding of word meaning resurfaced in his later and mature works.[114] Lastly, drawing from Müller's *Biography*, Huizinga's method revolved around a tragic understanding of the relation between word and meaning: words do not and cannot designate meaning, but their continued failure attests to a continued human longing for the idea. Taken together, these observations tie into Huizinga's more general

109 VW III: *Herfsttij der Middeleeuwen* (1919): 67, 67, 132, 252, 261.

110 VW I: *or letterkunde* (1898): 146. See e.g. *Autumntide of the Middle Ages* (1919), *America in Life and Thought* (1926), *In the Shadow of Tomorrow* (1935), *A World Betrayed* (1945).

111 Here I use Hayden White's analytical understanding of tragedy, see White, *Metahistory*, 9.

112 VW VII: *In de schaduwen van morgen* (1935): 322. The therapeutic and consoling purpose of Huizinga's historiography is also central to Léon Hanssen's depiction of Huizinga, though Hanssen does not stress the tragic element so central to my own argumentation on this point. Hanssen, *Huizinga en de troost der geschiedenis: Verbeelding en rede*, 355–60.

113 VW VII: *Over vormverandering der geschiedenis* (1941): 198.

114 As shall be explored in Chapters 5 and 6, Huizinga himself would most likely have objected to the designation 'existentialist'.

and Christian appreciation of cognitive modesty, and on numerous occasions this approach informed Huizinga's engagement with certain debates. For instance, as is explored in Chapter 3, Huizinga's epistemic antipathy for 'historical laws' can be seen to have influenced his critique of and his alternative to Lamprecht's historicism; and Chapter 5 argues that the same antipathy influenced Huizinga's analysis of Spengler's historical narratives. Like his moral sympathies, Huizinga's academic sympathies were developed and mobilized in tandem with the experiences and fashions of his time.

At this point, a crucial observation regarding the relations between Huizinga's personal habits, generational ethics and academic convictions can be tentatively discerned. Huizinga's value system has been depicted as conservative, Christian-humanist and universalist (in that order), and on these levels, Huizinga's world consisted of a slowly yet continuously receding past, leaving in its vacuum a liberal-industrial, instrumental and material world. 'The past', in this sense, symbolized to Huizinga not just bygone times or a mere reservoir of moral examples. 'The past' symbolized to him – on a personal, ethical and intellectual level – the awareness that the world is void of historical continuities, neatly defined periods, conceptual orders and foreseeable futures. Just as loved ones pass, historical cities will crumble, art will be commodified and genuine symbols will turn to platitudes, unless one labours consistently and dutifully towards their conservation by, say, drawing the dying, depicting the perished and describing the ends, however inevitable the ultimate demise of the present may be. Throughout his life, I shall contend, Huizinga held on to this three-dimensional conservative conviction and attitude, and these are of primary importance in understanding how Huizinga's experiences of loss shaped his historical works on the past.

Method and material

Against the background of the above reflections on Huizinga's person, generation and education, the upcoming chapters investigate Huizinga's response to 'experiences of loss' from 1900 until 1940. For this purpose, five particularly potent and drastic 'losses' are adduced in turn. The upcoming five chapters investigate, respectively, the effects on Huizinga's historical perspective through the experiences of:

1. The loss of historical heritage through the urban modernization projects of the 1900s.
2. The loss of confidence in historical reproduction and restoration following the Great War.

3. The loss of silence and spirituality through mechanized labour in the 1920s.
4. The loss of an internationalist way of life due to nationalism in the 1910s and '30s.
5. The loss of a 'democratic culture' to Dutch National Socialism in the 1930s.

These losses, I will show, not only shook and renegotiated fundamental historical categories such as the 'Middle Ages', 'Renaissance', 'modernity' and even 'culture.' The experiences of these losses mediated and enabled time and again, and under different circumstances, a new way of *perceiving and narrating* history and, in turn, conditioned a new answer to a central question: why should any human in her right mind be concerned with the investigation of the past? As the 1910s, '20s and '30s unfolded, Huizinga, along with many of his peer historians, came up with different answers to this question. However, in order to investigate the role of his experiences of loss in these transitions, a methodological question, briefly mentioned above, needs to be answered here first: *how* does one study 'experiences of loss' historically? *Where* does one find these 'experiences'? Answers to these questions shall be given by examining, in turn, (1) the recent methodological interest in 'experiences' in the present historical discipline and (2) the material used for this research.

Method

In order to tackle the tremendous amount of literature on 'historical experiences', it is helpful to distinguish two commonly adduced (yet often only implicitly alluded to) kinds of such experiences: (1) the experiences of history of the historical actors in question, and (2) the experience of history of the historians investigating historical actors. To be sure, these two kinds of 'historical experiences' do not necessarily exclude one another, but an already complicated term becomes even more complex if one does not make this distinction, especially in projects such as this book, which executes a historical investigation of a historian. For this reason, I shall restate this project's research object as succinctly as possible: *this book investigates Johan Huizinga's 'experiences of loss' and examines the role they played in his historical writings*. This book investigates 'experiences' in the first sense of the term listed above.

But what does that mean? In what sense, if any, can experiences be retrieved from historical material? And, more importantly, considering the term's methodological complexity: why? Why should historians be interested in experiences? To start with this last question, I cite with deep sympathy from the paper that brought 'experience' back to 'history' in the 1990s:

> [E]xperience is so much a part of everyday language, so imbricated in our narratives that it seems futile to argue for its expulsion. [...] Given the ubiquity of the term, it seems to me more useful to work with it, to analyse its operations and to redefine its meaning.[115]

These words are Joan Scott's, and, like her, I believe that 'experience' should be important precisely because it already is. Of course, 'experiences' have the ability to shape and change perspectives; of course, 'experiences' have the ability to shock, frighten and upset the way people view the world, the past and the future. For these reasons, any historiographical definition of 'experience' must be as intuitive as the term is urgent. In our everyday language, it makes perfect sense to use 'experiences' in such a way, and given the vernacular attraction and intuitive authority of the term, its inclusion in historical research is an imperative, especially when investigating humans in times as disruptive as the interwar period.[116] Yes, the war shocked Huizinga; yes, the evaporation of international exchange and life did alienate him from friends and loved ones; and yes, the erosion of democracy in the 1930s was frightening to him. Huizinga experienced his times as times of loss – the loss of a world in which men of his age had been heavily invested and that now seemed to be pulled from under their feet. By investigating such 'experiences of loss', one can learn about the effects of, say, urbanization, female emancipation and nationalism on the (male) historians of that period – how the past appeared and disappeared in their present. The issue, I argue, is not whether historians should be interested in 'experiences' but, as Scott implored, how we can define the word in such a way that it becomes accessible to historians.[117]

115 Scott, 'The Evidence of Experience', 797. For helpful overviews of the term's historiographic career in the 1970s and '80s, see C. Ireland, *The Subaltern Appeal to Experience: Self-Identity, Late Modernity, and the Politics of Immediacy* (Montreal/London/Ithaca: McGill-Queen's University Press, 2004), 3–26; E. Domanska, 'Frank Ankersmit: From Narrative to Experience', *Rethinking History* 13, no. 2 (2009): 181; J. Grethlein, 'Experientiality and "Narrative Reference", with Thanks to Thucydides', *History and Theory* 49, no. 3 (2010): 316.
116 This is the main and central point of Frits Boterman's latest published study. F. Boterman, *Tussen utopie en crisis* (Amsterdam: De Arbeiderspers, 2021), 12.
117 At roughly the same time, yet from an altogether different tradition, essays by Reinhart Koselleck were being translated to English, arguing why and how 'historical experience is to be transformed into historical science'. R. Koselleck, '"Space of Experience" and "Horizon of Expectation": Two Historical Categories', in *Futures Past: On the Semantics of Historical Time*,

Scott's call was to investigate 'experience' in the first sense of the term listed above. She called for an investigation of the 'operations' of 'experiences' in the history under consideration.[118] Yet, in the three decades after her paper appeared, most of the ensuing literature has taken her invitation to go into the second direction, more commonly defining 'experience' analytically rather than empirically investigating it.[119] However, in the recent Anglophone literature, I believe one exception exists that is significant to the present research. This exception takes up a definition of 'historical experiences' that centres around an element of 'loss' so as to appreciate the consequences that 'the virtuosity of destruction displayed in such a brilliant and shattering way over the last one hundred years or so' have had on how history has been experienced in modern times.[120] This account is Dariusz Gafijczuk's, and its understanding of experience centres around one term in particular, 'ruins', meaning not 'piles of rubble, but [something] that is intimate to our modern constitution: a process of ruination'. About this process, Gafijczuk writes:

> [I] will be speaking about ruins as phenomena that trigger a collapse in various dimensions of experience. Accordingly, I take ruins in the strictest meaning of *ruina* (from Latin, meaning collapse, collapsing) – as material structures and *perceptive textures* that undermine the boundaries between proximity and distance, presence and absence, inside and outside, past and present[.][121]

Two words stand out: 'experience' and 'ruins'. 'Ruins' here instil a 'temporal hesitation' in our world, whereby the past is present *because* it is

ed. K. Tribe (New York: Columbia University Press, 2004), 275. For a study on the reception of Koselleck in the Anglophone world, see S. L. Hoffmann, 'Koselleck, Arendt, and the Anthropology of Historical Experience', *History and Theory* 49, no. 2 (2010): 212–36; S. L. Hoffmann, 'Koselleck in America', *New German Critique* 44, no. 3 (2017): 167–87.

118 Scott, 'The Evidence of Experience', 797.

119 For an overview of Scott's reception in this ensuing literature, see e.g. J. H. Zammito, 'Reading "Experience": The Debate in Intellectual History among Scott, Toews, and LaCapra', in *Reclaiming Identity: Realist Theory and the Predicament of Postmodernism*, ed. P. M. L. Moya and M. R. Hames-García (Berkeley/Los Angeles/London: University of California Press, 2000), 279–311; Jay, *Songs of Experience*, 216–60; H. Mah, 'The Predicament of Experience', *Modern Intellectual History* 5, no. 1 (2008): 97–119. In English, two journals in particular facilitated these and related discussions: *History and Theory* and *Rethinking History*. The former launched a special forum edition in 2015 titled *After Narrativism*, exploring how 'experience' could lead the way out from the structuralism cultivated by narratological approaches.

120 D. Gafijczuk, 'Dwelling Within: The Inhabited Ruins of History', *History and Theory* 52, no. 2 (2013): 150.

121 Gafijczuk, 'Dwelling Within: The Inhabited Ruins of History', 151.

conspicuously absent, and as such, ‘ruins’ trigger ‘experiences’ of history in our present. I find Gafijczuk’s distinction both intuitive and helpful to historians and to anyone trying to understand Huizinga’s perspective in particular. Huizinga’s texts regularly explicitly addressed events of loss. These features, to be sure, are textual observations: they concern terms, metaphors, paragraph structures and narratives used by Huizinga to describe loss and make present a process of disappearance. Yet, in Huizinga’s case, these textual structures often first figured in passages wherein he described loss in his own lived world – Amsterdam’s changing cityscape, his deceased wife, his concern over his detained friend Pirenne – and these phrasings often appeared later in his technical reflections. Whether these ‘ruins’ triggered or were mediated by textually perceptible experiences of loss does not matter for the present task. What matters is that Gafijczuk’s method presupposes that experiences of collapse can be related to processes of ‘ruination’ outside the text, and this presupposition is helpful in mapping Huizinga’s historical writings (i.e. his textual ‘experiences of loss’) onto phenomena of loss in his world (the ‘processes of ruination’). For now, I adopt an intentionally loose operational understanding of experience, inspired by but not taken from Gafijczuk: I speak of textual manifestations of experiences of loss when a text *(a) narrates disappearance and (b) when its invocation can be related directly to external events of loss – ruination – in the author’s lived world.*

Against this background, all upcoming chapters follow the same principle to investigate the role played by Huizinga’s ‘experiences of loss’ in his academic output. Each chapter starts by presenting a ‘ruin’ and ‘experience’ – that is, the five events listed above – in the Gafijczukian sense. Each of these events figures as a moment of ‘ruination’ and can be directly related to Huizinga’s textually retrievable experiences in the aforementioned sense of the term. The sections of each chapter then present against this background the historical research Huizinga carried out during or soon after the experience(s) of loss in question and examine on a textual level whether, how and where these experiences can be seen to have reappeared in Huizinga’s more technical historical writings. Now, Huizinga was a prolific writer, and his books were frequently quite scattered – so where does one go looking for these ‘textual manifestations’ called ‘experiences’?

This book’s second central methodological principle now becomes explicit. To argue generally that Huizinga’s writings were affected by certain experiences is one issue, but to precisely distinguish the most powerful examples in his academic output is an entirely different point. For this latter relation – the relation between Huizinga’s ‘lived ruins’ and his cultural-historical writings

– this book privileges one particular dimension in Huizinga's writings: his most pronounced, lively and acknowledged disputes. At different stages of his career, Huizinga engaged (repeatedly) with various acknowledged antagonists, and the present investigation sets out to trace the effects of said experiences within the scope of these antagonisms. More specifically, this book examines the role played by Huizinga's experiences of loss in his disputes with other historians, investigating how certain experiences of loss informed Huizinga's position, tone and passion in the academic debates with which he engaged.

To be sure, alternatives to this strategy exist. For example, one could study how Huizinga's books or certain correspondence were informed by these experiences. However, it was on the level of antagonism and disagreement that the role of such experiences in Huizinga's historical outlook grew most pronounced. Huizinga used these antagonisms to navigate and find meaning in the experiences that upset him most, and in these academic collisions, his interpretative schemas (and the developments therein) often crystalized more clearly than in other parts of his written output. Huizinga's disagreements were often directly infused by a particular and experienced loss and ensuing fear and antipathy. This relation between experience and antagonism is, however, not a point I can (or wish to) argue for by theoretical means here; rather, it is the task of the upcoming chapters to *show* that this was the case. I can only *show*, for example, how Huizinga's confidence in the 'objective' retrieval of historical facts (vis-à-vis Karl Lamprecht) eroded as the Great War grew ever more destructive, or how his initial reservation towards democracy (vis-à-vis Carl Schmitt) withered as the threats of National Socialism to his world grew ever more pronounced. Ultimately, the epistemic privilege this project bestows on dispute and antagonism shall rest on an empirical argument.

Finally, I wish to make one last and crucial methodological point. By articulating the relation between experience and Huizinga's stated beliefs in the manner suggested above, I by no means intend to reduce the relation to a single causal arrow. Amsterdam's changing cityscape did not determine Huizinga's response to Jacob Burckhardt's *Renaissancebegriff*; the experience of the destruction of Ypres did not dictate Huizinga's position concerning Lamprecht's *kulturhistorische Methode*. The relations I present are neither reductionist nor causal. Rather, this book sets out to show how certain experiences mediated and returned in Huizinga's technical writings. This book shows that certain experiences influenced Huizinga's position and understanding of and attitude to 'the past' without arguing that these experiences themselves were sufficient conditions for these positions, understandings and attitudes. And still, these experiences today give us an impression of what kinds of concerns informed and made urgent the historical consciousness of interwar Europe.

In sum, each of the upcoming chapters examines, within the context of Huizinga's generational ethics and professional training, a given experience and its relation to Huizinga's position in a respective academic debate at the time. For this purpose, the two most central analytical terms are: (1) 'experience of loss' and (2) 'antagonism'. For now, what remains is a closer look at the material wherein these experiences and antagonisms are identified and investigated.

Material

In its investigation of Huizinga's 'experiences of loss' and his academic disputes, this research turns primarily to four different kinds of material: Huizinga's personal correspondence, his (lecture) notes, his lectures (both those meant for publication and those not) and his published works. His published lectures and works were collected by Leendert Brummel, Willem Rudolf Juynboll and Th. J. G. Locher and were published by Tjeenk Willink in nine volumes between 1948 and 1953.[122] His letters were collected by Léon Hanssen, Wessel Krul and Anton van der Lem, and a selection was published by Veen & Tjeenk Willink in three volumes that appeared between 1989 and 1991.[123] Lastly, Huizinga's lecture notes are to be found in the Huizinga Archives at Leiden University Library, and they were digitized by the library and published on their website in 2019.[124] The archive was inventoried by Anton van der Lem.[125] In addition, this research draws from a variety of materials from Huizinga's own times: newspapers, photographs, architectural drawings, sketches, paintings, literature, poems and buildings. Most newspapers have been mined from *Delpher*, the online Dutch national newspaper database; the photographs have typically been drawn from Amsterdam's city archive.[126] Some photographs have been used from Eindhoven's and Rotterdam's respective city archives.[127]

122 J. Huizinga, *Verzamelde Werken I–IX*, ed. L. Brummel, W. R. Juynboll, and Th. J. G. Locher (Haarlem: H. D. Tjeenk Willink & Zoon N.V., 1948–53).

123 Hanssen, Van der Lem, and Krul, *Johan Huizinga: Briefwisseling I–III.*

124 digitalcollections.universiteitleiden.nl/view/collection/Huizinga and Huizinga-online.nl

125 Van der Lem, *Inventaris van het archief van Johan Huizinga.*

126 Respectively, delpher.nl and archief.amsterdam/beeldbank

127 Respectively, www.rhc-eindhoven.nl and https://stadsarchief.rotterdam.nl/

Figure 1.8. **Huizinga commonly wrote his notes on strips of paper, usually on the back of paper that had already been written on, either by him or someone else. Next, he grouped and organized these strips in envelopes with particular designations. Sometimes the envelopes concerned certain concepts; at other instances, they concerned, say, other authors or particular periods. Above are some notes Huizinga used for Autumntide on medieval beggars. To the left (A) are his notes; on the right (B) is the back of the paper he used. The Huizinga archives are organized according to Huizinga's original logic.**

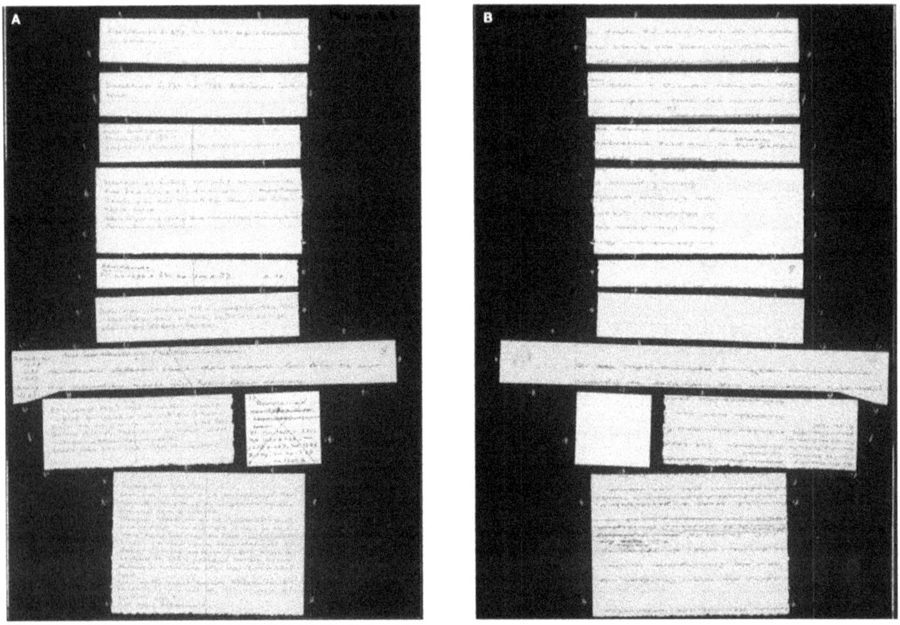

As discussed above, this book's method takes seriously Huizinga's acknowledged antagonists. Some of these were his contemporaries, but others were not. Regardless, I use Huizinga's antagonists as primary material, but mostly in terms of how this material would have appeared to Huizinga and in his milieu. His antagonists are thus read as historical material mostly (but not only) in the context of their Dutch reception rather than in the context of their inception. For this reason, material from, say, Burckhardt's Basel, Lamprecht's Leipzig and Schmitt's Prussia figure only where necessary to an understanding of Huizinga's response to them.

Lastly, a note on the translation work in this project is in order. Most translations – both of primary and secondary material and, in the former case, both of Huizinga's works and that of his interlocutors – are my own, unless stated otherwise in the footnotes. Two reasons underpin this decision. First, for the sake of readability, I have tried to not include more voices than necessary in this project. Second, many works by Huizinga available in English have been translated from the German translations prepared by Werner Kaegi (1901–1979). Only when translations from the original Dutch are available do I cite them alongside my own translations.

Structure

As was stated above, this book's main body follows a rather strict blueprint. Each chapter adduces an 'experience' and an 'antagonism' and investigates the relation between them. To begin, the next chapter examines Huizinga's critique of Jacob Burckhardt's *Renaissancebegriff* in *Die Kultur der Renaissance in Italien* (1860) against the background of Huizinga's experiences of Amsterdam's rapidly changing cityscape between 1903 and 1905; the third chapter does the same for Huizinga's critique of Karl Lamprecht's historical *Methode* against the background of Huizinga's experiences of the Great War. A fourth chapter takes up Huizinga's experiences of 'mechanization' and juxtaposes them with his technical critiques of Alexis de Tocqueville's version of *l'Amérique*. The fifth chapter distinguishes Huizinga's generational hopes, expectations and experiences of internationalist lifestyle and how the Great War suddenly challenged its former self-evidence; it uses this as a background for an investigation of Huizinga's critiques of Oswald Spengler's *Der Untergang des Abendlandes* (1918). The sixth chapter, finally, examines Huizinga's rejection of Carl Schmitt's *homo homini lupus* in the context of an increasingly fragile democratic culture in the 1930s.

Bibliography

Aerts, R. 'Op zoek naar een Nederlands fin de siècle'. *De Gids*, no. 156 (1993): 91–101.

Anchor, R. 'History and Play: Johan Huizinga and His Critics'. *History and Theory* 17, no. 1 (1978): 63–93.

Ankersmit, F. R. *Sublime Historical Experience*. Palo Alto: Stanford University Press, 2005.

Arnade, P., M. Howell, and A. van der Lem, eds. *Rereading Huizinga*. Amsterdam: Amsterdam University Press, 2019.

Berlage, H. P., W. M. Dudok, J. Gratama, A. R. Hulshoff, H. van der Kloot Meiburg, J. F. Staal, and J. Luthmann. 'Het doel van deze uitgave'. In *Moderne bouwkunst in Nederland*, 5–7. Rotterdam: W.L. & J. Brusse's Uitgeversmaatschappij, 1932.

Bosch, M. *Strijd! De vrouwenkiesrechtbeweging in Nederland, 1882–1922*. Hilversum: Verloren, 2019.

Boterman, F. *Tussen utopie en crisis: Nederland in het interbellum 1918–1940*. Amsterdam: De Arbeiderspers, 2021.

Brugmann, K. *Grundriss der Vergleichenden Grammatik der indogermanischen Sprachen*. Strasbourg: Karl J. Trübner, 1886.

Burke, P. *What Is Cultural History?* Cambridge: Polity Press, 2004.

Colie, R. L. 'Johan Huizinga and the Task of Cultural History'. *The American Historical Review* 69, no. 3 (1964): 607–30.

Daniel, U. *Kompendium Kulturgeschichte*. Frankfurt: Suhrkamp Verlag, 2001.

Daston, L. 'The History of Science and the History of Knowledge'. *KNOW* 1, no. 1 (2017): 131–54.

Doesberg, T., and P. Mondriaan. 'Manifesto I of De Stijl (1922)'. In *100 Artists' Manifestos: From the Futurists to the Stuckists*, edited by A. Danchev, 215. London: Penguin Books, 2011.

Domanska, E. 'Frank Ankersmit: From Narrative to Experience'. *Rethinking History* 13, no. 2 (2009): 175–95.

Dorléac, L. B. 'L'histoire de l'art ou l'éloge de la faiblesse'. In *L'histoire culturelle en France et en Espagne*, edited by B. Pellistrandi and J. F. Sirinelli, 77–86. Madrid: Casa de Velázquez, 2008.

Dorsman, L. 'Periodisering als integrale benadering: Nederlandse historici in het Fin-de-Siècle'. *Theoretische geschiedenis* 16, no. 3 (1989): 277–96.

Dunk, H. W. von der. 'Huizinga als kultuurpessimist'. *Groniek*, no. 26 (1973): 150–54.

Dupré, S., and G. Somsen. 'The History of Knowledge and the Future of Knowledge Societies'. *Berichte zur Wissenschaftsgeschichte* 42, no. 2–3 (2019): 186–99.

Edwards, J. 'Play and Democracy: Huizinga and the Limits of Agonism'. *Political Theory* 41, no. 1 (2013): 90–115.

Febvre, L. 'Un moment avec Huizinga'. *Annales. Histoire, Sciences Sociales* 6, no. 4 (1951): 493–96.

Gafijczuk, D. 'Dwelling Within: The Inhabited Ruins of History'. *History and Theory* 52, no. 2 (2013): 149–70.

Geyl, P. 'Huizinga als aanklager van zijn tijd'. *Mededelingen der Koninklijke Nederlandse Akademie van Wetenschappen, Afdeling Letterkunde* 24, no. 4 (1961): 137–77.

Geyl, P. 'Huizinga as Accuser of His Age'. *History and Theory* 2, no. 3 (1963): 231–62.

Gombrich, E. *In Search of Cultural History*. Oxford: Oxford University Press, 1969.

Gombrich, E. 'Huizinga's *Homo ludens*'. In *Johan Huizinga 1872–1972*, edited by W. R. H. Koops, E. H. Kossmann, and G. van der Plaat, 133–54. The Hague: Martinus Nijhoff, 1973.

Grethlein, J. 'Experientiality and "Narrative Reference", with Thanks to Thucydides'. *History and Theory* 49, no. 3 (2010): 315–35.

Hanssen, L. *Huizinga en de troost van de geschiedenis: Verbeelding en rede*. Amsterdam: Uitgeverij Balans, 1996.

Hoffmann, S. L. 'Koselleck in America'. *New German Critique* 44, no. 3 (2017): 167–87.

Hoffmann, S. L. 'Koselleck, Arendt, and the Anthropology of Historical Experience'. *History and Theory* 49, no. 2 (2010): 212–36.

Hollander, J. Den, H. Paul, and R. Peters. 'Introduction: The Metaphor of Historical Distance'. *History and Theory* 50, no. 4 (2011): 1–10.

Hübinger, G. 'Konzepte und Typen der Kulturgeschichte'. In *Geschichtsdiskurs Band 4: Krisenbewußtsein, Katastrophenerfahrungen und Innovationen 1880–1945*, edited by W. Küttler, J. Rüsen, and E. Schulin. Frankfurt: Fischer Taschenbuch Verlag, 1997.

Huizinga, J. *Erasmus of Rotterdam*. Edited by F. Hopman. London: Phaidon Press, 1952.

Huizinga, L. *Herinneringen aan mijn vader*. The Hague: H. P. Leopold's Uitgeversmij, 1963.

Ireland, C. *The Subaltern Appeal to Experience: Self-Identity, Late Modernity, and the Politics of Immediacy*. Montreal/London/Ithaca: McGill-Queen's University Press, 2004.

Jay, M. *Songs of Experience*. Berkeley/Los Angeles: University of California Press, 2005.

Jolles, A. *Einfache Formen*. Berlin: De Gruyter, 2006.

Kaegi, W. *Das historische Werk Johan Huizingas*. Leiden: Universitaire Pers Leiden, 1947.

Kantorowicz, E. 'Review of Johan Huizinga, "Geschichte und Kultur"'. *American Historical Review* 60, no. 4 (1955): 853–55.

Kelley, D. R. *Fortunes of History: Historical Inquiry from Herder to Huizinga*. New Haven/London: Yale University Press, 2003.

Kloos, W. *Nieuwere literatuur-geschiedenis. Deel 2*, of 5 vols. Amsterdam: L. J. Veen, 1925.

Koops, W. R. H., E. H. Kossmann, and G. van der Plaat, eds. *Johan Huizinga 1872–1972*. The Hague: Martinus Nijhoff, 1973.

Koselleck, R. '"Space of Experience" and "Horizon of Expectation": Two Historical Categories'. In *Futures Past: On the Semantics of Historical Time*, edited by K. Tribe, 255–75. New York: Columbia University Press, 2004.

Kossmann, E. H. 'Postscript'. In *Johan Huizinga 1872–1972*, edited by W. R. H. Koops, E. H. Kossmann, and G. van der Plaat, 223–34. The Hague: Martinus Nijhoff, 1973.

Kossmann, E. H. 'Romeins "Breukvlak" en de Nederlandse geschiedenis'. *BMGN – Low Countries Historical Review* 106, no. 4 (1991): 652–58.

Krul, W. E. 'Ter Braak contra Huizinga. Over de grenzen van de ironie'. *Groniek*, no. 100 (1988): 61–79.

Krul, W. E. 'Het leven der woorden. Taalkunde en geschiedenis in Huizinga's vroegste wetenschappelijke werk'. *BMGN – Low Countries Historical Review* 104, no. 3 (1989): 365.

Krul, W. E. *Historicus tegen de tijd: Opstellen over leven en werk van J. Huizinga*. Groningen: Historische Uitgeverij Groningen, 1990.

Krul, W. E. 'Nederland in het *fin-de-siècle*. De stijl van een beschaving'. *BMGN – Low Countries Historical Review* 106, no. 4 (1991): 581–94.

Krul, W. E. 'Huizinga in zijn brieven: Beeld en zelfbeeld'. *Tijdschrift voor Geschiedenis* 106, no. 1 (1993): 23–37.

Krul, W. E. 'Het *fin-de-siècle* als vertoning'. *BMGN – Low Countries Historical Review* 117, no. 4 (2002): 519–25.

Krumm, C. *Johan Huizinga, Deutschland und die Deutschen*. Münster/New York/Munich/Berlin: Waxmann, 2011.

Lecuppre-Desjardin, É., ed. *L'odeur du sang et des roses: Relire Johan Huizinga aujourd'hui*. Villeneuve d'Ascq: Presses universitaires du Septentrion, 2019.

Lem, A. van der. *Johan Huizinga: Leven en werk in beelden en documenten*. Amsterdam: Wereldbibliotheek, 1993.

Lem, A. van der. *Het eeuwige verbeeld in een afgehaald bed: Huizinga en de Nederlandse beschaving*. Amsterdam: Wereldbibliotheek, 1997.

Lem, A. van der. *Inventaris van het archief Johan Huizinga*. Edited by A. Th. Bouwman and J. J. Witkam. Leiden: Leiden University Library, 1998.

Locher, Th. J. G. 'Johan Huizinga'. *Jaarboek van de Maatschappij der Nederlandse Letterkunde* (1946): 88–109.

Mah, H. 'The Predicament of Experience'. *Modern Intellectual History* 5, no. 1 (2008): 97–119.

Meertens, P. J. 'Nederlandse filologie'. In *Geestelijk Nederland 1920–1940 II*, edited by K. F. Proost and J. Romein, 1–22. Amsterdam/Antwerp: Kosmos, 1948.

Müller, M. *Lectures on the Science of Language*. London: Longmans, Green & Co., 1864.

Müller, M. *Biographies of Words and The Home of the Aryans*. London: Longmans, Green & Co., 1912.

Noordegraaf, J. '"On Light and Sound": Johan Huizinga and Nineteenth-Century Linguistics'. In *The Dutch Pendulum. Linguistics in the Netherlands 1740–1900*, 130–58. Münster: Nodus Publikationen, 1996.

Noordegraaf, J. 'Uit het verleden van een historicus: De taalkundige ambities van de jonge Huizinga'. *Voortgang*, no. 13 (2011): 197–204.

Osthoff, H., and K. Brugmann. *Morphologische Untersuchungen auf dem Gebiete der indogermanischen Sprachen*. Cambridge: Cambridge University Press, 2013.

Östling, J., D. Larsson Heidenblad, E. Sandmo, A. Nilsson Hammar, and K. Nordberg. 'The History of Knowledge and the Circulation of Knowledge: An Introduction'. In *Circulation of Knowledge: Explorations in the History of Knowledge*, edited by J. Östling, E. Sandmo, D. Larsson Heidenblad, A. Nilsson Hammar, and K. Nordberg, 9–33. Lund: Nordic Academic Press, 2018.

Otterspeer, W. *Orde en trouw: Over Johan Huizinga*. Amsterdam: De Bezige Bij, 2006.

Otterspeer, W. *Reading Huizinga*. Edited by B. Jackson. Amsterdam: Amsterdam University Press, 2010.

Otterspeer, W. *De kleine Huizinga*. Amsterdam/Antwerp: Atlas Contact, 2019.

Pree, C. du. *Johan Huizinga en de bezeten wereld*. Leusden: ISVW Uitgevers, 2016.

Roland Holst, H. *Tusschen twee werelden*. Rotterdam: W. L. & J. Brusse, 1923.

Romein, J. *Geschiedenis*. Edited by K. F. Proost and J. Romein. *Geestelijk Nederland 1920–1940 II*. Amsterdam: Kosmos, 1948.

Romein, J. 'De geschiedschrijving in Nederland tijdens het interbellum'. In *Carillon der tijden: Studies en toespraken op cultuurhistorisch terrein*. Amsterdam: Querido, 1953.

Romein, J. *Op het breukvlak van twee eeuwen: De westerse wereld rond 1900*. Amsterdam: Querido, 1976.

Rooy, P. de. 'Een hevig gewarrel. Humanitair idealisme en socialisme in Nederland rond de eeuwwisseling'. *BMGN – Low Countries Historical Review* 106, no. 4 (1991): 625.

Rossum, V. van. 'De stad gebouwd: De oude binnenstad vernieuwd'. In *Amsterdam in de tweede Gouden Eeuw*, 36–58. Bussum: Thoth, 2000.

Santing, C. 'Het liefkozen van schone vormen. Johan Huizinga en het vrouwenvraagstuk'. *De Gids* 168, no. 2 (2005): 121–30.

Sas, N. C. F. van. 'Fin-de-siècle als nieuw begin. Nationalisme in Nederland rond 1900'. *BMGN – Low Countries Historical Review* 106, no. 4 (1991): 595–609.

Scott, J. W. 'The Evidence of Experience'. *Critical Inquiry* 17, no. 4 (1991): 773–97.

Spaanstra-Polak, B. 'De atmosfeer van het fin-de-siècle'. In *Het symbolisme in de Nederlandse schilderkunst 1890–1900*. Bussum: Uitgeverij Thoth, 2004.

Strupp, C. *Johan Huizinga: Geschichtswissenschaft als Kulturgeschichte*. Göttingen: Vandenhoeck & Ruprecht, 2000.

Thys, W. 'Huizinga en de beweging van negentig'. In *Johan Huizinga 1872 – 1972*, edited by W. R. H. Koops, E. H. Kossmann, and G. Van der Plaat, 29–52. The Hague: Martinus Nijhoff, 1973.

Vanwesenbeeck, B. 'Huizinga, Theorist of Lateness?' In *Rereading Huizinga: Autumn of the Middle Ages, a Century Later*, edited by P. Arnade, M. Howell, and A. van der Lem, 245–58. Amsterdam: Amsterdam University Press, 2019.

Veen, A. van. *Amsterdam 1900: Foto's van Olie, Breitner, Eilers en tijdgenoten*. Bussum: Thoth, 2016.

Velde, H. te. 'Fin de siècle in de Nederlandse politiek'. *Leidschrift* 1, no. 14 (1998): 45–55.

Verburgt, L. 'The History of Knowledge and the Future History of Ignorance'. *KNOW* 4, no. 1 (2020): 1–24.

Vree, F. van. 'De stad van het betere leven. Cultuur en samenleving in Nederland rond 1900'. *BMGN – Low Countries Historical Review* 106, no. 4 (1991): 641–51.

Vries, J. de, ed. *Nederland 1913: Een reconstructie van het culturele leven*. Haarlem: Meulenhoff/Landshoff, 1989.

Vugt, G. de. 'De droomtekeningen van Johan Huizinga'. *De Witte Raaf*, 209 (January–February 2021). dewitteraaf.be/artikel/de-droomtekeningen-van-johan-huizinga/.

White, H. *Metahistory*. Baltimore: John Hopkins University Press, 2014.

Woud, A. van der. *Het landschap en de mensen: Nederland 1850–1940*. Amsterdam: Prometheus, 2020.

Zammito, J. H. 'Reading "Experience": The Debate in Intellectual History among Scott, Toews, and LaCapra'. In *Reclaiming Identity: Realist Theory and the Predicament of Postmodernism*, edited by P. M. L. Moya and M. R. Hames-García, 279–311. Berkeley/Los Angeles/London: University of California Press, 2000.

Zweig, S. *Die Welt von Gestern: Erinnerungen eines Europäers*. Frankfurt: Fischer Taschenbuch Verlag, 2019.

2 'The Tyranny of the Present'[1]

Abstract

This chapter examines how Huizinga's experiences of loss following the urban modernization projects in Amsterdam around 1900 conditioned his critiques of Jacob Burckhardt (1818–1897) and Burckhardt's *Renaissancebegriff* in particular.

Keywords: Johan Huizinga; modern cityscapes; Jacob Burckhardt; Renaissance; *Autumntide of the Middle Ages*

Upon obtaining his doctoral degree in philology at the University of Groningen in 1897, Johan Huizinga moved to the city of Haarlem to teach history at a secondary school.[2] His relocation brought about many changes in his life, whether intellectually, socially, professionally and geographically. For one, Huizinga started visiting Amsterdam frequently, and there he experienced first-hand just how extensive and intrusive urban modernization had become. In his native Groningen, industrialization had been predominantly an agricultural development that had taken place outside of the city's borders on designated industrial sites; only later would the city be affected by industrialization within its own perimeters.[3] In Amsterdam, however, architects, engineers and philanthropists such as Samuel Sarphati (1813–1866), Jacobus G. van Niftrik (1833–1910), Hendrik P. Berlage (1856–1934)

1 This chapter includes a partially revised version of my article published by Taylor & Francis in the journal *History of European Ideas* on 27-11-2020, available online: https://doi.org/10.1080/01916599.2020.1842229. T. Rydin, 'Huizinga's "Heimwee": Responding to Burckhardt's "Die Kultur der Renaissance in Italien" in Times of Loss', *History of European Ideas* (2020): 732–47.

2 Formally, Huizinga's PhD was in 'Dutch philology' (*Nederlandsche letterkunde*). Petrus Blok advised Huizinga to obtain more teaching experience in the Dutch secondary school system before applying for an academic position. This was not uncommon at the time in the Netherlands. C. du Pree, *Johan Huizinga en de bezeten wereld* (Leusden: ISVW Uitgevers, 2016), 44; A. van der Lem, *Johan Huizinga: Leven en werk in beelden en documenten* (Amsterdam: Wereldbibliotheek, 1993), 55–56.

3 P. Kooy, 'Groningen en de industrialisatie van Nederland', *Groniek*, no. 64 (1979): 13–22.

Rydin, T., *The Works and Times of Johan Huizinga (1872–1945): Writing History in the Age of Collapse.* Taylor & Francis Group 2023
DOI: 10.5117/9789463724593_CH02

and Floor Wibaut (1859–1936) had been mapping out a city for the future for decades – and they had it built. From the 1900s onwards, Huizinga found that the industrial metropolitan developments had affected not only the sounds, shapes, speeds and smells of the world but also the very experience of the past. In a letter from November 1913, Huizinga wrote to his friend and author Willem G. C. Bijvanck (1848–1925) about the changing cityscapes in the Netherlands:

> [I] have always felt a certain nostalgia [*heimwee*] for the pre-Sarphatian Amsterdam and the Haarlem of 1860. How strange: whilst we are still busy condemning these times for their soberness and vandalism, they are already becoming romantic to us in the true and deep sense of the word.[4]

In Huizinga's eyes, the popular legitimacy of the 'creative' destruction of historical sites in Dutch and European cities around 1900 rested at least in part on an opportunistic misrepresentation of a particular period in European history: the Renaissance.[5] Some of the most prominent Dutch modernists, such as Berlage, had mobilized a Renaissance architecture and aesthetic to convey modern ideals and views of secularism, individualism, scientific reasoning and industrial ethics. This was no coincidence. Especially since the publication in 1860 of Jacob Burckhardt's *The Civilization of the Renaissance in Italy* (*Die Kultur der Renaissance in Italien*), the Renaissance had come to symbolize the human ability to create a world unrestrained by theological dogma, social conventions and historical authority – a world that

4 '[Naar] het "praesarphatisch" Amsterdam en het Haarlem van 1860 heb ik altijd met een zeker heimwee kunnen verlangen. Wat is het toch vreemd: terwijl wij nog bezig zijn, dien tijd om al zijn nuchterheid en vandalisme te verfoeien, begint hij toch tegelijkertijd voor ons al romantisch te worden in den waren en diepen zin des woords.' BW I: Huizinga–Bijvanck (1913): 107. See p. 11 for further details on the references to Huizinga's collected works and letters as well as to the Huizinga archives. Samuel Sarphati was a recognized Dutch physician. He had become particularly well known for his philanthropic efforts to improve the general public's hygiene in Amsterdam.

5 In 1913 observations such as Huizinga's, including his *Heimweh*, were not particularly uncommon. A certain hesitation and suspicion towards urban modernization, and an ensuing historical romanticism and *Heimweh*, was shared even among authors and artists more progressive than Huizinga. M. Wagenaar, 'De stad ontworpen: Stadsontwerp tussen wens en werkelijkheid', in *Amsterdam in de tweede Gouden Eeuw*, ed. M. Bakker et al. (Bussum: Thoth, 2000), 9–35. Willem Otterspeer traces this part of Huizinga's conservatism back to a certain nostalgia that had been present in the educated elite of Huizinga's Groningen during his childhood. W. Otterspeer, *Orde en trouw: Over Johan Huizinga* (Amsterdam: De Bezige Bij, 2006), 20. To me, however, it seems unlikely that the liberal household in which Huizinga grew up would have partaken in this conservatism.

actually had 'abolished' the past. As such, when Huizinga started arguing in 1907 that the Renaissance had in fact been a 'recreation' of medieval culture rather than a display of a furious anti-traditional 'creation', he was addressing not just historical accounts such as Burckhardt's, with which he was well acquainted.[6] Huizinga was also addressing his expressed *Heimweh* and what appeared to him as the tremendous carelessness with which his time dispensed with images of the past and the authority of tradition. In his eyes, the past and the creativity it enabled were being 'lost' in more than just one sense when cities were being modernized.

The argument Huizinga had started drafting in 1907 later grew into the central contention of his *Autumntide of the Middle Ages* (*Herfsttij der Middeleeuwen*), which first appeared in 1919.[7] Huizinga's conception of the late Middle Ages rested on an appreciation of the period's supposed ability to be creative not despite but because of tradition – its ability to be a *Re-naissance* in the literal sense of the term. In a letter from that year to the Belgian historian Henri Pirenne (1862–1935), Huizinga admitted that *Autumntide* had been first and foremost an argument against Jacob Burckhardt's canonical *Renaissancebegriff*. Still, the antagonism Huizinga perceived between his *Autumntide* and Burckhardt's *Civilization* is obscure: *Autumntide* dealt with Burgundian and French culture between 1300 and 1500; Burckhardt's *Civilization* considered northern Italian culture from 1350 to 1450.[8] So if not along temporal-spatial dimensions, how then was his historical image supposed to be a refutation of Burckhardt's analysis?

6 In his autobiographical essay *My Path to History* (1947), he writes that he had conceived of *Autumntide*'s main argument 'between 1906 and 1909, probably in 1907'. VW I: *Mijn weg tot de historie* (1947): 39.

7 For two helpful studies on how Huizinga went about finding and organizing his material for *Autumntide*, see G. Small, 'The Making of "The Autumn of the Middle Ages" I: Narrative Sources and Their Treatment in Huizinga's "Herfsttij"', in *Rereading Huizinga: Autumn of the Middle Ages, a Century Later*, ed. P. Arnade, M. Howell, and A. van der Lem (Amsterdam: Amsterdam University Press, 2019), 169–210; A. van der Lem, 'The Making of "The Autumn of the Middle Ages" II: The Eagle and His Pigeonholes: How Huizinga Organized His Sources', in *Rereading Huizinga: Autumn of the Middle Ages, a Century Later*, ed. P. Arnade, M. Howell, and A. van der Lem (Amsterdam: Amsterdam University Press, 2019), 211–26. All Dutch citations from *Autumntide of the Middle Ages* (*Herfsttij der Middeleeuwen*) have been taken from the fifth edition from 1941 unless stated otherwise. The differences from the first edition from 1919 are, however, either non-existent or for present purposes negligible. For the sake of maintaining an accurate timeline, I therefore continue to mention the year 1919 in the references to this work.

8 In his lifetime, Huizinga was often mentioned in terms of his similarity to Burckhardt. The National Socialist historian Christoph Steding (1903–1938), for example, introduced Huizinga as 'den man den Burckhardt der Niederlande nennt'. C. Steding, *Das Reich und die Krankheit der europäischen Kultur* (Hamburg: Hanseatische Verlagsanstalt Hamburg, 1942), 34.

This chapter argues that the criticism launched by Huizinga in *Autumntide* (1919) against Burckhardt's *Civilization* (1860) was mediated by Huizinga's experience of modernization in general, and in particular by the changing cityscapes and the *Heimweh* this transition inspired. This chapter's argument breaks down into two parts. First, it argues that Huizinga's rejection of Burckhardt's conception of the Renaissance drew from a moral anthropological disagreement.[9] Huizinga disagreed not with Burckhardt's particular reading of this or that historical example but objected more generally to the conceivability of Burckhardt's *uomo singolare*. Where Burckhardt had reconstructed the Renaissance around its ability to 'create' (*schöpfen*), Huizinga chose to depict the period in terms of its ability to 're-create' (*her-scheppen*). Secondly, this chapter argues that Huizinga's moral anthropological perspective in *Autumntide* was steeped in a certain experience of loss following, amongst other things, the urban modernization projects and the transition towards a modernist aesthetic. Huizinga's emphasis on the role and importance of 're-creation' in late medieval creativity was interwoven with the importance he attributed to historical re-creation in the cities of his own time. His interest in such diachronic narratives of re-creation and re-vitalization fitted into the *fin-de-siècle* literature and philosophy with which he had grown up, but the interest was not a bookish issue alone; it was lived and experienced.[10]

By making these two arguments – one concerning the moral anthropological claim of *Autumntide* and the other regarding this claim's relation to Huizinga's experiences of loss following urban modernization projects – the present chapter intervenes in two debates. First, I argue with Tollebeek, Guggisberg and Kaminsky, and against Gombrich and Kaegi, that Huizinga's historical analysis of late medieval culture is incommensurable with Burckhardt's Renaissance.[11] Unlike Tollebeek and Guggisberg, however, I argue

9 Strupp has argued that the book's purpose should be understood in terms of a methodological and a historical claim. By stressing the book's moral anthropological dimension, I look to add to Strupp's dissection of *Autumntide*. C. Strupp, *Johan Huizinga: Geschichtswissenschaft als Kulturgeschichte* (Göttingen: Vandenhoeck & Ruprecht, 2000), 141–42.

10 For this reason, Huizinga has recently been described as 'theorist of lateness'. Vanwesenbeeck, 'Huizinga, Theorist of Lateness?', 248. See also A. van der Lem, *Johan Huizinga: Leven en werk in beelden en documenten* (Amsterdam: Wereldbibliotheek, 1993), 34–37.

11 On this point, I side with J. Tollebeek, '"Renaissance" and "Fossilization": Michelet, Burckhardt, and Huizinga', *Renaissance Studies* 15, no. 3 (2001): 354–66; H. Kaminsky, 'From Lateness to Waning to Crisis: The Burden of the Later Middle Ages', *Journal of Early Modern History* 4, no. 1 (2000): 85–125. Conversely, on this point I side against E. Gombrich, 'Huizinga's *Homo ludens*', in *Johan Huizinga 1872–1972*, ed. W. R. H. Koops, E. H. Kossmann, and G. van der Plaat (The Hague: Martinus Nijhoff, 1973), 133–54; W. Kaegi, *Das historische Werk Johan Huizingas* (Leiden:

that this incommensurability cannot be properly conceptualized without considering the anthropological claims at stake. Secondly, by arguing that Huizinga's anthropological critique of Burckhardt was steeped in the (Dutch) *fin-de-siècle* culture of the 1900s, I buttress insights by Kaminsky and Krul more firmly in historical material. Yes, Huizinga's Middle Ages indeed followed 'a projection of late-nineteenth-century conceptions of culture into the age of Burgundy', but how exactly did this projection work, and how were the projected concepts related to Huizinga's lived world?[12] Where did their urgency come from? This chapter shows how the anthropological logic of Huizinga's critique of Burckhardt's *Renaissancebegriff* was steeped in *fin-de-siècle* experiences of disappearance and loss.[13] The modernization of cityscapes helped mediate new conceptions of 'past', 'present' and 'creativity.'

For these purposes, this chapter's main body has been divided into three parts. The first part examines Huizinga's experiences of loss due to urban modernization in the 1900s and 1910s and zooms in on the nostalgia Huizinga consequently reported. The second part sidesteps into Burckhardt's conception of the Renaissance, especially along those lines most relevant to an understanding of Huizinga's later criticism. The third part explores Huizinga's critique of Burckhardt along its moral anthropological dimensions and reads these dimensions against the background of the said modernization of Dutch and European cityscapes. Finally, a concluding section ties together these observations: Huizinga's late Middle Ages had known a *fin-de-siècle* of its own. Or perhaps, rather: in Huizinga's book, the *fin-de-siècle* had its own Middle Ages.

A modern city and its ruins[14]

That Huizinga should have taken an interest in the cultural importance of historical re-creation was not a coincidence. Between 1903 and 1905,

Leiden University Press, 1947). To be sure, especially Kaegi's position is nuanced and articulate. My present disagreement with Kaegi concerns his analysis leading to the idea that 'dort, wo Burckhardt aufhörte, hat Huizinga begonnen.' Kaegi, *Das historische Werk Johan Huizinga*, 14.

12 W. Krul, 'In the Mirror of Van Eyck: Johan Huizinga's "Autumn of the Middle Ages"', *Journal of Medieval and Early Modern Studies* 27, no. 3 (1997): 373; Kaminsky, 'From Lateness to Waning to Crisis: The Burden of the Later Middle Ages'.

13 Such as was, for instance, expressed through Huizinga's description of his period's supposed 'uprooting'. 'ontaarding', VW III: *Herfsttij der Middeleeuwen* (1919): 388.

14 This section includes a partially revised version of Rydin, 'Huizinga's "Heimwee": Responding to Burckhardt's "Die Kultur der Renaissance in Italien" in Times of Loss', 737–741. See footnote 1 of this chapter.

Huizinga lectured as a *privaat-docent* at the University of Amsterdam, and during this period, he witnessed the accelerated modernization of the Dutch urban landscape.[15] Huizinga mourned not only the loss of medieval canals and early modern architecture; he also regretted the seeming indifference with which heritage had been destroyed since at least the 1870s.[16] The city's concentric canals could not meet the demands of modern labour division and industrial logistics, which required straight streets and wide avenues, and new districts had to be built to relieve and expand the vastly overpopulated working-class neighbourhoods. As a result, and on a tremendous scale, crooked streets were straightened, building blocks replaced and canals drained. New areas were drafted in a modern aesthetic. The city's face was changing at an industrial speed, and these developments more in particular informed the 'nostalgia' Huizinga wrote about in his aforementioned letter to Bijvanck from 1913. Consider the following passage from an article on Amsterdam's 'rage of destruction', published in 1903 in the liberal newspaper *De Courant/Het Nieuws van den Dag*:

> In the city's heart, around the traffic vein of the Dam [Amsterdam's central square], one sees on a daily basis large trucks pass by filled with rubble and broken plaster; and on our beautiful canals, boats loaded with beams, frames and stones – the remnants of numerous demolished plots, are brought to places where the debris is sold. [...] Sturdy workers swing their sledgehammers so as to raze away once and for all that which no longer suffices, that which is old and deficient, sickly and decrepit.[17]

15 Later, Huizinga would also address in writing the urban modernization of Leiden, Delft and Haarlem. Huizinga's experience of and response to changing urban landscapes is discussed in A. van der Lem, *Verbreek nooit een nude rooilijn! Huizinga over stadhuis en stadsschoon in Leiden* (Leiden: Antiquariaat Klikspaan, 2021).

16 A recently published piece on Huizinga's reconstruction of historical urban cultures leaves his personal experiences of urban culture and urbanization strangely unmentioned. J. Dumolyn and É. Lecuppre-Desjardin, 'Huizinga's Silence', in *Rereading Huizinga: Autumn of the Middle Ages, a Century Later*, ed. P. Arnade, M. Howell, and A. van der Lem (Amsterdam: Amsterdam University Press, 2019), 65–84. For an overview of the industrial and urban transitions at this time in the Netherlands, see for example J. C. Kennedy, 'Vooruitgang en crisis, 1870–1949', in *Een beknopte geschiedenis van Nederland* (Amsterdam: Bert Bakker, 2017).

17 'In het hartje van de stad, rondom de verkeersader den Dam, ziet men dagelijks groote wagens met puin en verbrokkeld pleister en in onze fraaie stadsgrachten schuiten, zwaar beladen met balken, binten en steenen, de overblijfselen van tal van gesloopte perceelen, welke naar plaatsen afgevoerd worden, waar "afbraak" geveild wordt. [...] Het moet toch wel de aandacht van velen trekken, dat in verschillende drukke wijken van de stad de handen uit de mouwen worden gestoken, allerwegen gehamerd, gemetseld en gegraven wordt en men stoere arbeiders den moker ziet zwaaien, om voor goed weg te maaien wat niet meer deugde, wat oud en gebrekkig,

Old, defect, sickly and decrepit: in the eyes of the editors of *De Courant*, 'sledgehammers' had to 'once and for all' pave the way for a 'self-aware spirit of entrepreneurship'.[18] In a later article from 1916 titled *Destruction in Times of Peace*, the Dutch painter, art critic and public intellectual Jan Veth (1864–1925) commented on the effect of these 'ruthless sledgehammers'.[19] The destruction by modern urban planners of transgenerational points of aesthetic reference, Veth argued in his article, compromised the bedrock of human creativity. A 'tyranny of the present', he argued, cannot bring forth art, because it will destroy its fruits the moment they have been created; after all, 'the present will be history tomorrow.'[20] In this sense, art and architecture continued to need a dialogue with, or at least a recognition of the past, in order to 'create' (*scheppen*).[21] This historicism, Veth insisted, drew neither from mere antiquarianism nor ignorance of the 'instability of all things' but from a realization that creativity – in the past, present and future – relies on an independence from contingent fashion and temporary whims.[22] The connection Veth made between the ongoing modernization of Dutch urban space and the conditions of cultural creativity resonated with Huizinga, arguably one of Veth's most avid readers.

Huizinga had corresponded with Veth since at least 1895, when they both became involved in the newly founded social democratic weekly magazine *De Kroniek*.[23] Huizinga embraced Veth's suspicion of the ongoing urban modernization in the Netherlands. In 1915, a day after having heard Veth's presentation of an early draft of *Destruction in Times of Peace*, Huizinga wrote to Veth that the architectural transformations of Amsterdam had given him stomach aches.[24] 'You and I', Huizinga proceeded to write in this letter, share the same 'emotion of beauty.' The similarity of their conservative conception of cultural creativity – that is, of a creativity relying on a continued dialogue with the past – became apparent in *Autumntide*, which

ziekelijk en afgeleefd was.' 'Verbouwingen in Amsterdam', *Het Nieuws van den Dag*, 22 June, 1903, 14.

18 'zelfbewuste ondernemingsgeest'. 'Verbouwingen in Amsterdam', 14.

19 'meedogenloozen moker', J. Veth, 'Vredes-verwoestingen', *De Gids* 80, no. 1 (1916): 512.

20 'ook het heden zal morgen weer gisteren zijn en zoo min als dat heden de toekomst mag tyranniseeren, mag het rauwelijks wat achter ons ligt verdonkeremanen.' Veth, 'Vredes-verwoestingen', 524.

21 Veth, 'Vredes-verwoestingen', 524.

22 'onbestendigheid aller dingen'. Veth, 'Vredes-verwoestingen', 523.

23 Huizinga and Veth continued to correspond until Veth's death in 1925, and in 1927 a eulogical biography of Veth by Huizinga was published. VW VI: *Leven en werk van Jan Veth* (1927): 339–480.

24 BW I: Huizinga–Veth (1915): 181.

Huizinga had been working on since at least 1907 and which appeared a couple of years after this letter was written.[25] The opening passage of the book would be devoted to the virtues of medieval architecture and urban planning, and in the preceding foreword to the first edition, Huizinga reflected more generally on the need to appreciate the fertile soil of historical cultures:

> It is usually the origin of the new that our mind seeks in the past. [...] Yet in searching for the new life that was emerging, one easily forgets that, in history as in nature, dying and being born perpetually keep pace with one another. Old forms of civilization die off at the same time and in the same soil in which the new finds the nourishment to blossom.[26]

Historians, Huizinga argued, had tended to privilege the singular beginning of 'the new' and to forget that 'the new' can only grow on and from the soil of that which preceded it. On this seasonal soil, the new appropriates the material that is already present at its inception. In other words, there is no singular, original and independent raw creation; there is only recreation. By placing such reflections at the very outset of his book on the late Middle Ages, Huizinga addressed at least two different points at once: (1) on both a methodological and anthropological level, he addressed Burckhardt's understanding of the Renaissance as a unique, ahistorical and singular *Schöpfung* supposedly independent of the preceding medieval times, and (2) he continued Veth's arguments on the destruction of medieval urban planning and early modern architecture both in Amsterdam and beyond.[27] Regarding

25 Regarding the dating of Huizinga's book project, see e.g. BW I: Huizinga–Colenbrander (1906): 81; VW I: *Mijn weg tot de historie* (1947): 39.

26 J. Huizinga, *Autumntide of the Middle Ages*, trans. D. Webb, e.d. G. Small, and A. van der Lem (Leiden: Leiden University Press, 2020), 3. 'Het is meestal de oorsprong van het nieuwe, wat onze geest in het verleden zoekt. [...] Doch bij het zoeken naar het nieuwe leven, dat opkwam, vergat men licht, dat in de geschiedenis als in de natuur het sterven en het geboren worden eeuwig gelijke tred houden. Oude beschavingsvormen sterven af terzelfdertijd en op dezelfde bodem, waarin het nieuwe voedsel vindt om op te bloeien.' VW III: *Herfsttij der Middeleeuwen* (1919): 3.

27 In this period, Huizinga's political and aesthetic conservatism was entangled with the greatest personal loss of his life to that date: on 21 July 1914, his wife Mary V. Huizinga-Schorer died of cancer. In his personal correspondence of the time, Huizinga wrote of an 'inexpressible sadness', which appears to have loomed over him for decades rather than years. BW I: Johan Huizinga–Jakob Huizinga (brother) (1914): 165. The role of his grief over Mary's death in *Autumntide* is discussed in the next chapter.

both points, Huizinga drew the conclusion from his aforementioned predisposition towards a conservative aesthetics that human creativity had been (his historiographical claim) and must continue to be (his ethical claim) involved with the ideals of the past. The former point shall be expanded on in the upcoming sections; for now, it is the second point to which I turn.

The implicit role of Huizinga's experiences of modernity in his medieval historiography was embedded in a pronounced feature of Dutch historical discourse at the time. In the Netherlands, the historical discipline had grown, particularly since 1830, from an elaborate system of historical societies, journals and museums, each trying to explore and make sense of the many archives that had been made available to the public since the Batavian Revolution and the Napoleonic occupation that ensued.[28] From the 1840s onwards, several publications on Dutch national history by Dutch historians such as Guillaume Groen van Prinsterer (1801–1876), Reinier Bakhuizen van den Brink (1810–1865), Robert Fruin (1823–1899) and Petrus Blok (1855–1929) had contributed to the politicization of especially medieval and seventeenth-century Dutch history.[29] By emphasizing, downplaying or altogether neglecting either of these periods, historians could and did express sympathy for a variety of political and ethical positions regarding monarchism, republicanism, liberalism and the role of religious pluralism in Dutch identity. Huizinga was well aware of the ethical-political potential of historical research. To him and his peers, historical works could accommodate contemporary concerns alongside empirical diligence, impartiality and precision.

Huizinga's book on the 'autumn' and 'crisis' of the late Middle Ages was conceived and composed as the inner city's medieval anatomy and seventeenth-century architectural heritage receded from the urban landscape's character. The synchronization of these two 'crises' by Huizinga

28 The networks, practices and interests from which the Dutch historical discipline grew have been studied with admirable detail in J. Tollebeek, *De toga van Fruin. Denken over geschiedenis in Nederland sinds 1860* (Amsterdam: Wereldbibliotheek, 1996); P. Huistra, *Bouwmeesters, zedenmeesters: Geschiedbeoefening in Nederland tussen 1830 en 1870* (Nijmegen: Uitgeverij Vantilt, 2019). Huistra's book in particular is truly foundational and indispensable for any comprehensive understanding of Dutch nineteenth-century historical culture.

29 See e.g. Groen van Prinsterer's *Kort overzigt van de geschiedenis des vaderlands* (1841) and *Handboek der geschiedenis van het vaderland* (1846), Fruin's *Het voorspel van den Tachtigjarigen Oorlog* (1859) and Blok's *Geschiedenis van het Nederlandsche volk* (1892–1908). Each of these works addressed more or less explicitly the common political state of affairs through 'impartial' (*onpartijdig*) historical inquiry.

Figure 2.1. The canal along the Nieuwezijds Voorburgwal in Amsterdam had been dug in the fifteenth century and was drained in 1884 to accommodate traffic and the transportation of goods. As a consequence, the figure of Atlas, located on the roof of the royal palace, overlooked not (A) silent water but (B) bustling traffic.

took place on at least three levels:[30] (1) both the late Middle Ages and the early twentieth century experienced the destructive forces of 'decadence' and 'mechanization';[31] (2) the 'mechanization' and 'optimization' of the modern urban spaces brought about the gradual destruction of architectural references to the past; and in effect, (3) the medieval past was understood both in *Autumntide* and in early twentieth-century urban planning precisely in terms of its non-functional nature and value. In order to appreciate and conceptualize the relation between (1) these three levels of historiographic interconnection between modernity and the Middle Ages and (2) Huizinga's *Heimweh* for a past reduced to debris,[32] consider the following passage

30 Huizinga himself used the word 'crisis' to describe both his own and the late medieval period.

31 Huizinga's language, antipathies and medieval romanticism were typical of *fin-de-siècle* literary fashions in the Netherlands at the time. Van der Lem, *Johan Huizinga: Leven en werk in beelden en documenten*, 82–83.

32 BW I: Huizinga–Bijvanck (1913): 139.

Figure 2.2. (A) The draining of canals opened up the possibility of implementing new technologies underneath the city's skin. Here a sewage system was installed on the Nieuwezijds Voorburgwal in 1884. (B) Berlage and his peers introduced modern, straight streets to Amsterdam. This image shows the Hoofdweg in 1920.

from Dariusz Gafijczuk on 'the burden of the past' and 'ruins' as a form of 'historical awareness':[33]

> [Inhabited] ruins, weighed down by the 'burden of history' and the trauma of destruction, invent a past that 'lives as nervously and unpredictably as the present into which it protrudes its face'. Inhabited ruins, as forms of unique modern presence based on distance ontology, act as *frontier*

33 Gafijczuk, 'Dwelling Within: The Inhabited Ruins of History', 150. In this citation, Gafijczuk cites, respectively, Quignard's *The Roving Shadows* and Heidegger's *Building, Dwelling, Thinking*.

Figure 2.3. The modern world of commerce and technology was steeped in a Renaissance aesthetic. Berlage had been commissioned to build a new stock exchange in the 1885. The construction work started in 1898, and the building was revealed to the public in 1903.

formations, where something begins its presencing, not a boundary at which something ceases. This frontier is marked by a present that has partially collapsed under its own weight.[34]

Huizinga did not merely respond to the external object that was the waning of an architectural heritage. Rather, the experience of its loss – its transformation into debris and its subsequent transportation to dumping grounds – mediated and conditioned Huizinga's historical depiction of fifteenth-century Burgundian culture. In Gafijczuk's vocabulary: Huizinga's 'inhabited ruins', a Netherlands in an industrial aesthetic transition, conditioned the dramatic form wherein the Middle Ages were made urgent, were made present precisely through their disappearance. Under these conditions, the late Middle Ages appeared to Huizinga at the beginning of the twentieth

34 Gafijczuk, 'Dwelling Within: The Inhabited Ruins of History', 164.

century in terms of their failed resistance to the instrumental, industrial ethics of modernity. Huizinga's late Middle Ages (both in 1403 and 1903, so to speak) took shape in contrast to the malleable and instrumental world of industrial 'labour'.[35]

When Huizinga wrote of his *heimwee* for the pre-modern city, he did not write of a diachronic wish to access an inaccessible past; he expressed a synchronic sympathy for a frame of mind sensitive to non-instrumental considerations. Huizinga's *heimwee* and his depiction of the Middle Ages were not answers established independently of and in response to the question of how to feel at home in a modern world. The *heimwee* did not stand outside of the ruins it addressed. If anything, Huizinga's depiction of late medieval culture itself responded to the ruins or, in Gafijczuk's vocabulary, the 'inhabited ruins' of his world: Huizinga's appreciation and his account of late medieval culture were the dialectic counterparts of the modernization he witnessed, whose ability to transform revealed at once the need and responsibility to preserve and to re-create. In this capacity, Huizinga's inquiry into the Middle Ages and its call to remain sensitive to 'the eternally balanced pace' between past and future was always in antagonistic dialogue with 'the tyranny of the present' being experienced.[36] Huizinga's Middle Ages began in 1903 as a modern experience of loss.

When Jan Veth died in 1925, Huizinga took on the task of writing his biography. The book appeared in 1927 and dealt with both Veth's personal and academic life. Tellingly, however, Huizinga devoted significant attention to what Veth had taught his contemporaries about historical heritage and the need for sustained symbolic and aesthetic codes for cultural output to subsist and withstand mere whimsical fashion.[37] In this biography, Huizinga celebrated Veth as his cultural 'signpost', a 'guide' for authors in the modern world.[38] The continued value of Veth for modern times was clear to Huizinga:

35 A review of Huizinga's *Autumntide* from 1926 commented on this feature: 'One continually perceives an idealistic spirit trying to make its way through the contemporary climate of ideas, which appears to be ruled exclusively by political and economic considerations and values.' F. W. N. Hugenholtz, 'The Fame of a Masterwork', in *Johan Huizinga 1872–1972*, ed. W. R. H. Koops, E. H. Kossmann, and G. van der Plaat (The Hague: Martinus Nijhoff, 1972), 94.

36 My preferred translation of 'eeuwig gelijken tred'. VW III: *Herfsttij der Middeleeuwen* (1919): 3. Diane Webb's translation is 'perpetually keep pace with one another'. Huizinga, *Autumntide of the Middle Ages*, 3.

37 Whilst writing this book, Huizinga continued to argue for the preservation and restoration of historical landmarks in the Netherlands. Van der Lem, *Verbreek nooit een oude rooilijn! Huizinga over stadhuis en stadsschoon in Leiden*.

38 VW VI: *Jan Veth* (1924): 482.

> I remember exactly why, amidst my admiration for the passions of other [artists], Veth offered me the consolation and support I needed. He confirmed what my heart told me: that impressionism was not the only way towards renewal, and that it was not necessary to abandon the old in order to love the new.[39]

'It is not necessary to abandon the old in order to love the new': it is telling that precisely this conviction is highlighted by Huizinga, and I believe it illustrates the conceptions of historical temporality in the Dutch *fin-de-siècle*, a culture in which Huizinga had matured and been educated. The changes of the 1880s and 1890s had not resulted in the same experience of rupture in the Netherlands as elsewhere, and so the end of a century allowed for a greater sense of a persevering historicity. Sentiments of decay, decadence and historical dissolution were not alien to Dutch *fin-de-siècle* authors, but they were not alloyed with and further reinforced by, say, the French experience of a disastrous defeat in the Franco-Prussian War (1870–1871), the divisiveness of the German *Kulturkampf* or the Austro-Hungarian political disintegration.[40] The Dutch *fin-de-siècle* was paradoxically optimistic in its pessimism. Aerts has called the Dutch *fin de siècle* a time of 'critical optimism'; N. C. F. van Sas has labelled it 'happy'. Huizinga rejected conceptions of historical linear continuity, but he gravitated strongly towards the conviction that images of the past needed to be reckoned with in our cultural efforts.[41] The turn of a century in the Dutch context did not foster a cleavage from the previous century but rather a historical reworking of the past for contemporary needs.[42]

To conclude, Huizinga, like Veth, fell into neither a romantic idealization of the past nor a pessimistic rejection of change. Huizinga privileged, first and foremost, the need to engage with and not reject the past as one seeks avenues for aesthetic renewal. This conviction manifested itself first in Huizinga's response to the urban modernization projects across the Netherlands

39 'Ik weet nog nauwkeurig, waarom, bij al mijn bewondering voor de losse drift dier anderen, Veth mij een troost en een steun gaf, die ik behoefde. Hij bevestigde mij, wat het hart mij zei, dat het impressionisme niet de eenige weg was ter vernieuwing, en dat het niet noodzakelijk was, het oude af te zweren, om het nieuwe lief te hebben.' VW VI: *Artikel in NRC, 'Jan Veth'* (1924): 482.

40 W. E. Krul, 'Nederland in het *fin-de-siècle*. De stijl van een beschaving', *BMGN – Low Countries Historical Review* 106, no. 4 (1991): 581.

41 Contrast this sentence with the upcoming chapter, wherein Huizinga's experience of the Great War is discussed.

42 This attitude in Huizinga's writings would re-appear time and again in the 1920s and '30s. See images 29–32 and their captions in Van der Lem, *Johan Huizinga: Leven en werk in beelden en documenten*.

in the 1900s; this conviction figured in tandem with a stated experience of loss and nostalgia for a waning past. Huizinga feared that an outright rejection of historical architecture and aesthetic traditionalism would give way to a destructive and raging instrumentalism and an ensuing sense of homelessness, both in personal and urban life. Only through recreative play, Huizinga held, could human culture and the material world belong to each other, and for this purpose, one needs to hold on to the rules that are handed down by those images called history. This conviction emerged not only through his attitude towards modern metropolitanism but also through his appreciation of medieval culture in times hostile to medieval aesthetics. Huizinga's image of medieval culture reflected his own ethical identity: his medium of creativity was not instrumentalism but rule-abiding play. Burckhardt's image of Renaissance Italian culture could hardly have been more different.

Burckhardt's *uomo singolare*[43]

Huizinga first read Burckhardt's *The Civilization of the Renaissance in Italy* in 1893.[44] At this point, Burckhardt was still alive and widely read by historians throughout Europe.[45] No evidence exists of Huizinga's first impressions of Burckhardt's book, but from the archival material it appears that Huizinga started teaching and re-reading *Civilization* in a systematic and critical fashion in the early 1900s after having been appointed Professor in Dutch and General History at the University of Groningen in 1905.[46] That same year, Burckhardt's *Weltgeschichtliche Betrachtungen* was posthumously published, which Huizinga had read by 1919 at the latest. In Huizinga's ensuing understanding of Burckhardt, there had actually been two different 'Burckhardts': one had written *Civilization*, and the other had lectured on *Weltgeschichtliche Betrachtungen*. Huizinga preferred the latter to the former.[47] For the purpose of this chapter, however, Huizinga's relation

43 This section includes a partially revised version of Rydin, 'Huizinga's "Heimwee": Responding to Burckhardt's "Die Kultur der Renaissance in Italien" in Times of Loss', 734–35. See footnote 1 of this chapter.

44 W. E. Krul, *Historicus tegen de tijd: Opstellen over leven en werk van J. Huizinga* (Groningen: Historische Uitgeverij Groningen, 1990), 120. From now on, this book is referred to as *Civilization*.

45 R. Chickering, *Karl Lamprecht: A German Academic Life (1856–1915)* (Atlantic Highlands, NJ: Humanities Press, 1993), 89.

46 HA: 110.2.8; 110.2.32.

47 BW I: Huizinga–Pirenne (1919): 269.

to *Civilization* takes centre stage.[48] To understand (1) Huizinga's critical attitude to *Civilization* and (2) the role Huizinga's experience of loss played in it, a careful look at the inception and main argument of Burckhardt's book is in order.

Upon its publication in 1860, Burckhardt's *Civilization* had entered into an intricate and volatile set of debates in the post-revolutionary German Confederation, Switzerland and beyond.[49] By the 1860s, Renaissance, Reformation and medieval studies had become arenas for debates on large political themes such as secularism, nationalism, democracy and even German cultural identity as a whole. Philosophers and historians alike could express sympathy or antipathy for, say, emerging nationalist narratives or state centralization by arguing for or against the interrelation between the Renaissance and the Reformation, the existence of a northern Renaissance or the dependence of Renaissance art on late Middle Age culture.[50] In *The Renaissance in Historical Thought* (1948), which to this day remains a central point of reference in studies of Burckhardt, Wallace K. Ferguson argued that the success of Burckhardt's *Civilization* rested in significant part on its ability to tie together and address all major political themes of his day into a single book on the Italian Renaissance.[51] On the canvas of Renaissance culture, Burckhardt projected the origins of modernity – 'the mother of our own [civilization]'[52] – as well as an alternative trajectory.[53]

In order to set the stage for Huizinga's positive appreciation of Burckhardt's depiction of the Renaissance, what matters is not Burckhardt's political alliances per se; what matters more is Burckhardt's opposition in *Civilization* to Hegelian and Rankean conceptions of the Renaissance as a transitional stage within a larger, continuous and progressive historical space.[54] In his *Lectures on the History of Philosophy* (*Vorlesungen über die*

48 Huizinga's relation to the second Burckhardt – the Burckhardt of *Weltgeschichtliche Betrachtungen* – shall be discussed in the next chapter regarding Huizinga's critiques of Karl Lamprecht's notion of *Gotik*.

49 M. A. Ruehl, *The Italian Renaissance in the German Historical Imagination, 1860–1930* (Cambridge: Cambridge University Press, 2015), 61–70.

50 Ruehl, *The Italian Renaissance in the German Historical Imagination*, 69–70.

51 W. K. Ferguson, *The Renaissance in Historical Thought* (New York/Chicago/Dallas/Atlanta/San Francisco: The Riverside Press Cambridge, 1948), 195–252.

52 'Mutter der unsrigen'. J. Burckhardt, *The Civilization of the Renaissance in Italy* (Oxford: Phaidon Press, 1981), 1; J. Burckhardt, *Die Kultur der Renaissance in Italien* (Vienna: Phaidon-Verlag, 1934), 1.

53 Ruehl, *The Italian Renaissance in the German Historical Imagination, 1860–1930*, 66.

54 See for Hegel's understanding of the Reformation as (1) a result of the Middle Ages and (2) as the source of modern humanity's self-recognition.

Geschichte der Philosophie), first published in 1832, Hegel described the Renaissance period within a larger 'awakening' (*Wiederauflebung*) of 'the selfhood of spirit [*Geist*]'. In *History of the Popes, their Church and State* (*Die römischen Päpste in den letzten vier Jahrhunderten*), first published in 1834–1836, Ranke described the period as the 'bending [of] refractory spirits [*Geistern*] [to] the pure laws of Christian truth'.[55] Moreover, both Hegel and Ranke found that the Renaissance had gone hand in hand with (1) the Reformation, (2) the emergence of the modern state and (3) the revival of classicism. Though Burckhardt, too, explored the Renaissance in terms of *Geist*, his understanding of the term was fundamentally different, and Burckhardt came to challenge both Ranke and Hegel on most if not all major points:[56] he argued that the Renaissance concerned not synchronic relations but a diachronic cross-section, and that the period was marked not by a continuously progressing state formation but by everyday, mundane practices.

Its criticism of state-centred historiography notwithstanding, Burckhardt's account of the Italian Renaissance commenced from the political balances between the northern Italian city-states 1350–1550, especially Florence and Venice. The capricious political circumstances among and within these city-states in the fourteenth and fifteenth centuries were both the cause and result of a new interest in 'calculation' (*Berechnung*).[57] Leaders were not given but 'possible' (*möglich*), events were not catalogued but 'calculated' (*berechnet*), a given situation was not fixed but 'uncertain' (*unsicher*). Action, in short, was not taxonomical but statistical, and its calculation was not principled but instrumental.[58] This new political uncertainty was performative of a new experience of unpredictability and possibility of history and future, and it amounted to the removal of the medieval 'veil [of] faith, illusion, and childish prepossession'.[59] Uncertainty and transformation, in yet other words, had become the new fabric of the state:

55 G. W. F. Hegel, *Lectures on the History of Philosophy*, ed. E. S. Haldane and F. H. Simson (London: Kegan Paul, Trench, Trübner & Co., 1895), 108; L. Ranke, *The History of the Popes During the Last Four Centuries* (London: G. Bell and Sons, 1913), 26.

56 In *In Search of Cultural History* (Oxford: Oxford University Press, 1969), Ernst Gombrich (1909–2001) argued that Burckhardt's interest in 'reconstructing' the spirit of an age drew from Hegelian philosophy of history, but this seems wrong to me. Gombrich, *In Search of Cultural History*, 14–25.

57 Burckhardt, *The Civilization of the Renaissance in Italy*, 47; Burckhardt, *Die Kultur der Renaissance in Italien*, 44.

58 Burckhardt, *The Civilization of the Renaissance in Italy*, 2, 2, 79; Burckhardt, *Die Kultur der Renaissance in Italien*, 2, 2, 75.

59 Burckhardt, *The Civilization of the Renaissance in Italy*, 81.

> The wondrous Florentine spirit, at once keenly critical and artistically creative, was incessantly transforming [*umgestaltet unaufhörlich*] the social and political condition of the State, and as incessantly describing and judging the change. Florence thus became the home of political doctrines and theories, of experiments and sudden changes, but also, like Venice, the home of statistical [science].[60]

The political uncertainty fed into an altogether different kind of politics whose business it was to 'incessantly transform' the conditions (*Zustand*) of the state according to whimsical circumstance. The state became 'a calculated and conscious outcome' rather than a given authority; it became itself a 'work of art' (*als Kunstwerk*) rather than its mere commissioner.[61] The instrumental conception of politics belonged to the defining features of the Renaissance, according to Burckhardt, and on numerous occasions in *Civilization*, he emphasized the unprecedentedness of this new 'living thing' (*ein neues Lebendiges*) in history:[62] it embodied nothing less than 'a new position' (*auf einem neuen Boden*), 'a wholly different foundation' (*die Basis ist eine andere*) from which personhood, agency, creativity and the entire natural world were experienced anew and 'essentially differently from the Middle Ages' (*von der des Mittelalters wesentlich verschieden*).[63] But how, then, did conceptions of the state alter the very borders of human individuality? How did political circumstance translate into the emergence of a 'spiritual individual' (*geistiger individuum*)?[64] Consider the following passage:

> Despotism [fostered] in the highest degree the individuality not only of the tyrant or Condottiere himself, but also of the men whom he protected

60 Burckhardt, *The Civilization of the Renaissance in Italy*, 48. The original reads: 'Der wunderbare florentinische Geist, scharf räsonierend und künstlerisch schaffend zugleich, gestaltet den politischen und sozialen Zustand unaufhörlich um und beschreibt und richtet ihn ebenso unaufhörlich. So wurde Florenz die Heimat der politischen Doktrinen und Theorien, der Experimente und Sprünge, aber auch mit Venedig die Heimat der Statistik und allein und vor allen Staaten der Welt die Heimat der geschichtlichen Darstellung im neuern Sinne.' Burckhardt, *Die Kultur der Renaissance in Italien*, 45.

61 '[Der] Staat als berechnete, bewußte Schöpfung, als Kunstwerk'. Burckhardt, *Die Kultur der Renaissance in Italien*, 2.

62 Burckhardt, *Die Kultur der Renaissance in Italien*, 2. The Phaidon Press edition has translated this as 'a new fact'; I suggest 'a new living thing' instead.

63 Burckhardt, *The Civilization of the Renaissance in Italy*, 5, 217, 279; Burckhardt, *Die Kultur der Renaissance in Italien*, 4, 205, 261.

64 Burckhardt, *The Civilization of the Renaissance in Italy*, 81; Burckhardt, *Die Kultur der Renaissance in Italien*, 76. Huizinga discusses this term coined by Burckhardt in VW IV: *Het Probleem der Renaissance*: 245 (1926).

> or used as his tools – the secretary, minister, poet and companion. These people were forced to know all the inward resources of their own nature [*seine innern Hilfsquellen*], temporary or permanent; and their enjoyment of life was enhanced and concentrated by the desire to obtain the greatest satisfaction from a possibly very brief period of power and influence.[65]

The instrumental logic of the modern city-state affected both despots and subordinates, as both found themselves answering to new and perpetually changing conditions, and ultimately it fed into a new experience of personality and individuality. The decisive feature of one's life was no longer the general group to which one belonged – *Rasse, Volk, Partei, Korporation, Familie*[66] – but one's ability to respond to and manipulate changing circumstances. Medieval interest in the preservation of traditional social taxonomies had been replaced by images of incessant transformation, in whose storm only images of 'immortality' (*Verewigung*) and 'greatness' (*historische Größe*) offered direction.[67] Burckhardt called this new individuality *uomo singolare* or *uomo unico*.[68] This new human was prepared to dispense at its convenience with any tradition, promise or alliance in its pursuit of survival, whether through worldly subsistence, immortal legacy or both. This transformation of the political world affected not only the images of moral guidance and the conceptions and practices of individuality; it also affected, according to Burckhardt, the temporal framework formerly offered by Christian doctrine. That is, it was truly and fully invested in the *saeculum* (world time). This temporal reorganization suggested by Burckhardt later became of utmost importance to Huizinga's critique of him:

> While the men of the Middle Ages look on the world as a vale of tears, which Pope and Emperor are set to guard against [*hüten müssen*] the coming of the Antichrist; while the fatalists of the Renaissance oscillate between seasons [*zwischen Zeiten*] of overflowing energy and seasons of superstitions or of stupid resignation, here, in this circle of chosen spirits, the doctrine is upheld that the visible world was created by God in love,

65 Burckhardt, *The Civilization of the Renaissance in Italy*, 82; Burckhardt, *Die Kultur der Renaissance in Italien*, 77.

66 Burckhardt, *The Civilization of the Renaissance in Italy*, 81; Burckhardt, *Die Kultur der Renaissance in Italien*, 76.

67 Burckhardt, *The Civilization of the Renaissance in Italy*, 160, 262; Burckhardt, *Die Kultur der Renaissance in Italien*, 151, 245.

68 Burckhardt, *Die Kultur der Renaissance in Italien*, 336. These terms appear in the notes to *Die Kultur*; they did not make it into the translation by Phaidon Press.

> that it is the copy of a pattern pre-existing in Him [*ein Abbild des in ihm präexistierenden Vorbildes*], and that He will ever remain its eternal mover and restorer [*dauernder Beweger und Fortschöpfer*].[69]

The transition from the Middle Ages to the Renaissance had been accompanied by new perceptions of human agency, and the reorganization of agency corresponded directly to a reorganization of time. The temporality of the Middle Ages could be captured in terms of guarding, delaying, slowing down and securing a given state in the face of a continuously impending doom, judgment and salvation – that is, the simultaneously horrendous and blissful arrival of the anti-Christ. Here agency was understood in terms of cataloguing, organizing and preserving. The temporality of the Renaissance, on the other hand, was fundamentally different: the world was not awaiting its salvaging *release*; the world was an orchestra of furious *creation* and 'overflowing energy'.[70] Time was not measured relative to an awaited moment of judgment; time answered to the creative force of a divine human, driven by a hunger only for historical greatness and monumental honour.

Burckhardt's interest in the cultural rather than political history of Renaissance Italy drew precisely from the supposed Renaissance interest in *saeculum*. Not lofty ideas but everyday practice had become the object of both the Renaissance mind and Burckhardt's historiography. As 'the need of salvation became felt more and more dimly', political practice, art, feasts, an emerging scientific method and ruptures with tradition merged with the Renaissance *Geist*. Here political history became *Kulturgeschichte*.[71] The categories of cultural practices now came to approximate, better than formal diplomatic correspondence, the human soul's 'flashes, expansions, and pauses' (*ihr plötzliches Aufblitzen, ihre Verbreitung, ihr Innehalten*).[72]

69 This passage is worth citing in full in the original German: 'Während die Menschen des Mittelalters die Welt ansehen als ein Jammertal, welches Papst und Kaiser hüten müssen bis zum Auftreten des Antichrist, während die Fatalisten der Renaissance abwechseln zwischen Zeiten der gewaltigen Energie und Zeiten der dumpfen Resignation oder des Aberglaubens, erhebt sich hier, im Kreise auserwählter Geister, die Idee, daß die sichtbare Welt von Gott aus Liebe geschaffen, daß sie ein Abbild des in ihm präexistierenden Vorbildes sei, und daß er ihr dauernder Beweger und Fortschöpfer bleiben werde.' Burckhardt, *The Civilization of the Renaissance in Italy*, 341; Burckhardt, *Die Kultur der Renaissance in Italien*, 322.

70 'gewaltige Energie'. Burckhardt, *Die Kultur der Renaissance in Italien*, 322; Burckhardt, *The Civilization of the Renaissance in Italy*, 341.

71 Burckhardt, *The Civilization of the Renaissance in Italy*, 304; Burckhardt, *Die Kultur der Renaissance in Italien*, 284.

72 Burckhardt, *The Civilization of the Renaissance in Italy*, 281; Burckhardt, *Die Kultur der Renaissance in Italien*, 263.

Sixty years later, in 1920, Huizinga asked rhetorically and in celebration of Burckhardt: 'who before [him] had considered including social life, fashion, dilettantism and feasts in their cultural-historical significance?'[73]

In sum, Burckhardt's *The Civilization of the Renaissance in Italy* presented a new historiographical approach by which an entire epoch could be treated as a whole and on its own terms, rather than as a transitional stage within a larger progressive development. Commencing from (1) a new political reality in fourteenth- and fifteenth-century Italy, Burckhardt showed how (2) the ensuing political practices, artistic norms, scientific endeavours and rituals of feast conditioned (3) the release of a new kind of morality and humanity – a new kind of individual. This newly released individual was not merely a spiritual epiphenomenon of political and cultural reality; this spiritual individual permeated all levels at once but was also revealed through them. The defining feature of this *Zeitgeist* was its utter and complete contempt for moral principle and the promise of salvation. The only temporal orientation that mattered was the immortality of this-worldly fame, not the universality of other-worldly holiness. Burckhardt's Renaissance was through-and-through secular, that is, drafted in *saeculum* (world time). A veil had been lifted from medieval eyes.

In the 1860s, Burckhardt's image of *uomo unico* had figured as a powerful critique of and alternative to the contemporary fashions of romantic nationalism, political centralization and mass politics of his times. In 1905, against the background of municipalities and architects eagerly calling for the dismissal of the past and its traditions in favour of modern invidualism and commerce, Huizinga found that this *uomo singolare* had lost its historical *and* moral anthropological appeal.

Huizinga's medieval *homo ludens*[74]

Huizinga's *Autumntide of the Middle Ages* (1919) is commonly read as a direct and explicit critique of Burckhardt's *Civilization* (1860).[75] Though

73 'Wie had tevoren eraan gedacht, het gezelschapsleven, de mode, het dilettantisme, de feesten in haar cultuurhistorische beteekenis te beschouwen?' VW IV: *Het probleem der Renaissance* (1926): 247.

74 This section includes a partially revised version of Rydin, 'Huizinga's "Heimwee": Responding to Burckhardt's "Die Kultur der Renaissance in Italien" in Times of Loss', 735–37. See footnote 1 of this chapter.

75 For a helpful overview of the reception of *Autumntide* and Huizinga's position in Renaissance historiography, see Tollebeek, '"Renaissance" and "Fossilization": Michelet, Burckhardt, and

I do believe this antagonism to be the correct backdrop for a proper understanding of Huizinga's book, it is by no means as evidently the case as many previous accounts have implied.[76] On face value, this omission of references to Burckhardt may not seem particularly puzzling: Burckhardt wrote about Renaissance Italy, while Huizinga wrote about late medieval France, the Low Countries and the Burgundian Netherlands. In the entirety of *Autumntide*, Huizinga mentions Burckhardt only five times. He expresses agreement with Burckhardt four times,[77] and only once does he explicitly critique him, and even then, only within the span of a single paragraph.[78] How could these books collide in the face of such different material? As was briefly mentioned at the beginning of this chapter, Huizinga himself admitted to Pirenne shortly after the publication of *Autumntide* that his book had been meant to expose 'the severe shortcomings [*graves défauts*] of [Burckhardt]'.[79] But where, then, did this opposition lie? What precisely did it concern? How do two works of history, dealing with different materials, periods and locations, meet?

In order to answer this question, the following subsections explore the two works in which Huizinga most explicitly addressed Burckhardt's conception of Renaissance culture: the second chapter of *Autumntide* (1919) and a slightly later essay, *The Problem of the Renaissance* (1920). This essay is seldom mentioned in relation to Huizinga's appreciation of Burckhardt's

Huizinga'. For a helpful overview of the book's earliest reception in both the Netherlands and abroad, see Strupp, *Johan Huizinga: Geschichtswissenschaft als Kulturgeschichte*, 142–49; L. Hanssen, *Huizinga en de troost van de geschiedenis: Verbeelding en rede* (Amsterdam: Uitgeverij Balans, 1996), 192–200; Van der Lem, *Johan Huizinga: Leven en werk in beelden en documenten*, 149–51.

76 For example, good reason exists to read *Autumntide* in terms of Huizinga's opposition to Otto Oppermann (1873–1946), a German medieval historian at Utrecht University at the time. P. Raedts, *De ontdekking van de Middeleeuwen: Geschiedenis van een illusie* (Amsterdam: Wereldbibliotheek, 2014), 263–65; Van der Lem, *Johan Huizinga: Leven en werk in beelden en documenten*, 117. At the time, medieval history was rather unpopular in the Dutch historical discipline. Through their research, Opperman and Huizinga laboured together, despite their differences, to effectuate a new historical sensitivity in their field. Huistra, *Bouwmeesters, zedenmeesters: Geschiedbeoefening in Nederland tussen 1830 en 1870*, 112; Raedts, *De ontdekking van de Middeleeuwen: Geschiedenis van een illusie*, 263. Also, the same year *Autumntide* was published, a book by the Dutch historian B. Heijmans was published under the title *The Renaissance in Italy* (*De Renaissance in Italië*). This book proposed an image of the relation between the Middle Ages and the Renaissance that was diametrically opposed to Huizinga's. Huizinga did not discuss Heijmans's work, but Heijmans's book points to a common opinion at the time to which Huizinga was responding. B. Heijmans, *De Renaissance in Italië* (Zutphen: W. J. Thiemen & Cie, 1919).

77 VW III: *Herfsttij der Middeleeuwen* (1919): 19, 47, 79, 180, 280.

78 VW III: *Herfsttij der Middeleeuwen* (1919): 79.

79 BW I: Huizinga–Pirenne (1919): 271.

Civilization. Yet its inclusion in the matter carries significant potential; the essay served as one of the few occasions in which Huizinga spelled out his understanding of contemporary debates in medieval and Renaissance studies and his position on them. The essay, in fact, accommodated the most extensive discussion of Burckhardt in Huizinga's entire oeuvre.

Autumntide of the Middle Ages (1919)

Like Burckhardt's *Civilization*, Huizinga's *Autumntide* was steeped in a political dimension upon its publication.[80] In a narrow but not unimportant sense, it repeated a line of argumentation that had appeared in one of Huizinga's earlier publications on the history of the University of Groningen, 1814–1914. Here, Huizinga had discussed at length the Dutch Higher Education Act of 1876, which had delivered a decisive blow to the medieval organization of Dutch universities and their pedagogical ideals. From 1876 on, Dutch university curricula were organized along the boundaries of new disciplines, effectively wiping out the traditions of a universal liberal arts education after the medieval model. Philosophy was no longer mandatory, nor was Latin; ideals of a humanistic education made way for those of professional training and disciplinary depth. To top it off, these modern developments were accommodated in newly erected faculty buildings, often designed in a 'quasi-historical' style of the 'Hollandic Renaissance', thus further solidifying the late nineteenth-century alliance between liberalism, bureaucratization and the Renaissance aesthetic in the Netherlands.[81] Read against this background, *Autumntide* – and its appreciation of late medieval culture as a precedent of Renaissance civilization – figured as a way of criticizing such university reforms.

When placed against the background of this chapter's first section, this particular political investment fits into a wider, extra-political experience. *Autumntide* not only addressed political projects of modernization, whether or not they had to do with urban geography or education reforms; *Autumntide* addressed the value of the past as such, and by the 1900s, few periods of European history symbolized the 'pastness' of the past as potently

80 See e.g. L. Hanssen, *Huizinga en de troost der geschiedenis: Verbeelding en rede* (Amsterdam: Uitgeverij Balans, 1996), 230–33; N. C. F. van Sas, 'Fin-de-Siècle als nieuw begin. Nationalisme in Nederland rond 1900', *BMGN – Low Countries Historical Review* 106, no. 4 (1991): 595–609.
81 For an overview of Huizinga's understanding of the modern university reforms in the Netherlands of the 1870s, see VW VIII: *Geschiedenis der universiteit gedurende de derde eeuw van haar bestaan* (1914): 291–93, 324.

as the images of medieval culture.[82] When Huizinga researched the forms of medieval life and thought and when he took issue with Burckhardt's contention that the Renaissance had marked a complete departure from its 'past', Huizinga was not only addressing political issues. In a wider, extra-political, moral and perhaps unconscious sense, he was also giving expression to a more general experience of loss and of longing for the past. Huizinga's medievalism was political, but not only political; it belonged to a moral and existential perspective on the value of 'the past' to life. Before the extra-political dimension of this opposition is explored, first consider the only passage from *Autumntide* in which Huizinga takes explicit issue with Burckhardt's reading. In this passage, Huizinga addresses the continued importance of medieval values in Renaissance times:

> It appears to me that this is one of the points where Burckhardt has exaggerated the distance between the Middle Ages and the Renaissance, between Western Europe and Italy. The longing for honour and glory in the Renaissance is essentially like the chivalrous longing for honour from earlier times in France, an expansion of caste honour towards general urgency, ridded of the feudal sentiment and fertilized with an ancient idea.[83]

Against the background of the political reforms of his time, Huizinga's opposition stated in this passage seems fairly straightforward: Burckhardt had (grossly) exaggerated the difference between western and southern Europe as well as the difference between the Middle Ages and the Renaissance. Why should this matter? Well, it undermines central tenets of Dutch and European culture, whose continued recognition matters in times of a rapidly modernizing state. Medieval curricula, Latin, philosophy and medieval urban geography might still be of value to the modern mind. A closer look at the antagonism between *Autumntide* and *Civilization*, however, reveals a more fine-grained moral and perhaps even anthropological tension, directly

82 This perspective offers an alternative answer to a question recently asked in an edited volume on *Autumntide*: why does the book disregard medieval urban life as much as it does? This was not only a methodological statement; it also reflected an antipathy for the fetishism of modern cosmopolitan life. Dumolyn and Lecuppre-Desjardin, 'Huizinga's Silence', 66.

83 'Het schijnt mij toe, dat dit een der punten is, waarop Burckhardt den afstand tussen Middeleeuwen en Renaissance, tusschen West-Europa en Italië te groot gezien heeft. Die roemliefde en eerzucht der Renaissance is in haar kern de ridderlijke eerzucht van vroeger tijd en Fransche herkomst, de standseer uitgebreid tot wijder gelding, ontdaan van het feodale sentiment en bevrucht met antieke gedachte.' VW III: *Herfsttij der Middeleeuwen*: 79 (1919).

related to Huizinga's experiences of loss discussed above. This tension becomes particularly salient in the second chapter of *Autumntide*, wherein Huizinga put forth an anthropological theory of human consciousness as his book's key historiographical tool. The particular political conditions explain a great deal (in a sense, 'the form') of Huizinga's opposition to Burckhardt, but not its content. The content, I will argue, relates to an experience of time, or an experience of time as loss. But to unpack this claim, a look at the second chapter of *Autumntide* is in order.

The second chapter of *Autumntide* is titled 'The longing for a more beautiful life', and in this chapter, Huizinga provided a taxonomy of three basic 'spiritual attitudes' or 'fundamental mentalities', each corresponding to a particular mode of 'longing'.[84] The first mode amounts to outright asceticism: a Christian removal of thought from world. This attitude is driven by a 'nostalgic longing for eternal blissfulness' in the nostalgic image of a supposedly uncorrupted, pristine state of being.[85] This frame of mind ultimately feeds into utter 'complacency' towards the world at present and 'world-renouncement'.[86] The second mode of longing and striving, on the other hand, is that of 'improvement' and 'completion' of 'vocation': here longing is captured in terms of an impending improvement of the sensuous world through labour and production.[87] The presupposed possibility of an idea's future materialization through labour is what Huizinga calls 'optimism'.[88] In Europe this infusion of life with labour was, according to Huizinga, to be found first among eighteenth-century thinkers.[89] Labour redirected longing from an ascetic life-beyond to a life-future through its virtues.

But neither of these mentalities, Huizinga argued, could capture the dynamics of late medieval culture. The first mentality is ultimately a subjection to mere cerebral asceticism; the second is a subjection through labour to an idealized ever-impending future. In *Autumntide*, however, Huizinga set to show how late medieval culture had erected vast systems of social rituals and symbolic communitarian behaviour (not ascetic), which, in turn,

84 Respectively 'de zucht naar schooner leven', 'geesteshoudingen' and 'zucht.' VW III: *Herfsttij der Middeleeuwen* (1919): 42, 35, 34.

85 'Heimwee naar een eeuwig heil'. VW III: *Herfsttij der Middeleeuwen* (1919): 42.

86 Respectively 'onverschilligheid' and 'wereldverzaking'. VW III: *Herfsttij der Middeleeuwen* (1919): 42, 152.

87 Respectively 'verbetering', 'volmaking', 'beroep', 'arbeid' and 'productie'. VW III: *Herfsttij der Middeleeuwen* (1919): 40, 40, 251, 41, 42.

88 VW III: *Herfsttij der Middeleeuwen* (1919): 41.

89 'levensarbeid'. VW III: *Herfsttij der Middeleeuwen* (1919): 41–42.

were altogether independent from any notion of 'progress' (not labour). Thus, in order to understand late medieval culture, Huizinga invoked a third mentality, one of 'aesthetic longing'. This mentality took shape in a conception of 'life as art' and centred around the concept of 'play'.[90] 'Play' was defined here by Huizinga as the activity of spirit that 'recreates' a given suffering into a directed suffering towards beauty.[91] This beauty, in turn, resides not in the world's achievement of any particular empirical quality but in the directedness itself. Before we turn to Huizinga's understanding of beauty, in whose orientation suffering is recast, we explore the dynamic of this orientation, that is, play:

> How does this third approach affect life: the longing for a more beautiful life according to a dreamt ideal? She [the third approach] recreates [*herscheppen*] the forms of life into forms of art. Yet it is not only the works of art as such wherein she expresses her understanding of beauty, she wants to refine life itself with beauty, and she complements society with game and forms. Here, the highest demands are made to precisely the personal art of life, demands that can only be pursued by an elite in an artistic game of life.[92]

Central to Huizinga's third mode of life is one of 'recreative longing'.[93] This longing is not a longing for something beyond itself; it is not, for example, a longing aching for its dissolution upon the improvement of living conditions, increased status, happiness or even an impending moment of utter blissfulness – that is, 'works of art themselves'.[94] The longing does not work towards its dissolution. Rather, it exhausts itself, and can be conceived of as an end in itself, as an 'art of life' or 'life-play'. Here the 'ethical ideal' becomes an 'ideal of life' – that is, an ideal to which one can strive and, through that pursuit, give meaning to romantic, political, political and professional

90 'levenskunst'. VW III: *Herfsttij der Middeleeuwen* (1919): 48.

91 'spel' and 'herscheppen'. VW III: *Herfsttij der Middeleeuwen* (1919): 42.

92 'Hoe werkt nu op het leven de derde houding: de zucht naar het schoonere leven volgens een gedroomd ideaal? Zij herschept de vormen van het leven in kunstvormen. Maar het zijn niet enkel de kunstwerken als zoodanig, waarin zij haar schoonheidsdroom uitdrukt, zij wil het leven zelf veredelen met schoonheid, en vult de samenleving zelf met spel en vormen. Hier worden juist aan de persoonlijke levenskunst de hoogste eischen gesteld, eischen, die alleen kunnen worden nagestreefd door een élite, in een kunstig levensspel.' VW III: *Herfsttij der Middeleeuwen* (1919): 43.

93 'zucht'. VW III: *Herfsttij der Middeleeuwen* (1919): 43.

94 VW III: *Herfsttij der Middeleeuwen* (1919): 43.

practice.[95] In this sense, the ideals 'recreate the forms of life into forms of art'; they are inherently creative.[96] This is not to say, however, that longing resides in an abstract existential domain, for then it would spiral into the world-renouncement of Huizinga's first mode of being. No: longing occurs in forms of life, and these forms are ideal only in terms of their self-sufficiency, for in every other respect they are delineated by the shape of fabric, the walls and ceiling of a seminar room, the river or forest along a battlefield. The longing takes shape not in the art actually produced but in the recreative play of life, and it gives meaning – cultural meaning – to it.

Huizinga's anthropology was directly reflected by the anatomy of *Autumn-tide*. Roughly, this book can be subdivided into four parts: (1) chapters 1–2 deal with the structure of longing; (2) chapters 3–9 deal with the ideals and forms for which medieval culture strove; (3) chapters 10–21 deal with the resulting empirical expressions – the 'images' (*beelden*) – of this striving; and lastly, (4) chapter 22 deals with the revitalization potential dormant within a collapsing culture. The subdivision of the book's main body (the second and third part in the subdivision above) mirrors Huizinga's anthropological theory of longing: ideals constitute and regulate the domain of meaningfulness, and the activity within this regulated domain recreates results in their images. The conviction that cultural behaviour was always regulated by cultural ideals was thus paramount both to the content and structure of Huizinga's book, and it is on this level that Burckhardt's Renaissance comes in.

Huizinga contrasted his conception of the 'medieval dream' and its culture of 'recreation' with Burckhardt's conception of Renaissance individualist instrumentalism. Whilst Burckhardt had conceived of the Renaissance as an amoral longing for honour and monumental immortality, Huizinga understood these conceptions of honour and immortality as guiding principles in whose images individual lives achieved meaningfulness. In other words, Huizinga objected to the very notion and even the very possibility of amoral instrumentalism, for what, other than a certain idealization, could direct the employment of instruments? Fifteenth-century images of honour were not amoral. On the contrary, they were idealizations taken to their most extreme significance: nothing mattered more than the image of honour. The 'Renaissance Mensch', as the very term itself states, did not conceive of itself as an isolated body; it recast, reimagined and rebirthed itself in the image of an ideal.

95 Respectively 'ethisch ideaal' and 'levensideaal'. VW III: *Herfsttij der Middeleeuwen* (1919): 78, 43.
96 VW III: *Herfsttij der Middeleeuwen* (1919): 43.

Interlude: Van Eyck's mirror

This anthropological disagreement regarding the possibility of an amoral frame of mind discussed above – the contrast between Huizinga's 'dream' and Burckhardt's *uomo singolare* – was not a matter of mere theoretical dispute. It trickled down to disagreements on an empirical level, too. A particularly good example of how their anthropological perspectives shone through in their readings of historical materials can be found in their respective interpretations of works by Jan van Eyck (1390–1441).[97] Huizinga found his discussion of Van Eyck's *The Arnolfini Wedding* (1434) in *Autumntide* to be so important that he initially considered calling his book on the medieval ages *In Jan van Eyck's Mirror* ('In den spiegel van Jan van Eyck').[98] In *Civilization*, Burckhardt, too, ascribed great significance to Van Eyck's work:

> In the fifteenth century, the great masters of the Flemish school, Hubert and Jan van Eyck, suddenly lifted the veil from nature. Their landscapes are not merely the fruit of an endeavour to reflect the real world in art, but have, even if expressed conventionally, a certain poetical meaning – in short, a soul.[99]

Two features of Burckhardt's account are of interest here: the art produced by Hubert and Jan van Eyck mark a rupture from previous art – a 'veil' is lifted, similar to the medieval 'veil' removed by the Renaissance – and this rupture consists in part of a newfound ability to 'reflect the real world'. New geometrical insights and technologies had enabled a depiction of the world more true to nature.[100] Within the parameters of Burckhardt's amoral, instrumental *uomo singolare*, Van Eyck's painting achieves significance through the expanded instrumental ability it displayed: its bolstered ability

97 VW III: *Herfsttij der Middeleeuwen* (1919): 324; Burckhardt, *Die Kultur der Renaissance in Italien*, 395.

98 This appears from the different covers Huizinga designed for his work on late medieval culture, see G. Small, "Epilogue," in *Autumntide of the Middle Ages* (Leiden: Leiden University Press, 2020), 542; Van der Lem, *Johan Huizinga: Leven en werk in beelden en documenten*, 138. See for another, interesting discussion of the title's history Krul, 'In the Mirror of Van Eyck: Johan Huizinga's "Autumn of the Middle Ages"', 370.

99 Burckhardt, *The Civilization of the Renaissance in Italy*, 181; Burckhardt, *Die Kultur der Renaissance in Italien*, 170.

100 The convex image in the mirror depicted in the painting is commonly held to be the fruit of new geometrical technologies in Renaissance art, see e.g. D. Hockney and C. M. Falco, 'Optical Insights into Renaissance Art', *Optics and Photonics News* 11, no. 7 (2000): 52–59.

to project, form and create. This could not have been further from Huizinga's understanding of Jan van Eyck's work:

> Through Van Eyck's art, the pictorial representation of things holy has reached a degree of detail and naturalism that art historically could be called a beginning, yet cultural-historically means an end. The most extreme tension in the earthy representation of the divine had been reached here; the mystical content of its imagination stood on the verge of escaping its images and leaving behind a mere lust for colourful form.[101]

In this passage, Huizinga argues that the 'pictorial representation' in Van Eyck's painting had at its time indeed reached an unprecedented level of detail and natural sensitivity, which art historically could be 'called a beginning'. But – and this is crucial – cultural-historically speaking, the painting 'signifies [rather] an end'.[102] The painting's technical form may have entered a new stage, but its content and semantics had not. In fact, Van Eyck's naturalism may be a beginning in the domain of artistic technique, but culturally, it is the final 'unfolding of the late medieval spirit'.[103] Van Eyck's creativity lay in his feverish attempt to 'recreate' in even greater detail the traditional ideals and symbolic codes. In Huizinga's account, Van Eyck's sharpening of form is a final and most decadent attempt to express in the greatest of detail the medieval idea of love, and thus 'in Van Eyck's art the content remains entirely medieval.'[104] Finally, it is precisely this ability to narrate scenes in the light of ideals, striving and recreation (the medieval peak of which is found in Van Eyck's naturalism) that Huizinga equates with medieval 'play'.

In this difference between Burckhardt's and Huizinga's appreciation of Van Eyck's art, what one sees expressed is not a head-on collision of arguments but two altogether different ways of looking, quite literally. It is not so much Burckhardt's argument on the Renaissance's supposed

101 'Met de kunst der Van Eyck's heeft de picturale uitbeelding der heilige dingen een graad van détailleering en naturalisme bereikt, die misschien strikt kunsthistorisch een begin kan heten, maar cultuurhistorisch een einde beduidt. De uiterste spanning in het aardsch verbeelden van het goddelijke was hier bereikt; de mystische inhoud dier verbeelding stond gereed om uit die beelden te ontvlieden en enkel den lust aan den bonten vorm achter te laten.' VW III: *Herfsttij der Middeleeuwen* (1919): 330.

102 VW III: *Herfsttij der Middeleeuwen* (1919): 330.

103 VW III: *Herfsttij der Middeleeuwen* (1919): 331.

104 '[In] de kunst der Van Eyck's is de inhoud nog volkomen middeleeuwsch.' VW III: *Herfsttij der Middeleeuwen* (1919): 331.

individualism that Huizinga contradicts but rather his very standard of proof. Both authors here proposed a synchronic cross-section of a cultural object, but Huizinga's emphasis on ideals operating through rituals, metaphors and allegories fundamentally desensitized his reasoning to Burckhardt's emphasis on technological and instrumentalist developments of the time. Huizinga understood technology, such as Van Eyck's geometrical method of projection, as meaningful only in so far as it expressed the ideals underpinning Van Eyck's culture. Instrumentalism in itself, Huizinga argued in stark opposition to Burckhardt, is inert and historically unintelligible. According to Huizinga, the historical question is always: instrumental towards *what*?

So far, four key steps have been made in this section: (1) Huizinga understood *Autumntide* as an argument against Burckhardt's *Civilization*; (2) this perceived opposition did, however, not take place on an empirical level, as *Autumntide* and *Civilization* are empirically too distant from one another to uphold such a claim; (3) only on an anthropological level – in which Huizinga was deeply invested, as both the content and form of *Autumntide* show – can this opposition be perceived clearly and meaningfully. In *Civilization*, Burckhardt appreciated creativity in terms of 'creation' (*schöpfen*); in *Autumntide*, Huizinga understood creativity in terms of its ability to 'recreate' (*her-scheppen*).[105] Lastly, (4) this anthropological difference was reflected in the single most important empirical overlap between *Autumntide* and *Civilization*: Van Eyck's *The Arnolfini Wedding*. In a later essay, Huizinga further developed his opposition to Burckhardt's *Renaissancebegriff* along similar lines, but by different means. In this work, not anthropological but philological considerations served to methodologically buttress historical ideals.

These observations allow an intermediate reflection on the debate spelled out at this chapter's outset regarding Huizinga's relation to Burckhardt. Two general positions on the matter have been presented: either (1) Huizinga's cultural history is read as an exponent of Burckhardt's cultural-historical project (Kaegi, Gombrich), or (2) Huizinga's cultural history is understood to be qualitatively different from Burckhardt's (Guggisberg, Tollebeek, Kaminsky). The four observations listed above offer support for the second position: Huizinga's conceptual apparatus was, I have argued, fundamentally different from Burckhardt's, and this becomes particularly apparent in their respective anthropological convictions. Huizinga had a teleological and

105 Burckhardt, *Die Kultur der Renaissance in Italien*, 2; VW III: *Herfsttij der Middeleeuwen* (1919): 43.

Figure 2.4. **(A) Jan van Eyck's *The Arnolfini Wedding* (1434) is shown. On the right, two images show geometrical features of primary importance to the painting's art historical status. (B) A non-aligned, three-dimensional spatial orientation of the chandelier. (C) A convex geometry displayed in the mirror. Both these geometrical features are commonly understood to be emblematic of significant technical advances in Renaissance art.**

idealistic understanding of human action; that is, he understood action as a pursuit of meaningfulness. Burckhardt's *uomo singolare*, on the other hand, rests on the conviction that human culture can under certain circumstances be utterly non-idealistic and instrumental. In the next section, Huizinga's anthropological convictions are further explored through their philological investment as they appeared in his essay *The Problem of the Renaissance* from 1920.

The Problem of the Renaissance (1920)

In order to understand in greater detail Huizinga's objection to Burckhardt's analysis of the Italian Renaissance, it is helpful to read *Autumntide* in conjunction with his essay *The Problem of the Renaissance* (*Het probleem der Renaissance*, 1920). In most of his writings, Huizinga's explicit discussion of other authors was fairly limited; he rarely spelled out contemporary debates in his publications. Yet in this essay, Huizinga went to considerable lengths to develop his understanding of and position in the wider fields of medieval and Renaissance studies. The essay, in fact, accommodated the most extensive discussion of Burckhardt in Huizinga's entire oeuvre. This time around, Huizinga invoked a philological perspective unto the tenability of Burckhardt's understanding of the concept of 'Renaissance'.

Huizinga's essay begins by arguing that the 'Renaissance' has not always been a historical category but has become one, and in order to understand the Renaissance, one must understand the conditions under which the term transitioned from the theological to the historical vernacular. In a nutshell, this is the central contention of Huizinga's *The Problem of the Renaissance*, and it was meant as an argument against those historians who since Burckhardt had treated the Renaissance as a predicate-carrying substance.[106] The question 'what is the Renaissance?' had to be equated with 'what has 'Renaissance' meant?'[107] Huizinga the historian had not shed his philological ways, and, as shall become clear, they fundamentally affected both the method and outcome of his historical inquiry.

The essay consisted of two parts: the first traced the development of the term 'Renaissance' from the sixteenth century onwards and showed how it had transitioned from modifier (the Renaissance *of*) to distinct period (of the *Renaissance*) in the late eighteenth and early nineteenth centuries; the second part juxtaposed four historical concepts – Middle Ages, Renaissance, Reformation, modernity – and contended that recent arguments by Konrad Burdach (1859–1936) and Ernst Troeltsch (1865–1923) made urgent a fundamental revision of their traditionally perceived interrelation.[108] In both parts, Burckhardt was central: (1) Burckhardt had singlehandedly introduced a Renaissance understood solely on its own terms rather than

106 Huizinga targeted historians such as Carl Neumann (1832–1925), Louis Courajod (1841–1896) and Hyppolite Fierens-Gevaert (1870–1926). VW IV: *Het probleem der Renaissance* (1920): 253.
107 'Daarom is het probleem der Renaissance, de vraag: wat is zij geweest? niet los te maken van het groeien van den term, die haar aanduidt.' VW IV: *Het probleem der Renaissance* (1920): 232.
108 VW IV: *Het probleem der Renaissance* (1920): 261.

as a transitional phase in a larger historical progression. In so doing, he had (2) used his conception of the Renaissance to secure secular origins for modernity independent of both the Middle Ages and the Reformation. Huizinga described the consequences and aftermath of Burckhardt's *Civilization* as follows:

> One falsely imagined the Renaissance type of culture being a free, genius personality, elevated beyond doctrine and moral, a haughty frivolous hedonist, who seizes the power to live according to his own norms in a pagan lust for beauty. The artists of the late nineteenth century found in this imagination of historical life the echo of their own wishes.[109]

'Elevated beyond doctrine and moral', the Burckhardtian *Renaissance Mensch* 'seizes the power to live according to his own norms'. Burckhardt's human, Huizinga argued, was truly 'demonic in its unwavering pride, complacency and boldness';[110] this human's contempt for conscience and empathy – in fact, for anything that did not serve the purpose of assertion and creation – was captured in the name Burckhardt gave this being: *uomo singolare*.[111] This human that had lain dormant during the Middle Ages was uncovered, discovered when a veil (*sluier*, referring to Burckhardt's *Schleier*) was lifted. This was, in Huizinga's conception, the image of the Renaissance inspired by Burckhardt. And it was wrong, both in an anthropological and a philological sense.

Burckhardt could not be blamed for the later remediation and vulgarization of his *Renaissance Mensch*, Huizinga held, but he could have known that any depiction relying so strongly on one conception is 'necessarily one-sided'.[112] Burckhardt had overestimated the moral paganism and individualism in Renaissance culture, and ironically, this inaccuracy was nowhere more salient than in the word 'Renaissance' itself. The term had been employed already in the fourteenth and fifteenth centuries as a self-reflexive category of identity. Through its subsequent historical reification, the term had become unduly insensitive to the conditions under which it

109 'Men waande het type van de Renaissance-cultuur te zien als de vrije, geniale persoonlijkheid, verheven boven doctrine en moraal, een hooghartig frivool genotmensch, die in heidenschen schoonheidslust de macht grijpt om naar eigen norm te leven. Het artisticisme van de eindigende negentiende eeuw vond in die inbeelding van historisch leven den weerklank van zijn eigen wensch.' VW IV: *Het probleem der Renaissance* (1920): 247.

110 VW IV: *Het probleem der Renaissance* (1920): 247.

111 VW IV: *Het probleem der Renaissance* (1920): 247.

112 VW IV: *Het probleem der Renaissance* (1920): 248.

entered the discourse of identity at all. The term 'Renaissance', along with related concepts such as '*renovatio, restitutio, restauratio*', Huizinga argued, were consciously taken directly from the New Testament:[113]

> The origin of this entire growth of ideas lies in the concept of rebirth in the New Testament, which, in turn, has its roots in the images of revival of the Psalms and Prophets. The Gospels and the Letters had acquainted the spirit with concepts of revival, rebirth and regeneration, which in one part concerned the working of sacraments, especially baptism and communion, in another part concerned the expectation of eventual salvation and in a final part concerned the reception of a saving grace required to convert the living *mensch*.[114]

The term Renaissance drew from tropes and images of 'revitalization, recovery and rebirth' and as such perpetuated the Christian 'expectation of eventual salvation'.[115] The 'Renaissance', understood in its original usage as a historical artefact itself, was not pagan; on the contrary, it was through-and-through Christian. This insight, Huizinga argued, necessitated a fundamental revision of the relation between the Middle Ages, the Renaissance and the Reformation. In this sense, Huizinga understood himself to have answered the call of Ernst Troeltsch and Konrad Burdach, who, each in their own way, had argued for continuity among the three terms.

To conclude, what matters for now is this: in this essay, Huizinga understood Burckhardt's depiction of the *Renaissance Mensch* in terms of its supposed and self-perceived ability to create according to its own norm. Huizinga's critique was meant to debunk the supposed creative thunderbolt that was *uomo singolare*. He argued that the Renaissance, as the very word and its theological heritage itself implied, was meant a spiritual restorative endeavour. That is, the *Renaissance Mensch* did not create but *recreate*. It did not lash out independent of principle but answered to an image it

113 VW IV: *Het probleem der Renaissance* (1920): 265.

114 'De oorsprong van dien geheelen ideeëngroei ligt in het wedergeboortebegrip van het Nieuwe Testament, dat reeds wortelt in vernieuwingsvoorstellingen van Psalmen en Profeten. De Evangeliën en de Brieven hadden den geest vertrouwd gemaakt met begrippen van vernieuwing, herboorte, regeneratie, die ten deele betrekking hadden op de werking der sacramenten, met name doop en avondmaal, ten deele op de verwachting van het uiteindelijk heil, en ten deele op de omkeering van den levenden mensch tot het ontvangen der genade.' VW IV: *Het probleem der Renaissance* (1920): 262.

115 Respectively 'herleving', 'herstel', 'herboorte' and 'heilsverwachting'. VW IV: *Het probleem der Renaissance* (1920): 262.

perceived to be worth restoring. Huizinga's philological approach did not only problematize neatly defined historical periods, but it made urgent an analysis of Christian appreciations of the soul and their lasting heritage in secular times. As *fin-de-siècle* Netherlands underwent an aesthetic makeover towards secular ideals, inquiries into the creative potential of recreation, restoration and traditional ideals seemed more urgent than ever to Huizinga.

A symmetry of central importance to the present argument now emerges. Huizinga's understanding of his own times, exemplified by his appreciation of Jan Veth, is altogether congruent with his description of late medieval culture. In moments of tremendous development, of cultural decay and decadence, new ideas of universality and creativity will be found not through a renunciation of the past but through a re-appropriation and recreation of historical images. Just as the twentieth century would not require 'an abandonment of the old' but rather its reorganization for new means, the Renaissance was not merely the end of the Middle Ages; the Renaissance was the culmination of the Middle Ages, and from within the crises it experienced, it would have to reinvent the spiritual forces from which culture had originally sprung. This symmetry extends beyond Kaminsky's observation: the *fin-de-siècle* character of Huizinga's Middle Ages is not only discursive, but it also carries a similar temporal-historical orientation through a narrative of tragic perseverance. The demolition of medieval and early modern architecture in the 1900s for industrial purposes elicited an image of medieval culture in opposition to instrumental logic: Huizinga's image of medieval culture and his critique of Burckhardt's *Renaissancebegriff* were responses to industrialism in 1919.

Conclusion

Huizinga's reconstruction of the 'medieval mind' and corresponding arguments against Burckhardt's *Civilization* in *Autumntide* were mediated by his experience of loss following the transitions of Dutch cityscapes around 1900. To this effect, this chapter has made two points in particular: (1) the disagreement Huizinga expressed in *Autumntide* with Burckhardt's *Civilization* was moral anthropological in nature. Ultimately, it centred around Burckhardt's celebration of the creative potential of 'creation' (*Schöpfung*) and Huizinga's celebration of the creative potential of 're-creation' (*her-schepping*). (2) This disagreement was established retrospectively from within Huizinga's moral framework, a framework that was mediated by experiences of loss following the tremendous aesthetic transition of Dutch and other European societies around 1900. The 'tyranny of the present' called for new ideals of timelessness, creativity and restoration, and Huizinga's medieval mind and critique of Burckhardt were drafted in terms of this renegotiation. The steps taken towards these conclusions in this chapter can be reconstructed by means of three distinct steps:

1. Burckhardt's *Renaissance Mensch* had no conscience and cared for nothing beyond pride and immortality through honour. This amoral frame of mind was expressed on numerous levels: a calculating, political logic of state; a new conception of individuality independent from categorical structures; a new conception of time, invested not in a preservation of the world in the face of a looming final judgment but in creation. Burckhardt understood all aspects of Renaissance life as imbued with this single *Zeitgeist*, that of the *uomo singolare*, and it was because of this supposed holistic ubiquity of spirit that new historical material became of interest to Burckhardt: feast, food, dress and laughter were turned into historical objects, meaning objects telling of a wider, historically unique truth. Burckhardt held that only cross-sectional observations, not continuous historical narratives, could make the cultural historian sensitive to the period's spirit.
2. Huizinga's reconstruction of the late Middle Ages had little empirical overlap with Burckhardt's *Civilization*. Still, Huizinga conceived of *Autumntide* as a response to Burckhardt's analysis of Italian culture between 1350 and 1550. Why? Huizinga understood the late Middle Ages as a 'Renaissance' only in the term's traditional Christian capacity – that is, as the term was used by Renaissance authors themselves, according to Huizinga. Huizinga understood the late Middle Ages as a cultural epoch obsessed with symbolism. In the Middle Ages, each and every cultural practice had become drenched in ritual codes so as to give birth anew (*Renaissance*) to original, uncorrupted ideals. What had appeared to Burckhardt as a new, egocentric pursuit of honour was in fact the frantic climax of a medieval longing for *realization*, the creation of idols and ideals. This conviction held by Huizinga was not supported by historical facts alone but was part of what I call Huizinga's moral anthropology: he understood paradigmatically human culture and creativity as the pursuit of recreation of traditional codes and ideals. Huizinga's anthropology appreciated creativity as re-creativity.
3. This moral anthropological disagreement over the creative potential of re-creation repeated Huizinga's experience of loss in response to the transformation of Dutch cityscapes. First, it was only in relation to these metropolitan developments and new aesthetics that Huizinga

> started expressing conservative ethics. In fact, in his student days, Huizinga would have been identified as a progressive youth, invested in cosmopolitan ideals and an international scientific community. Huizinga never let go of these ideals, yet during and after the 1910s, they were laced with more conservative and traditionalist elements. These elements took shape in opposition to the speed with which historical architecture and urban planning disappeared from Dutch cities and are fruitfully understood in this light. The seeming disappearance of a world enabled a new experience of 'the past' as a ruin, as an artefact that stood in opposition to the instrumental logic under which it crumbled. Through the urban modernization projects by which medieval and early modern aesthetics disappeared, the Middle Ages achieved their dialectic significance as a world and time that was driven not through instrumental creation but through a non-instrumental, devoted relation to traditional ideals and their continued recreation.

These findings put the reader in a good position to evaluate anew various statements made in previous literature on the relation between Huizinga's Middle Ages and Burckhardt's Renaissance. Werner Kaegi held that 'wo Burckhardt aufhörte, hat Huizinga begonnen' (Burckhardt ended where Huizinga began); Gombrich called Burckhardt '[Huizinga's] great predecessor'.[116] Now, to be sure, Burckhardt's work played a tremendous role in both Huizinga's historiographical and medievalist analyses. But this role, I have argued, was one neither of continuation (Kaegi) nor succession (Gombrich). Rather, the role was one of a retrospectively constructed dialogue and antagonism, wherein the parameters of discussion, argumentation and moral urgency were set by Huizinga's *fin-de-siècle* perspective and experiences, not by a logic inherent to Burckhardt's texts. For this reason, I agree with Guggisberg's conclusion, which in turn is similar to Tollebeek's statement 'daß [es] falsch wäre, [eine] Art Lehrer – Schüler Verhältnis zu konstruieren' (that it would be wrong to construct a kind of teacher-student relationship) between Burckhardt and Huizinga.[117] Similar to Kaminsky, I believe that 'even before [Huizinga] found the imagery of decline in the late-medieval texts he read, he had already found it closer to hand, in the *fin-de-siècle* cult of decadence and modernization that occupied his youthful years.'[118] Other than Kaminsky, however, I stress how Huizinga found these images not

116 Gombrich, 'Huizinga's *Homo ludens*', 146.
117 Guggisberg, 'Burckhardt und Huizinga – Zwei Historiker in der Krise ihrer Zeit', 172.
118 Kaminsky, 'From Lateness to Waning to Crisis: The Burden of the Later Middle Ages', 106.

only in literature and 'high culture' but in his actual and lived experiences of loss as well.

The next chapter, however, shall show that *Autumntide* was laced with another kind of experience of loss, too. Whilst this chapter has shown the role played by 'destructions in times of peace' in Huizinga's medievalism, to speak with Veth, the next chapter turns to the role played by destruction in times of war. After all, when *Autumntide* appeared in 1919, an entire continent, including Huizinga, had been forced to reckon with losses of an altogether different kind and scale for years. The collapses and catastrophes of the 1910s were of various kinds and thus resonated in more than one way in the 'Middle Ages' perceived by Huizinga.

Bibliography

Burckhardt, J. *Die Kultur der Renaissance in Italien*. Vienna: Phaidon-Verlag, 1934.

Burckhardt, J. *The Civilization of the Renaissance in Italy*. Oxford: Phaidon Press, 1981.

Chickering, R. *Karl Lamprecht: A German Academic Life (1856–1915)*. Atlantic Highlands, NJ: Humanities Press, 1993.

Dumolyn, J., and É. Lecuppre-Desjardin. 'Huizinga's Silence'. In *Rereading Huizinga: Autumn of the Middle Ages, a Century Later*, edited by P. Arnade, M. Howell, and A. van der Lem, 65–84. Amsterdam: Amsterdam University Press, 2019.

Ferguson, W. K. *The Renaissance in Historical Thought*. New York/Chicago/Dallas/Atlanta/San Francisco: The Riverside Press Cambridge, 1948.

Gafijczuk, D. 'Dwelling Within: The Inhabited Ruins of History'. *History and Theory* 52, no. 2 (2013): 149–70.

Gombrich, E. *In Search of Cultural History*. Oxford: Oxford University Press, 1969.

Gombrich, E. 'Huizinga's *Homo ludens*'. In *Johan Huizinga 1872–1972*, edited by W. R. H. Koops, E. H. Kossmann, and G. van der Plaat, 133–54. The Hague: Martinus Nijhoff, 1973.

Guggisberg, H. R. 'Burckhardt und Huizinga – Zwei Historiker in der Krise ihrer Zeit'. In *Johan Huizinga 1872–1972*, edited by W. R. H. Koops, E. H. Kossmann, and G. van der Plaat, 155–76. The Hague: Martinus Nijhoff, 1973.

Hanssen, L. *Huizinga en de troost van de geschiedenis: Verbeelding en rede*. Amsterdam: Uitgeverij Balans, 1996.

Hegel, G. W. F. *Lectures on the History of Philosophy*. Edited by E. S. Haldane and F. H. Simson. London: Kegan Paul, Trench, Trübner & Co., 1895.

Heijmans, B. *De Renaissance in Italië*. Zutphen: W. J. Thiemen & Cie, 1919.

Hockney, D., and C. M. Falco. 'Optical Insights into Renaissance Art'. *Optics and Photonics News* 11, no. 7 (2000): 52–59.

Hugenholtz, F. W. N. 'The Fame of a Masterwork'. In *Johan Huizinga 1872–1972*, edited by W. R. H. Koops, E. H. Kossmann, and G. van der Plaat, 91–103. The Hague: Martinus Nijhoff, 1972.

Huistra, P. *Bouwmeesters, zedenmeesters: Geschiedbeoefening in Nederland tussen 1830 en 1870*. Nijmegen: Uitgeverij Vantilt, 2019.

Kaegi, W. *Das historische Werk Johan Huizingas*. Leiden: Leiden University Press, 1947.

Kaminsky, H. 'From Lateness to Waning to Crisis: The Burden of the Later Middle Ages'. *Journal of Early Modern History* 4, no. 1 (2000): 85–125.

Kennedy, James C. 'Vooruitgang en crisis, 1870–1949'. In *Een beknopte geschiedenis van Nederland*. Amsterdam: Bert Bakker, 2017.

Kooy, P. 'Groningen en de industrialisatie van Nederland'. *Groniek*, no. 64 (1979): 13–22.

Krul, W. E. *Historicus tegen de tijd: Opstellen over het leven en werk van J. Huizinga*. Groningen: Historische Uitgeverij Groningen, 1990.

Krul, W. E. 'Nederland in het *fin-de-siècle*. De stijl van een beschaving'. *BMGN – Low Countries Historical Review* 106, no. 4 (1991): 581–94.

Krul, W. E. 'In the Mirror of Van Eyck: Johan Huizinga's "Autumn of the Middle Ages"'. *Journal of Medieval and Early Modern Studies* 27, no. 3 (1997): 353–84.

Lem, A. van der. *Johan Huizinga: Leven en werk in beelden en documenten*. Amsterdam: Wereldbibliotheek, 1993.

Lem, A. van der. 'The Making of "The Autumn of the Middle Ages" II: The Eagle and His Pigeonholes: How Huizinga Organized His Sources'. In *Rereading Huizinga: Autumn of the Middle Ages, a Century Later*, edited by P. Arnade, M. Howell, and A. van der Lem, 211–26. Amsterdam: Amsterdam University Press, 2019.

Lem, A. van der. *Verbreek nooit een oude rooilijn! Huizinga over stadhuis en stadsschoon in Leiden*. Leiden: Antiquariaat Klikspaan, 2021.

Otterspeer, W. *Orde en trouw: Over Johan Huizinga*. Amsterdam: De Bezige Bij, 2006.

Pree, C. du. *Johan Huizinga en de bezeten wereld*. Leusden: ISVW Uitgevers, 2016.

Raedts, P. *De ontdekking van de Middeleeuwen: Geschiedenis van een illusie*. Amsterdam: Wereldbibliotheek, 2014.

Ranke, L. *The History of the Popes During the Last Four Centuries*. London: G. Bell and Sons, 1913.

Ruehl, M. A. *The Italian Renaissance in the German Historical Imagination, 1860–1930*. Cambdrige: Cambridge University Press, 2015.

Rydin, T. 'Huizinga's "Heimwee": Responding to Burckhardt's "Die Kultur der Renaissance in Italien" in Times of Loss'. *History of European Ideas* (2020): 732–47.

Sas, N. C. F. van. 'Fin-de-siècle als nieuw begin. Nationalisme in Nederland rond 1900'. *BMGN – Low Countries Historical Review* 106, no. 4 (1991): 595–609.

Small, G. 'The Making of "The Autumn of the Middle Ages" I: Narrative Sources and Their Treatment in Huizinga's "Herfsttij"'. In *Rereading Huizinga: Autumn of the Middle Ages, a Century Later*, edited by P. Arnade, M. Howell, and A. van der Lem, 169–210. Amsterdam: Amsterdam University Press, 2019.

Small, G. 'Epilogue'. In *Autumntide of the Middle Ages*, 538–79. Leiden: Leiden University Press, 2020.

Steding, C. *Das Reich und die Krankheit der Europäischen Kultur*. Hamburg: Hanseatische Verlagsanstalt Hamburg, 1942.

Strupp, C. *Johan Huizinga: Geschichtswissenschaft als Kulturgeschichte*. Göttingen: Vandenhoeck & Ruprecht, 2000.

Tollebeek, J. *De toga van Fruin. Denken over geschiedenis in Nederland sinds 1860*. Amsterdam: Wereldbibliotheek, 1996.

Tollebeek, J. '"Renaissance" and "Fossilization": Michelet, Burckhardt, and Huizinga'. *Renaissance Studies* 15, no. 3 (2001): 354–66.

Vanwesenbeeck, B. 'Huizinga, Theorist of Lateness?' In *Rereading Huizinga: Autumn of the Middle Ages, a Century Later*, edited by P. Arnade, M. Howell, and A. van der Lem, 245–58. Amsterdam: Amsterdam University Press, 2019.

'Verbouwingen in Amsterdam'. *Het Nieuws van den Dag*. 22 June 1903.

Veth, J. 'Vredes-verwoestingen'. *De Gids* 80, no. 1 (1916): 507–27.

Wagenaar, M. 'De stad ontworpen: Stadsontwerp tussen wens en werkelijkheid'. In *Amsterdam in de tweede Gouden Eeuw*, edited by M. Bakker, R. Kistemaker, H. van Nierop, W. Vroom, and P. Witteman, 9–35. Bussum: Thoth, 2000.

3 An Irretrievably Lost Past

Abstract

This chapter examines how Huizinga's experiences of loss ensuing from the destruction of the Belgian city of Ypres during the Great War informed and changed his critique of the German cultural historian Karl Lamprecht (1856–1915).

Keywords: Johan Huizinga; World War I; Karl Lamprecht; historical determinism

> Forty is probably a critical age; one is no longer young, one notices that one's own future is no longer the general future, but only one's own. You have to live your life to the end – a life that has already been overtaken by the course of the world. [...] When the turning point in an individual life coincides with the thundering of a turning point in the world, the awareness of this becomes terrifying.[1]
>
> – Thomas Mann in *Reflections of a Non-Political Man* (1918)

In the spring of 1913, Huizinga's wife Mary (née Schorer, 1877–1914) was operated on for a 'lump in the chest'.[2] At first the intervention seemed to have delivered; Mary was lively and in good spirits soon thereafter and quickly returned to her former routines. A few months later, however, her health declined again, and periodic disruptions to both her physical and mental well-being started taking over her life. In January 1914, she was

1 T. Mann, *Reflections of a Nonpolitical Man*, ed. Morris W. D. (New York: New York Review of Books, 2021), 9. 'Vierzig Jahre sind wohl ein kritisches Alter, man ist nicht mehr jung, man bemerkt, daß die eigene Zukunft nicht mehr die allgemeine ist, sondern nur noch – die eigene. Du hast dein Leben zu Ende zu führen, – ein vom Weltlauf schon überholtes Leben. [...] [Wenn] die Wende des persönlichen Lebens von den Donnern einer Weltwende begleitet und dem Bewußtsein furchtbar gemacht wird.' T. Mann, *Betrachtungen eines Unpolitischen* (Frankfurt: Fischer, 1974), 14.

2 BW I: Huizinga–Van Anrooy (1913): 101. See p. 11 for further details on the references to Huizinga's collected works and letters as well as to the Huizinga archives.

Rydin, T., *The Works and Times of Johan Huizinga (1872–1945): Writing History in the Age of Collapse.* Taylor & Francis Group 2023
DOI: 10.5117/9789463724593_CH03

diagnosed with a brain tumour.[3] That summer, Mary spent a stretch of weeks hallucinating, talking at a rapid pace but without much coherence. Then, suddenly, she fell silent: 'she has become most calm; she is still confused, but does not say much, her eyes have dimmed. Still, she recognizes us and seems happy when we come,' Huizinga wrote to his brother Jakob on 18 July.[4] The silence persisted, and three days after this letter was posted, it continued indefinitely: 'tonight the end came, without consciousness, without suffering.'[5] Five children lost their mother. Huizinga's sorrow overwhelmed him, and when the Great War erupted a week later, Huizinga's mind was clouded by grief: 'I cannot write about the war.'[6] In the hope of getting back on his feet, Huizinga moved to Leiden with his children in 1915 to take up a new position.[7] His grief travelled with him, but gradually, over the course of a few years, Huizinga grew increasingly able to occassinally lift his gaze beyond himself and his children. By 1917 he *could* write about the war. Coming from the Professor of General History at Leiden University, Huizinga's reflections could count on an audience.

During the first couple of months of 1918, and in a somewhat premature but not entirely unwarranted anticipation of the Great War's conclusion, the Belgian architect Huib Hoste (1881–1957) sent out letters to at least eighty recipients with the following question: should the Flemish city of Ypres be restored from an aesthetic, art historical, national or international point of view – if at all?[8] Since the war's outbreak, three major battles had been fought in and around Ypres. The most recent one had taken place in the autumn of 1917 and had become known as the Battle of Passchendaele. Those months, hellish forces of destruction had roared over the city. Ypres was not just damaged, compromised or scarred; it was obliterated. Hoste sent his questionnaire mainly to architects, artists and academics, and included among the recipients was Johan Huizinga, who at that point had held his chair at Leiden University for nearly three years. Throughout the 1900s and most of the 1910s, Huizinga had been an outspoken advocate of the preservation and restoration

3 A. van der Lem, *Johan Huizinga: Leven en werk in beelden en documenten* (Amsterdam: Wereldbibliotheek, 1993), 128.

4 BW I: Huizinga–Jakob Huizinga (1914): 142.

5 BW I: Huizinga–Van Anrooy (1913): 143.

6 BW I: Johan Huizinga–Jakob Huizinga (1914): 146.

7 'The hope to find new courage and a renewed ability to work weighs heavily in my decision [to take this role],' Huizinga wrote to his brother in September 1914. BW I: Johan Huizinga–Jakob Huizinga (1914): 149. Huizinga took up his position in Leiden in January 1915. BW I: Johan Huizinga–Jakob Huizinga (1915): 157.

8 Hoste's questionnaire was also discussed in Dutch newspapers, see e.g. *Algemeen Handelsblad* 19-02-1918: 6.

of historical landmarks, yet Huizinga's reply to Hoste was as negative as it was short.[9] Concerning the medieval city of Ypres, Huizinga judged:

> Let us accept the catastrophe as irreparable.[10]

To Huizinga and countless others, the Great War symbolized a discontinuity, a breaking-away of the past from the present in more than one sense. Against the background of the gripping grief over Mary's death, the war not only shook Huizinga's confidence in the ability to preserve historical sites in a world of industrial means but also affected his appreciation of historical inquiry itself. Before the war, Huizinga had considered historical inquiry to be ultimately the process of 'objectively' recovering past 'truths'.[11] After the war, Huizinga recast the historical discipline's value in another, existential language: historical inquiry answered to a basic 'need of life' to reconcile with the past; it could 'cure [us] of egocentrism', allow us to 'discard the borders of [our] narrow [personalities]', enable 'serene resignation', and teach us how to become 'stoic'.[12] Already during the initial aftermath of Mary's passing and the war, Huizinga had started appreciating the potency of 'stoicism' in the face of 'incomparable evils'.[13] Huizinga never discussed how his historiographic interests transitioned from the recovery of historical 'truth' to the therapeutic engagement with the 'needs of life', and it has gone unnoticed in Huizinga scholarship. This development, however, offers a window into the relation between Huizinga's experiences of loss and his historical thought.[14] Huizinga's history was in a state of shock, and historical reflection soon became a way of life amidst loss.

The present chapter examines this transition in Huizinga's historiography against the background of those losses he experienced between 1914 and 1918. For this purpose, the development of Huizinga's historiographic position during this period is teased out in terms of his opposition to the methodological works of Karl Lamprecht (1856–1915). Huizinga's dialogue with Lamprecht's works formed a backbone of his historiographical output,

9 For examples of Huizinga's ethics of preservation before the Great War, see e.g. VW VIII: *Geschiedenis der universiteit uedurende de derde eeuw van haar bestaan, 1814–1914* (1914): 321; (2) VW II: *De beteekenis van 1813 voor Nederland's geestelijke beschaving* (1913): 528; VW I: *Over de oudste geschiedenis van Haarlem* (1907): 365–66. See the first section of the previous chapter for a more exhaustive account of Huizinga's ethics of preservation.

10 BW I: Huizinga–Hoste (1918): 205.

11 VW VII: *Het aesthetische bestanddeel van geschiedkundige voorstellingen* (1905): 26–27.

12 VW VII: *De taak der cultuurgeschiedenis* (1929): 41, 65.

13 'stoïcisme', 'maux incomparables', BW I: Huizinga–Pirenne (1917): 195.

14 For an impression of how other Dutch historians responded to the Great War, see C. du Pree, *Johan Huizinga en de bezeten wereld* (Leusden: ISVW Uitgevers, 2016), 42–55.

and for this reason, the antagonism functions as a helpful proxy for the former's historiographic development.[15] In short, this chapter argues that: (1) there was, indeed, a discursive shift in Huizinga's historiography during the Great War; (2) after this shift, Huizinga's critiques of Lamprecht became increasingly concerned with the cultivation of a 'virtuous self'; and (3) this shift tied directly into an experience of tremendous upheaval caused by the war and Huizinga's personal grief. Mary's death and the outbreak of the war fed into a new awareness and, later, conviction: the past was such that it could be irretrievably lost. This sense of loss came as a shock to Huizinga. He and his peers had grown up with narratives of progress and a sense of excitement concerning what the future might bring. Now, images of tragedy, loss and uncertainty lay before his mind's eye.

By committing to a virtue-centred reading of Huizinga's later historiography, this chapter draws from and critically adds to the arguments of Christoph Strupp, who has offered the most detailed account to date of Huizinga's 'experience of war' and its relation to his 1920s historiography.[16] 'In the capacity of historian,' Strupp argues, 'Huizinga tried to keep the relativity of actual events before his mind's eye. [...] One did not need four thousand years before all this would seem as inconsequential as the Assyrian wars. Precisely Huizinga's intense historical consciousness offered him an escape [route].'[17] This chapter sides with Strupp's virtue-ethical perspective on Huizinga's post-war historiography, but only so as to give it more content. Strupp correctly argues that the value of history to life stood left, right and centre in Huizinga's historiography. Yet, Strupp does not go

15 E.g. VW VII: *Het aesthetische bestanddeel van geschiedkundige voorstellingen*: (1905): 10; *De taak der cultuurgeschiedenis* (1927): 47; *De wetenschap der geschiedenis*: (1934): 118. Alongside these publications, Huizinga produced one review of Lamprecht's *Einführung in das historische Denken. Ordentliche Veröffentlichung der pädagogischen Literatur* (1912), VW VII: *Boekbespreking van K. Lamprecht's* Einführung in das historische Denken (1913): 233.

16 'Erfahrung des Weltkriegs'. C. Strupp, *Johan Huizinga: Geschichtswissenschaft als Kulturgeschichte* (Göttingen: Vandenhoeck & Ruprecht, 2000), 56.

17 'Huizinga versuchte, sich als Historiker die Relativität der aktuellen Ereignisse vor Augen zu führen. [...] Es werde keine viertausend Jahre dauern, dann werde dies alles der Menschheit ebenso gleichgültig sein wie die Kriege Assyriens. Gerade Huizinga intensives historisches Bewußtsein verstellt ihm aber den Weg der [Flucht].' Strupp, *Johan Huizinga: Geschichtswissenschaft als Kulturgeschichte*, 54–55. Strupp uses Jörgen Rüsen's concept of *Daseinsorientierung* to denote the development of this virtuous dimension in Huizinga's historiography. Strupp, *Johan Huizinga. Geschichtswissenschaft als Kulturgeschichte*, 44. For Rüsen's original invocation of *Daseinsorientierung* as a historiographical category, see J. Rüsen, *Historische Vernunft. Grundzüge einer Historik I: Die Grundlagen der Geschichtswissenschaft* (Göttingen: Vandenhoeck & Ruprecht, 1983), 29. See the conclusion for a more critical analysis of this category's supposed value to understanding Huizinga's historiography.

beyond carefully chosen references to Huizinga's writing; he does not show *how* this was actually the case for Huizinga. This chapter jumps precisely into this gap, illustrating how Huizinga's experiences surfaced in and were mediated by his historiographic works.

For these purposes, the chapter's main body has been subdivided into four sections. The first section explores Huizinga's Dutch *Weltkriegserfahrung* through his personal correspondence and lectures from 1914 to 1919.[18] The second section then turns to Karl Lamprecht's *Methode* and presents those features of Lamprecht's work that occupied Huizinga most before and after the war. The third section, in turn, examines how Huizinga's experiences of loss from 1914 to 1918 returned in his *Autumntide of the Middle Ages* (1919) and in its critique of Lamprecht's *Deutsche Geschichte* (1891–1909) in particular, marking a move away from Huizinga's earlier objections to Lamprecht. Here it is argued that Huizinga's 'experiences of loss' during the war led to the emergence of an existential virtue-ethical concern in his historiographic perspectives. A fourth section picks up where the third ended by showing how a number of these experiences and convictions returned time and again in Huizinga's later discussions of both Lamprecht's work and historiography more generally.

Ypres and the 'irreparable' disappearance of the past

On 22 February 1918, Huizinga replied to Hoste's aforementioned questionnaire on the restoration of the Belgian city of Ypres. Alongside the questionnaire, Hoste included photographs of what Ypres had become after its third battle, and only after seeing the visuals did Huizinga realize precisely how unfathomable the destruction of Ypres and other Belgian cities had been. Whilst Huizinga had generally been in favour of the restoration of historical cities, as has been discussed in the previous chapter, he responded to Hoste: 'I have no doubts about my answer. Let us accept the catastrophe as irreparable, and not try to create the appearance that this evil could be

18 Here I will draw from Schulin's concept of a generational *Weltkriegserfahrung*. See E. Schulin, 'Weltkriegserfahrung und Historikerreaktion' (Frankfurt: Fischer Taschenbuch Verlag, 1997). In the past two decades, Dutch historians have argued for the analytical value of a Dutch *Weltkriegserfahrung*, see e.g. F. Boterman, *Tussen utopie en crisis* (Amsterdam: De Arbeiderspers, 2021), 12; A. van der Woud, *Het landschap en de mensen* (Amsterdam: Prometheus, 2020), 134–40; J. M. J. Sicking, 'Woede en verontwaardiging: De Eerste Wereldoorlog in de Nederlandse literatuur', *Armada* 14, no. 52 (2008): 44–52; P. Moeyes, *Buiten schot: Nederland tijdens de Eerste Wereldoorlog* (Amsterdam/Antwerp: Arbeiderspers, 2001).

made undone through reconstruction, which is bound to disappoint.'[19] Up until this point in time, Huizinga had defended the idea that culture rested on an inherent activity of re-creation and re-production, and he must have realized that his cultural-historical output so far could barely have been more strongly at odds with his reply to Hoste.[20] Later, Huizinga would continue to defend the idea that culture rested on the ability to reproduce codes from the past, yet his attitude and tone on the topic had changed after he had received Hoste's photographs of Ypres. Some damage could be irreparable, some parts of the past could be lost, and some components of history could be shattered to dust and irretrievably dispersed in the winds.

Huizinga's response to the photographs of Ypres fitted into a wider transition of both his personal and academic self.[21] Before the war, Huizinga had not only been a keen advocate of historical renovations in the architectural domain; he believed the historian, too, had an intellectual obligation to pursue an 'objectively' truthful reconstruction of past events.[22] In his piece on the history of Haarlem from that same year, Huizinga for instance insisted on appreciating the 'infinite multiplicity' of 'facts' resisting a 'synthetic' reconstruction in terms of 'real' relations and 'real objects'.[23] Huizinga continued to understand the historian's task roughly along these lines until at least 1915. That year, Huizinga delivered his inaugural lecture at Leiden University, and he used the occasion to address the expectations of his audience directly:

> Perhaps you have come here with the idea that only one single topic could possibly occupy anyone meditating on world history at this moment in time: the historical background to the world war. If this is the case, I

19 BW I: Huizinga–Hoste (1918): 205.

20 See the previous chapter's first section.

21 The most extensive account of Huizinga's experience of and relation to the Great War has been given in L. Hanssen, *Huizinga en de troost van de geschiedenis: Verbeelding en rede* (Amsterdam: Uitgeverij Balans, 1996), 75–80. Hanssen's account could be described as rather impressionistic; the present section hopes to give a more directed understanding of how one *could* study this relation.

22 This was a common attitude among the Dutch historians of his day. P. Huistra, *Bouwmeesters, zedenmeesters: Geschiedbeoefening in Nederland tussen 1830 en 1870* (Nijmegen: Uitgeverij Vantilt, 2019), 145; H. Paul, 'The Scholarly Self: Ideals of Intellectual Virtue in Nineteenth-Century Leiden', in *The Making of the Humanities Volume II: From Early Modern to Modern Disciplines*, ed. R. Bod, J. Maat, and T. Weststeijn (Amsterdam: Amsterdam University Press, 2012), 397–412; J. Tollebeek, 'Wetenschap en waardering: Nederlandse historici in verzet tegen het positivisme (1890–1910)', *Groniek*, no. 124 (1994): 70–73.

23 VW I: *De opkomst van Haarlem* (1905): 206–7

> will disappoint you. I will not speak about it. If I happen to call upon the stormy winds in your thoughts exalted by the war, it will be unintentional. It is not the historian's task to speak like Demosthenes in the storm, and that clear day, when he can observe peoples and states floating along the course of time as white clouds on a summer's day, remains far removed.[24]

This lecture was given against the background of the million Belgian refugees who had come to the Netherlands in 1914, the mobilization of Dutch forces, the intensification of Dutch military service, the destruction of Leuven and the publication of the German *Manifest der 93*, wherein German academics, including Karl Lamprecht, justified the levelling of Leuven and its university library.[25] With these and other events in mind, Huizinga concluded that the war and its development were, their blatant horror notwithstanding, still beyond the historian's reach. The Greek statesman Demosthenes (384–322 BC) had practised and developed his oratory capacities on stormy seashores; the historian, on the other hand, was not in a position to speak during stormy times. Historical reflection, Huizinga maintained, could not travel through storms of passions, for passions lead to partial perspectives and, ultimately, to self-deceit (*zelfbedrog*).[26] For the sake of intellectual and moral integrity, historical inquiry into the Great War would have to wait for the storm to pass.[27]

The storm, however, proved too violent and too enduring to await its dying down in mere silence, even for a historian such as Huizinga. At this point,

24 'Misschien zijt gij hier gekomen met de gedachte, dat er voor iemand, die op dit tijdstip inzichten omtrent wereldgeschiedenis te belijden heeft, slechts één onderwerp mogelijk is: de historische achtergrond van den wereldoorlog. Is dit zoo, dan zal ik u teleurstellen. Ik zal niet daarover spreken. Wanneer ik ten slotte nog even den stormwind mocht oproepen, dien de oorlog door uw denken jaagt, dan zal het mijns ondanks zijn. Het is niet het werk van den historicus, om als Demosthenes in den storm te spreken, en de heldere dag, waarin hij volken en staten langs de baan van den tijd kan zien voorbij drijven als witte wolken aan een zomerhemel, is thans verre.' VW IV: *Over historische levensidealen* (1915): 411.

25 The manifesto was published and discussed in both regional and national Dutch newspapers, see e.g. *Hoornsche Courant* 08-10-1914, *Het Volk* 16-10-1914 and *Arnhemsche Courant* 01-12-1914.

26 VW IV: *Over historische levensidealen* (1915): 423.

27 But what, then, should the historian do? Huizinga addressed this question too, however briefly, towards the lecture's end: if the historian is to continue to carry 'the burden of history' in times of turbulence, s/he will have to resign her own 'selfhood'. Only by removing one's observational perspective from one's selfhood, by turning to the duties of life, professional or other, can one resist the intellectually corrupting forces of passion. And though history might offer 'only a beautiful dream of peaceful perfection', it also offers a relieving forgetfulness. And, Huizinga was quick to add, this is more than one could say of the future in 1915. VW IV: *Over historische levensidealen* (1915): 431.

Huizinga had fallen into deep grief after the premature death of Mary just before the Great War erupted.[28] Now a widower with the responsibility of five surviving children resting on his shoulders, Huizinga's interest in topics of an academic and political nature evaporated.[29] When the war broke out, Huizinga's world was consumed by the care of his children, and only a few years after Mary's death did Huizinga's interest in the war start to take an articulate shape, not least through personal correspondence with his friend and Belgian colleague Henri Pirenne, whose work had become an important point of reference to Huizinga's research on the late Middle Ages. Huizinga and Pirenne had started corresponding in 1908, and as the war grew more violent, eventually leading to Pirenne's detention by the Germans, Huizinga became deeply invested in Pirenne's fate. Whilst imprisoned, Pirenne was still in a position to receive and send letters, and their exchange of 1917 especially left a deep impression on Huizinga. The grief over his wife and the concern over Pirenne tied into a sense of vulnerability that came to symbolize to him the ongoing war in its totality:

> I have been touched by everything you say in your letter of the happiness, neutrality and stoicism that war must engender in people's minds. It is all true, all of this. As for the first, I always felt it during the terrible years that we have been living in, and I was often surprised when we experienced cases of evils that these were still incomparable to those of the present war. This winter will probably be a little harsher, but it can only be useful if we also suffer a little of what the whole world is suffering.[30]

In this letter, Huizinga for the first time used 'stoicism' in an appreciative manner while applying it to his own life and times. Later on in the correspondence, but still that same year, Huizinga repeated his appreciation again: 'stoic virtues made you strong enough to endure the physical ailments (of long captivity).'[31] As shall be explored in this chapter's fourth section,

28 The effects of Mary's death on Huizinga have been vividly described in Van der Lem, *Johan Huizinga: Leven en Werk in Beelden en Documenten*, 124–28.

29 E.g. BW I: Johan Huizinga–Jakob Huizinga (1914): 146, 147, 150.

30 'J'ai été bien touché de tout ce que vous dîtes dans votre lettre du bonheur des neutres et du stoïcisme que la guerre doit engendrer dans les esprits. C'est bien vrai, tout cela. Quant au premier, je l'ai senti toujours durant les années terribles que nous passons, et je me suis étonné souvent quand on faisait beaucoup de cas des maux incomparables à ceux de la guerre actuelle. Cet hiver ce sera un peu plus rude probablement, mais il ne peut qu'être utile si nous souffrons aussi un peu de ce que souffre le monde entier.' BW I: Huizinga–Pirenne (1917): 195.

31 BW I: Huizinga–Pirenne (1917): 195, and later in e.g. BW I: Huizinga–Pirenne (1918): 240.

this appreciation of 'stoicism' later migrated to Huizinga's technical historiographical work. History teaches us how to become 'stoic', Huizinga wrote in the 1920s; its investigation reveals the way to 'serene resignation', he defended in the 1930s.[32] For now, what matters is not this later migration but that the experiences of war and loss coincided with Huizinga's new appreciation of a stoically virtuous character formation, and his appreciation of serenity and resignation more specifically. His most outspoken appreciation of such 'peaceful' virtues would come just after the war's conclusion. During the opening lecture of the 1919/20 academic year, Huizinga said to his students:

> I do not know whether you feel the need for lectures on history, but I need them in order to hold on to history – for a long time I have felt history escape me. Despair [vertwijfeling]: how could I continue to teach history? I no longer understand it and I no longer care for it. [...] The consequence of the war? Initially, I also felt that all the past had lost its importance. [What] now?' [...] The medicine – to work and have hope, to not despair of humanity, society and civilization. [...] To see how everything oscillates back neither to an ideal state nor to mere deplorability, but to a tolerable earthly imperfection and relative and deficient expediency, wholeness and happiness. [...] Then peace with history returns, and the love for people and the things of the past.[33]

In 1919 Huizinga's patience of 1915 had eroded and disappeared. To be sure, he held on to the same ideals of impartiality and moral selflessness, but by 1919 these ideals had been transformed into an altogether different creature. Historical inquiry had explicitly become the means of cultivating ideals in one's character, of practising virtues. The tone in this later lecture is fundamentally different from the 1915 lecture: an existential stake shines through.

32 VW VII: *De taak der cultuurgeschiedenis* (1929): 65. VW VII: *De wetenschap der geschiedenis* (1937): 157.

33 'Ik weet niet of gij 't gevoel hebt, college over geschiedenis te behoeven, maar ik heb ze noodig, om mij vast te houden aan geschiedenis, – want lang gevoeld alsof geschiedenis mij ontzonk. Vertwijfeling: hoe moet ik verder geschiedenis onderwijzen? Ik begrijp het niet meer en het kan mij niet meer schelen. [...] Het gevolg van den oorlog? In 't begin ook dat gevoel: al dat oude heeft zijn belangrijkheid verloren. [En] nu? [...] Het geneesmiddel – werken en hoop hebben, niet wanhopen aan de menschheid, de maatschappij en de beschaving. [...] Zien hoe alles terugschommelt, niet naar een ideaaltoestand, ook niet naar volstrekte verwerpelijkheid, maar naar een dragelijke aardsche onvolmaaktheid en betrekkelijke en gebrekkige doelmatigheid, heilzaamheid en gelukkigheid. [...] Dan keert ook de vrede met de historie terug, en de liefde voor de menschen en de dingen van 't verleden.' See HA 9.I.2 (1919), and Van der Lem, *Johan Huizinga: Leven en Werk in Beelden en Documenten*, 154–55.

In this lecture, historical inquiry is concerned not with the safeguarding of objective judgment but with the stability of inner life amidst turmoil, uncertainty, vulnerability and even despair. Huizinga had lost interest in historical minutiae ('I no longer care for it'), which, he now held, had rested on 'the illusion of progress'. Nevertheless, he still felt the need to 'hold onto history' and to re-establish a 'peace' of mind. The 'history' he wished to hold onto was not one of facts and disengaged reflection but a mode of historical thought and investigation that was able to cultivate certain frames of mind, character traits and virtues: hope and love for a world that simply is what it *is* – neither perfect nor abominable – so as to dispense with despair. In this lecture, history had ceased to be a reservoir of impartial statements; it was a reservoir of practices from which historians could attain an impartial and moral frame of mind in turbulent times.

To be sure, Huizinga's invocation and usage of 'stoicism' in the 1910s and later on should not be read in an overly technical capacity. When Huizinga adduced the value of a 'stoic' frame of mind, he was not mobilizing the ins and outs of an ancient ethic. Instead, he employed the word 'stoic' in a rather loose capacity alongside celebrations of virtues such as calmness, stability, modesty and an insensitivity to violent ruptures of passion. In a more critical fashion, one might say that Huizinga used 'stoicism' as a secularized term for the Christian virtues of ascetism which he had been brought up with and which he had cultivated as a young student in the 1890s.[34] In his 1924 biography of Erasmus, Huizinga lauded him for '[expressing] the meaning of philosophy in his life and virtues' rather than in 'theses'.[35] In 'stoic' fashion, Erasmus 'learned to scorn treasure not through artificial syllogisms but through disposition, through his facial expressions and eyes, through life itself. To live according to that norm is what Christ himself calls "Renascentia"'.[36] Huizinga was well aware that Christian virtues could pass through secular customs unnoticed when dressed in classical attire.

The textual testimonies discussed so far carry the features of Gafijczuk's historical experiences: in 1914–18 Huizinga found in personal loss and political catastrophe a new sense of an irreversibly lost past ('the catastrophe is irreparable') in the present. Huizinga's history was in ruins; or rather,

34 See Chapter 5 for a more detailed discussion of the generational virtues of the students of the 1890s.

35 VW IV: *Erasmus* (1924): 108.

36 'Hij is in waarheid theoloog, die niet door kunstige syllogismen, maar door zijn gezindheid, door zijn gelaat en zijn oogen, door zijn leven zelf leert dat men schatten moet versmaden. Naar die norm te leven dat is wat Christus zelf 'Renascentia' noemt.' VW VI: *Erasmus* (1924): 108.

Figure 3.1. (A) An undated photograph of Ypres's Cloth Hall from before the war. (B) It is not known which photographs of Ypres Hoste added to the questionnaire he sent to Huizinga. Most likely, they looked something similar to the bottom image, which was taken immediately after the Battle of Passchendaele.

'ruins' had now become his history.[37] At the age of forty-two, ruptures of different kinds shook the shackles that had connected his past to his future, and a sense of irreparable, irreversible damage had become central to his experience of life, time and history itself. The events of 1914–19 – funerals, the routines and challenges of raising children, photos, letters and newspapers – had revealed themselves as markers of history's discontinuity. From these experiences Huizinga induced an altogether new relation to the

37 Gafijczuk, 'Dwelling Within: The Inhabited Ruins of History', 150.

Figure 3.2. (A) A group of professors from the University of Leiden receive military training in the summer of 1915. Johan Huizinga is the fourth person from the left, just left of the standing lieutenant. (B) An undated photograph taken during the Great War in Amsterdam, most likely from 1914. It shows a stream of Belgian refugees exiting the central train station. Over one million Belgian refugees crossed the Belgian-Dutch border during the first months of the war.

'past' and 'history'. He no longer solely or even primarily looked to extract objective facts from historical materials and memories. Rather, historical investigations figured as the resource for virtues such as modesty, resilience and wisdom, a means of character formation in stormy times. Past ideals no longer applied to the present or the future the way they had, and this insight itself became a virtue-ethical resource. The repeatability of conduct on which the categorical imperative rested had been defused by time's apparent discontinuity.

In sum, this section has set the stage for the upcoming discussion of one of the most significant transitions in Huizinga's oeuvre: his turn towards a virtue-ethical, stoic understanding of the purpose of historical inquiry in the course of 1914 to 1919. Against the background of the loss of his wife, the ensuing grief and the responsibility of raising five children alone, as well as the horrific destruction of war to which his friends abroad were also subjected and whose consequences soon reached Dutch society, a particular development in Huizinga's correspondence and lectures becomes salient: historical inquiry transitioned from a more or less disengaged mode of investigation into a way of life, a way of developing a character able to withstand the experiences of rupture and uncertainty that flooded his mind and that of others. When history no longer appeared to be continuous the way it had seemed before, when it appeared as irretrievably lost, the value of historical inquiry had to be reimagined. Throughout this transition – before, during and after – Huizinga read and reflected on the methodological works of Karl Lamprecht. Before the role of Huizinga's experiences of loss in his critiques of Lamprecht are examined, Lamprecht's *kulturhistorische Methode* shall take centre stage: who was Huizinga's methodological antagonist and imagined partner in conversation throughout this development?

Lamprecht's laws

No evidence exists of any correspondence between Huizinga and Lamprecht, and it seems probable that they never spoke in person, but as shall be discussed later, their paths likely crossed in Leipzig in 1895.[38] Regardless, like most European historians in the 1900s and 1910s, Huizinga was aware of Lamprecht's works. Huizinga first came across Lamprecht's work around 1900 whilst he prepared his research on the early medieval history of the city of Haarlem, and by 1905 he had become particularly familiar with Lamprecht's then partially published *Deutsche Geschichte*, which was mentioned above, as well as Lamprecht's *Die kulturhistorische Methode* (1900). In his inaugural lecture at the University of Groningen in 1905, Huizinga discussed Lamprecht's latter publication to a significant extent. After the war, as shall soon be discussed, Huizinga's interest in deflating Lamprecht's historiographic method only increased. In order to appreciate the developments of Huizinga's historiography during the Great War as well as his critiques of Lamprecht more specifically, the central questions

38 Lamprecht visited the Netherlands at least once, in the 1890s. A. van der Lem, *Het eeuwige verbeeld in een afgehaald bed: Huizinga en de Nederlandse beschaving* (Amsterdam: Wereldbibliotheek, 1997), 406.

are now: to Huizinga, what were the most salient and upsetting features of Lamprecht's works? What did Lamprecht's theories of history symbolize to him? For this purpose, I turn to Lamprecht's methodological writings and the lines along which Huizinga read and understood them.

As the son of a Lutheran pastor in a small Saxon village, Karl Lamprecht (1856–1915) had been rocketed straight into what is now commonly called the nineteenth-century culture of German *Bildungsbürgertum*: a middle class with the spiritual ideals of 'high culture' and the professional ideals of state officials.[39] As a trained historian within this social-cultural space, however, Lamprecht endlessly frustrated his peers, most of whom researched in the tradition of Ranke and the Prussian School. Lamprecht was less interested in the Rankean details of history than he was in the 'laws of history' which he discussed in his writings on German history and historiography and taught starting in 1909 at his *Institut für Kultur- und Universalgeschichte* in Leipzig. In Roger Chickering's words, 'Karl Lamprecht's impulsive energies were not disciplined by patience, caution or a disposition to self-criticism. [The] signs of carelessness and haste that marred his early scholarship betrayed the price he paid.[40]

In an essay titled *The Cultural-Historical Method* (*Die kulturhistorische Methode*) from 1900,[41] Lamprecht rejected a number of historical schools and programmes; he dismissed what he called the 'theory of ideas', 'philosophical world history' and 'economic social-political history'.[42] More particularly, he criticized Ranke and that generation of 'so-called neo-rankeans'.[43] Each of these programmes, schools and authors, Lamprecht argued, had succumbed by different means to the same error: they had falsely assumed history to be 'synchronic' (*synchronistisch*).[44] By this, Lamprecht meant that they – even Ranke – had assumed certain principles (about which Lamprecht remained

39 Chickering, *Karl Lamprecht: A German Academic Life (1856–1915)*, 22.

40 Chickering, *Karl Lamprecht: A German Academic Life (1856–1915)*, 108.

41 Lamprecht had been a professor at the University of Leipzig since 1891, where he succeeded Georg Voigt (1827–1891). Chickering, *Karl Lamprecht: A German Academic Life (1856–1915)*, 113–14.

42 'Ideenlehre', 'philosopische Weltgeschichte' and 'wirtschaftlich-sozialpolitische Geschichte', K. Lamprecht, *Die kulturhistorische Methode* (Berlin: R. Gaertners, 1900), 22, 45, 26.

43 'die sogennante Jungrankianer', Lamprecht, *Die kulturhistorische Methode*, 24. Here Lamprecht would most likely have had the likes of Max Lenz (1850–1932) and Georg von Below (1858–1927) in mind.

44 Lamprecht stated: 'Zunächst mußte der holde Wahn aufgegeben werden, daß das historische Geschehen auf der Welt synchronistisch verläuft, etwa wie die geologischen Zeitalter, die, überall auf dieser Erdrinde von denselben großen astronomischen und atmosphärischen Einflüssen abhängig, im Tiefsten einen vollen Synchronismus aufweisen.' Lamprecht, *Die kulturhistorische Methode*, 42.

vague) to be outside of history, atemporal, resisting the diachronic maturation of culture. In order to become truly inductive, the historical discipline had to turn to the natural sciences, and to biology in particular.[45] After all, in terms of inductive sensitivity, he argued, 'there is no doubt that the natural sciences even today have developed further than the historical science.'[46] The natural sciences, and physics, chemistry and biology in particular, Lamprecht held, had accepted change and movement (*Bewegung*) rather than static laws to be the most fundamental categories of investigation, and it was in precisely this sense that history could learn from them:

> Are there parallels in the humanities to the position that mechanics, physics and chemistry and finally the biological disciplines have in natural science? The humanities encompass the sciences of the mental processes, mental movements and mental energy in the same way as the natural sciences encompass those of the physical. It is now obvious that an inductive psychology, which does not depend on any particular deductive, metaphysical system, must be the basic science of mental phenomena in the same way as mechanics is for physical phenomena.[47]

The natural sciences had progressed by diverting their attention from continuity to change and difference, and history ought to do the same, Lamprecht argued in this passage. History's object should be the 'mental movements' (*seelische Bewegungen*): a study of the 'mental energy' (*psychischen Energie*). Consequently, the historical sciences were to become the 'sciences occupying themselves with the historical development of mental lives'.[48] The natural sciences studied movement (*Bewegung*), so the historical science had to study development (*Entwicklung*), the human soul and its maturation through time. For this purpose, Lamprecht found, 'the sciences' (the natural sciences)

45 Chickering, *Karl Lamprecht: A German Academic Life (1856–1915)*, 43–45.

46 'es unterliegt keinem Zweifel, daß die Naturwissenschaften noch heute verhältnismäßig weiter entwickelt sind als die Geschichtswissenschaft.' Lamprecht, *Die kulturhistorische Methode*, 8.

47 'Gibt es in den Geisteswissenschaften Parallelen zu der Stellung, welche die Mechanik, die Physik und Chemie, welche schließlich die biologischen Disziplinen in der Naturwissenschaft einnehmen? Die Geisteswissenschaften umfassen die Wissenschaften der seelischen Vorgänge, der seelischen Bewegungen, der psychischen Energie, wie die Naturwissenschaft die der physischen. Da liegt nun auf der Hand, daß eine induktive, nicht von irgendwelchen besonderen deduktiven, metaphysischen Systemen abhängige Psychologie ebenso die Grundwissenschaft der geistigen Erscheinungen sein muße wie die Mechanik der physischen.' Lamprecht, *Die kulturhistorische Methode*, 12–13.

48 'Wissenschaften, die sich mit der geschichtlichen Entwicklung des Seelenlebens beschäftigen.' Lamprecht, *Die kulturhistorische Methode*, 13.

had to be introduced to the soul rather than leaving the soul to be studied by an insular and autonomous historical science.

Lamprecht used a host of terms to describe his 'natural' historical position as well as its alliances: 'natural history', 'developmental psychology', 'ethno-psychology', 'sociology', 'historical ethnology', 'scientific world history' and 'the biology of the historical spirit of life'. Yet the most important term was 'cultural history' (*Kulturgeschichte*).[49] Its central task (*Aufgabe*) was to develop 'a theory of typical cultural stages', which Lamprecht, in turn, understood in terms of 'types'.[50] These 'types' referred to 'natural' cause-and-effect relations, whose existence the historical discipline simply had to presuppose (*Voraussetzung*) 'in the spiritual field' in order to expel ahistorical metaphysics.[51] For this purpose, the 'types' had to be composed of concepts implying causal relations between events. The only theoretical continuity that historians may speak of, Lamprecht argued, was an 'uninterrupted connection of cause and effect'.[52] By identifying history's object – 'spiritual movement' and 'development' – and adopting a causal methodology, the historian could uncover laws similar to natural laws. In this capacity, Lamprecht wrote of 'psychological' (*psychologische*) and 'mental laws' (*seelische Gesetze*), lauding insights by Marquis de Condorcet (1743–1794) and Auguste Comte (1798–1857).[53]

With these analytical tools in hand, Lamprecht's aim was to compose nothing less than a new historical periodization of German and European history. Unhappy with what he found to be the anachronistic categories of traditional periodization – Antiquity, Middle Ages, Renaissance, Enlightenment, etc. – Lamprecht used his vocabulary to compose new categories in *Deutsche Geschichte* (German history): *Symbolismus* (<350 AD),

49 'Naturgeschichte', 'Psychologie der Entwicklung', 'Völkerpsychologie', 'Soziologie', 'historische Ethnologie', 'wissenschaftliche Weltgeschichte', 'die Biologie des geschichtlichen Geisteslebens', Lamprecht, *Die kulturhistorische Methode*, 7–8, 14, 32, 32, 30, 21, 16.

50 '[Eine] Lehre der typischen Kulturstufen', Lamprecht, *Die kulturhistorische Methode*, 45. Later Huizinga, too, would use the word, long before he first read Weber. Possibly, he drew from Burckhardt's usage of 'Typ'. See e.g. VW VII: *Het aesthetische bestanddeel van geschiedkundige voorstellingen* (1905): 5; *De wetenschap der geschiedenis* (1937): 132.

51 'auf geistigem Gebiete', Lamprecht, *Die kulturhistorische Methode*, 34.

52 'ununterbrochenen Zusammenhang von Ursache und Wirkung', Lamprecht, *Die kulturhistorische Methode*, 34.

53 Lamprecht was met by tremendous opposition in Germany, but his reception in France, especially among the later members of the *Annales* school, was much friendlier. The fact that Lamprecht never delivered on his promises (that is, never spelled out the laws of his *wissenschaftliche Weltgeschichte*) apparently did not diminish its appeal, see Chickering, *Karl Lamprecht: A German Academic Life (1856–1915)*, 81. For Lamprecht's references to Condorcet and Comte, see Lamprecht, *Die kulturhistorische Methode*, 33.

Typismus (350–1050 AD), *Konventionalismus* (1050–1450 AD), *Individualismus* (1450–1700 AD) and *Subjektivismus* (>1700 AD).[54] To be sure, Lamprecht's interest in labels did not obstruct his interest in detail. His revision of German history 'presented simply the richest history of Germany that had ever been written', according to Chickering. The next section shall show how Huizinga's *Autumntide* objected especially to Lamprecht's understanding of the second and third categories, *Typismus* and *Konventionalismus*, and how this opposition revealed a new and (post-)war historiography on Huizinga's side. For present purposes, however, what matters is that Lamprecht suggested a new historical typography that was scientific in its supposed ability to lay bare the psychological dynamics of human history.

Lamprecht's typological ideals were crucial to his understanding of what constituted decent historical research. From his conception of 'type', for one, Lamprecht deduced that the role Rankeans had ascribed to aesthetics in historiography had to be fundamentally reconsidered. In fact, Lamprecht was stricter than most on this point; he categorically denounced aesthetics in historical writings. According to him, aesthetics concerned only the description of a particular case, and considering his 'historical science' to be one of lawfulness, 'every attempt to describe the individual, be it rough or precise, belongs not to science, but to art.'[55] To Lamprecht, then, aesthetics was not only a matter of, say, extra-factual niceties of either a rhetorical or moral nature. To him, 'aesthetics' referred to something more general, namely arbitrariness: the arbitrariness of any given historical fact and its details. History, he found, concerned law, structures and types; that was what 'history', properly understood, referred to. The 'aesthetic' facts themselves were meaningless when not accommodated by patterns.

Before we return to Huizinga and the world in which he read Lamprecht, three features of Lamprecht's historiography in particular need to be kept in mind: (1) the historical object concerns the causes of the soul (*das Psychische*) and its respective development through natural time (*Entwicklung*); (2) this causal development is captured in terms of identifying 'types' (*Typen*) that facilitate this causal interrelation, so, as such, Lamprecht attributed a significant role to language in the historical method; and lastly, (3) history's newly acquired scientific status categorically ruled out any role for the aesthetic imagination. Throughout his historiographic output, Huizinga criticized Lamprecht's method, but the central target of these critiques

54 Chickering, *Karl Lamprecht: A German Academic Life (1856–1915)*, 120.

55 '[Jeder] Versuch der Umschreibung des Individuellen, er sei roh oder fein, gehört daher nicht der Wissenschaft an, sondern der Kunst.' Lamprecht, *Die kulturhistorische Methode*, 6.

developed over time. Before the war, Huizinga took issue with Lamprecht's anti-aestheticism; after the war, Huizinga objected to Lamprecht's understanding of causality. The war resurfaced in Huizinga's perspective on historical inquiry and its value. But how? How did Huizinga's budding interest in stoic virtues resurface in his historical outlook?

Two perspectives on a church

By the time Huizinga became Professor in Dutch and General History at the University of Groningen in 1905, he had become well acquainted with Lamprecht's *German History*.[56] Like Lamprecht, Huizinga had serious reservations about what both dismissively considered Rankean 'antiquarianism', but he had even greater reservations about Lamprecht's 'scientific history'.[57] First and foremost, Huizinga faulted Lamprecht for an alleged empirical insensitivity. Lamprecht's method fed into artificial periods and laws, and his rationality was harmful rather than conducive to the pursuit of 'objective truth', Huizinga argued in his 1905 inaugural lecture.[58] Instead, and drawing from a range of recent neo-Kantian works, Huizinga suggested an 'aesthetic' alternative to both the Rankean particularism and Lamprecht's

56 Huizinga's appointment was not without controversy: see e.g. BW I: Huizinga–Brom (1919): 256. An overview of the controversy has been given in Van der Lem, *Johan Huizinga: Leven en werk in beelden en documenten*, 108–12. After Bussemaker's death, Blok successfully vouched for Huizinga to the Dutch Department of Internal Affairs (which administered the appointment of academic positions at the time). For a description of how this appointment came about, see Van der Lem, 108–12. For references to Lamprecht in Huizinga's *The Emergence of Haarlem*, see VW I: *De opkomst van Haarlem* (1905): 205, 210, 216–17, 257.

57 Like his predecessor Bussemaker, and despite his deep sympathies towards Blok, Huizinga harboured reservations about the Rankean tradition. At that point, Blok was the face of the Dutch Rankean tradition together with his own predecessor Robert Fruin (1823–1899). Huistra, *Bouwmeesters, zedenmeesters: Geschiedbeoefening in Nederland tussen 1830 en 1870*, 151. For an impression of Huizinga's appreciation of Fruin and Blok, see VW VI: *Prof. Dr. P. J. Blok* (1909); VW VI: *Robert Fruin* (1940). For a discussion of the wider anti-Rankean tendencies in the Netherlands at the time, see P. Raedts, *De ontdekking van de Middeleeuwen: Geschiedenis van een illusie* (Amsterdam: Wereldbibliotheek, 2014), 265–66; Tollebeek, 'Wetenschap en waardering: Nederlandse historici in verzet tegen het positivisme (1890–1910)'; L. J. Dorsman, 'Periodisering als integrale benadering: Nederlandse historici in het Fin-de-Siècle', *Theoretische geschiedenis* 16, no. 3 (1989): 287. On the other hand, Huizinga held onto the Rankean disdain for rationalized historical speculation. H. Paul, 'Performing History: How Historical Scholarship Is Shaped by Epistemic Virtues', *History and Theory*, no. 50 (2011): 1–19.

58 'objectieve waarheid'. VW VII: *Het Aesthetische Bestanddeel van Geschiedkundige Voorstellingen* (1905): 27.

Gesetze.[59] On this occasion, Huizinga's opposition to Lamprecht was a moral one steeped in generational identities: his aestheticism was typical for an author navigating the tensions between *De Tachtigers* and *De Negentigers*, and in 1915 Huizinga repeated similar convictions during his inaugural lecture in Leiden. However, by the time Huizinga's *Autumntide* came out in 1919, something had changed; his opposition to Lamprecht's historiography had acquired an entirely new dimension.

This new dimension is helpfully illustrated through a comparison of the appreciation of aesthetic 'Gothicism' in Lamprecht's *German History* and Huizinga's *Autumntide*. The differences between their respective depictions of the period reveal a substantial change in Huizinga's historiographic convictions between 1915 and 1919. As has been mentioned above, Lamprecht understood the word in terms of a taxonomical category to be fitted in a larger historical lineage. Before, say, 1915, Huizinga would have objected to the ontological status Lamprecht ascribed to the term 'Gothicism': historical periods did not allow exhaustive and *real* definitions. In *Autumntide*, however, a new opposition shone through. Huizinga's understanding of

59 *'Aesthetische ontvankelijkheid'*. VW VII: *Het Aesthetische Bestanddeel van Geschiedkundige Voorstellingen* (1905): 21. More particularly, Huizinga referred to Simmel's *Probleme der Geschichtsphilosophie* (1892), Windelband's *Geschichte und Naturwissenschaft* (1894), Rickert's *Die Grenzen der naturwissenschaftlichen Begriffsbildung* (1896–1902) and Spranger's *Die Grundlagen der Geschichtswissenshaft* (1905). VW VII: *Het aesthetische bestanddeel van geschiedkundige voorstellingen* (1905): 5. Huizinga's appreciation of neo-Kantian authors has understandably led numerous commentators to frame Huizinga as a neo-Kantian historian, see e.g. D. G. Shaw, 'Huizinga's Timeliness', *History and Theory* 37, no. 2 (1998): 245–58; Anchor, 'History and Play: Johan Huizinga and His Critics'. This deduction is, however, unwarranted: it does no justice to Huizinga's prose and textual techniques. Neither his lecture nor his later books on the matter employed transcendental-critical argumentation. In fact, his arguments are often passionate, his prose is often loose and his terminology is predominantly empirical rather than analytical. These features of the text are by no means peripheral to the text's function and meaning and ought to be appreciated on their own terms. As shall be explored shortly, Huizinga's antipathy towards a 'system' was performative of the virtues he ascribed to historical inquiry. To this effect, a similar argument can be found in Strupp, *Johan Huizinga. Geschichtswissenschaft als Kulturgeschichte*, 49. What drew Huizinga to neo-Kantian authors, I hold, was not their vocabulary, argumentative strategies or deontological moral laws; what attracted him to neo-Kantianism was the space it prepared to reason from within subjectivity and its epistemic confinements. By recognizing the inherent unknowability of certain events, the historical discipline could be, as it were, freed from the yoke (and illusion) of having to truthfully restore past events. That Huizinga gravitated towards these points had less to do with the persuasiveness of neo-Kantianism than with the Christian epistemic virtues they allowed to be perpetuated. This was not uncommon at the time; neo-Kantianism was commonly understood as first and foremost a theory of value-ethics, H. Schnädelbach, *Philosophy in Germany 1831–1933*, trans. E. Matthews (Cambridge: Cambridge University Press, 2009), 161–62.

'Gothicism' in *Autumntide* reflected the therapeutic potential he had recently started ascribing to historical inquiry: the late medieval Gothic aesthetic had to be understood in terms of its ability to balance, guide and stabilize the 'passions'. 'Gothicism' was a 'style' of practising and directing the soul. Huizinga's 'Gothicism' showcased a new historiographic purpose. In order to appreciate this new purpose, we turn first to Huizinga's antagonist: Lamprecht's understanding of *Gotik*.

In 1891 Lamprecht obtained the position of Professor of Medieval History at the University of Leipzig, where he would remain for the rest of his career and life, founding the aforementioned *Institut für Kultur- und Universalgeschichte* along the way. From this appointment onwards, until 1909, he worked on the twelve-volume work on German history that was mentioned above, and by 1892 he was in the process of completing the third volume on the period 1050–1450, which he dubbed the 'age of conventionalism' (*Zeitalter des Konventionalismus*).[60] The period's Gothic aesthetic, he argued, was exemplary of how the taste and style at the time had been subjected to narrow social constraints. The architects of the period, Lamprecht argued, had been looking solely to maximize the symbolic density of surface areas available through the aesthetic norms in place: 'virtue now means social property, [and] morality its rules.'[61] An obsession with codes and rules had crippled the period's ability to appreciate individual occurrences; events and ornaments achieved significance only in terms of predetermined roles, categories and taxonomies.

Within the grander scheme of his *Deutsche Geschichte*, Lamprecht's 'age of conventionalism' functioned as the historical springboard for what was in fact the central object of both Lamprecht's moral and historical interests: the Lutheran Reformation and the pure individualism it had supposedly enabled. The church's Gothic aesthetic was only a chapter in a succession of ages, an announcement of times to come within a more exhaustive diachronic chain of events. Roger Chickering has for this reason described Lamprecht's *Deutsche Geschichte* as 'an immense exercise in collection and classification. Every detail of the past was to be fitted into a hierarchy of categories, at the top of which stood Lamprecht's five historical periods.'[62] The last historical period, to which all previous periods had amounted, was that of a Protestant 'subjectivism', and in this sense, the project was 'an attempt to ground the cultural authority of the imperial German *Bildungsbürgertum* in the context

60 Chickering, *Karl Lamprecht: A German Academic Life (1856–1915)*, 126.
61 Chickering, *Karl Lamprecht: A German Academic Life (1856–1915)*, 125.
62 Chickering, *Karl Lamprecht: A German Academic Life (1856–1915)*, 130.

of the nation's full historical experience'.[63] Lamprecht, Chickering argues, was invested in exposing the historical authority and perhaps even the historical necessity of Protestantism.

In Leipzig Lamprecht would have hardly been able to find a more potent example of Protestantism's succession over Gothicism than the Thomaskirche (St. Thomas Church), which had been built in the fifteenth century in the typical Gothic style of its times. The church, however, symbolized at once the cultural epochs that were later to overcome and condemn its aesthetic. Luther had preached here in 1539, and from the 1720s onwards, *Kapellmeister* Johann Sebastian Bach (1685–1750) had composed the sound of what Lamprecht held to be the pinnacle of Protestant subjectivism. In the 1890s the Leipzig cultural elite congregated at the Thomaskirche every Saturday to hear renditions of Bach performed by the church's boys' choir, and though he did not discuss the Thomaskirche in his *Deutsche Geschichte*, it seems more than likely that Lamprecht, a Bach enthusiast, would have visited this church on one such occasion. One could then imagine how Lamprecht may have closed his eyes, if even for just a moment, so as to heighten his senses and experience how a Gothic aesthetic gave way to baroque tones. The Thomaskirche would have been the stage for his beloved Bach, and only just that – a stage. And if, by any chance, he had attended in late 1895, he might have, upon reopening his eyes, seen a young, lost-looking student tucked away in one of the church's corners.

In the autumn of 1895, just after obtaining today's equivalent of a master's degree in philology, Huizinga spent one winter term in Leipzig to study with the linguist Karl Brugmann.[64] Huizinga was young – twenty-three when he arrived and twenty-four when he left – and from his correspondence letters at the time as well as his later recollections, it seems that he had a rather lonely experience in Leipzig. Both socially and academically, Huizinga was ill at ease in Leipzig, and he thus had ample time to head over to the Thomaskirche each Saturday to hear the boys' choir perform, which he did.[65] Whilst seated in front of its central altar, a somewhat desolate and still boyish Huizinga heard modern renditions of Bach's baroque interpretations of a late medieval genre of music: the *motet*, a polyphonic vocal composition that had been particularly popular in late medieval and early Renaissance

63 Chickering, *Karl Lamprecht: A German Academic Life (1856–1915)*, 137.

64 For a discussion of Huizinga's term in Leipzig, see Van der Lem, *Johan Huizinga: Leven en werk in beelden en documenten*, 44–48.

65 Huizinga lived on Kurprinzstraße, today called Grünewaldstraße, only a few blocks away from the Thomaskirche. BW I: Huizinga–Veth (1895): 2.

times. These Saturdays, different times morphed into each other, and a young Huizinga would have had what he would later call 'a historical sensation': a sense of traversing temporal boundaries. If only for an hour, the choir dismissed temporal limitations, catapulting a nineteenth-century world into a baroque world of Gothic tones. It is not unlikely that Huizinga's and Lamprecht's paths crossed one such Saturday afternoon.

The concerts at the Thomaskirche made a tremendous impression on Huizinga. And as the Great War reached its conclusion, Huizinga had not forgotten about the concerts he attended at this church in 1895.[66] The experiences of secession following the war would have contributed to the temporal complexity of the Thomaskirche for Huizinga: the church's Gothic architecture symbolized at once a world before the Reformation, a world before revolution, a world before maturity, a world before world war and now a world after it.[67] In this temporally layered, contemporary sense, the Gothic aesthetics were performative of the symbolic saturation Huizinga attributed to the aesthetic of Gothicism in his *Autumntide*:

> The flamboyant Gothicism is like an endless organ play: she dissolves all forms through self-disintegration, gives to each detail its continued effect, to each line its counter line. It is an unfettered proliferation of form over the idea; the decorated detail affects all surfaces and lines. In this art, the *horror vacui* reigns, which perhaps characterizes all spiritual ages that are reaching their conclusion.[68]

66 Huizinga ended up recounting his appreciation of the concerts he had attended at the Thomaskirche in his autobiographical essay *My Path to History* from 1943. This period of Huizinga's life figures prominently in Chapter 4.

67 Huizinga's prose in *Autumntide* is heavy with metaphors, allegories and ornamental descriptions of passions. This observation has, in turn, led recent authors to argue that Huizinga's *Autumntide* ought to be appreciated either (1) as a product of *fin-de-siècle* symbolism or (2) as a performative investigation of the supposed decadence of the late medieval mind. The observations in the present section – regarding Lamprecht and the memories of the Thomaskirche – add a third and complementary alternative to the menu: the prose of *Autumntide* reflected the contemporary experiences of a world in seemingly continuous transition, a world in which a single object, such as a church, had multiple temporal vectors, symbolizing different times at once replete and exhausted with histories. This third alternative treats *Autumntide* not only as a textual object but as an object of experience: as the Great War tore away 'a world before the war' so violently and so drastically from 'the present', the world was augmented by another temporal layer, one highlighting rupture, decline and 'autumnal' tones. Huizinga's Middle Ages, thus perceived, were the Middle Ages of the Great War. I return to this point in this section's conclusion.

68 'De flamboyante gothiek is als een eindeloos orgelnaspel: zij lost alle vormen op in zelfontbinding, geeft aan elk détail zijn voortgezette doorwerking, aan elke lijn haar tegenlijn. Het is

Figure 3.3. (A) A drawing of the Thomaskirche from 1749, by Joachim Ernst Scheffler. (B) A postcard image of the Thomaskirche displayed from the other side from 1918. The church's outer construction underwent a number of modifications during the nineteenth century.

One can only speculate whether Huizinga had those Saturday afternoons he spent amidst the Gothic aesthetics of the Thomaskirche in mind when he wrote on Gothicism in terms of an 'endless organ play'. Either way, in the passage above, Huizinga recounted a medieval experience along the lines discussed in the previous chapter: the late medieval mind sought to exhaust its life with forms, ornaments and symbols in its pursuit of a meaningful existence amidst an otherwise meaningless vacuum. A *horror vacui* whipped and jolted the soul into symbolic thought and practice. In its over-heated symbolic activity, the late medieval mind was confronted with its impotence. The more symbols it cooked up, the more they turned into meaningless platitudes, and on this particular point, Huizinga felt a close proximity to this period. This stage of symbolic exhaustion was, according to Huizinga, by no means particular to the fourteenth and fifteenth centuries. They belonged to all 'spiritual ages reaching their conclusion', and to Huizinga it seemed that the 1910s had achieved just that: as empires fell, political systems changed, and new uncertainties took central stage, the nineteenth century had properly been exhausted – '[what] now?'[69]

To Huizinga, late medieval Gothicism and his own times were contemporaries through their congruence in crisis, and this fact, in turn, reveals how Huizinga's antagonism with Lamprecht's historiography had

een ongebonden woekeren van den vorm over de idee; het versierde détail tast alle vlakken en lijnen aan. Er heerscht in deze kunst die horror vacui, die misschien een kenmerk van eindigende geestesperioden mag heeten.' VW III: *Herfsttij der Middeleeuwen* (1919): 312.

69 HA: 9.I.2 (1919).

an immediate historical counterpart. As was examined earlier on in this chapter, Huizinga had come to believe over the course of the Great War that historical investigations had the ability and duty to inspire a 'love' for the world's imperfections. Regarding this duty, the 'self-disintegration' effectuated by the baroque aesthetic was, if not ideal, at least of interest. Here, mediated by a supposed anthropological continuum, the late medieval experience of crisis could be shared by a contemporary mind looking to accommodate a similarly overwhelming sense of *horror vacui*. A moral-pedagogical potential could be released by those instances when a spiritually exhausted mind had reached rock bottom. Thus, Huizinga's historiographic aims were part and parcel of the actual history he wrote: history and historians were not dictated by laws of development and progression, as Lamprecht had contested. Rather, the historian had to identify points of contact between historical periods, and these points of contact lay in crisis. At these temporal intersections, Huizinga held, the historian could put into practice the virtues of historical inquiry. Concerning the late Middle Ages, Huizinga wrote:

> The passionate and violent mind, hard yet at the same time rich in tears, always staggering between a black despair towards the world and a revelling in her varied beauty, could not live beyond the strictest forms of life. It was necessary that the affections were contained in a fixed window of calibrated forms; only thus could society acquire order at least as a rule. Consequently, the events of one's own life and those of others turned into a beautiful play for the mind; one enjoyed the pathetic outfits of suffering and joy in the light of art.[70]

The Gothic medieval mind had found a way to deal with violent passions, despair, affections and pathetic outfits amidst conditions of crisis and suffering. How? Through forms: by sticking to the forms of custom, play and rituals, the medieval mind could distance itself from its most violent afflictions and perceive forms in terms of beauty in 'the light of art'. These forms, Huizinga held, introduced a 'noble harmony' and brought a 'rhythm'

70 'De hartstochtelijke en gewelddadige geest, hard en tevens tranenrijk, altijd wankelend tusschen de zwarte vertwijfeling aan de wereld en het zwelgen in haar bonte schoonheid, kon niet buiten de strengste vormen van het leven. Het was noodig, dat de aandoeningen waren gevat in een vast raam van geijkte vormen; zoodoende kreeg het samenleven althans in den regel orde. Zoo werden de eigen levensgebeurtenissen en die van anderen tot een schoon schouwspel voor den geest; men genoot de pathetische uitmonstering van leed en geluk onder kunstlicht.' VW III: *Herfsttij der Middeleeuwen* (1919): 57.

to pain.[71] As Huizinga raked over his manuscript for *Autumntide*, carefully preparing it for publication in 1917–19, the contemporary times were on his mind. In fact, in the prologue of the first edition, Huizinga wrote that the project had functioned as 'the mirror of the spirits of times': 'whilst writing this book, my gaze has been directed to the depths of a nocturnal sky – one drenched in blood, heavy and furious with the haunting grey of lead, replete with false copper appearances.'[72] Huizinga's take on the medieval culture of play and rituals was a pedagogical device for a particular way of accepting and experiencing the affections of suffering without becoming a mere ball for the violent play of passions. And this was exactly what set his appreciation of Gothicism apart from Lamprecht's:

> In the reduction of everything to the general lies that quality Lamprecht has called typism, which he has elevated into being the more distinctive feature of the medieval mind. She [the medieval mind] is, however, rather a consequence of that subordinating need of the mind that springs from an embedded Idealism. It is not so much an inability to see the special in things as it is a conscious intent to see the meaning of things in relation to the highest, their virtuous idealness, their general meaning.[73]

Here, in relation to his interpretation of late medieval Gothicism, Huizinga put into practice the convictions he had started developing during the Great War: history was primarily neither a matter of establishing an inventory of events nor of carving out hierarchic periodizations. Instead, history had a pedagogical duty to answer to what Huizinga would later start calling the 'needs of life'. In his take on the 'flamboyant Gothicism', Huizinga opposed both Lamprecht's historiographic and his medieval historical perspectives: the Gothic aesthetic was not a diachronic phase within a more exhaustive development. Rather, it was a synchronic slice of a way of life that had sought to accommodate *horror vacui*, a set of symbolic practices answering

71 VW III: *Herfsttij der Middeleeuwen* (1919): 64, 74.

72 'De blik is bij het schrijven van dit boek gericht geweest als in de diepten van een avondhemel, – maar van een hemel vol bloedig rood, zwaar en woest van dreigend loodgrijs, vol valschen koperen schijn.' VW III: *Herfsttij der Middeleeuwen* (1919): 3.

73 'In die herleiding van alles tot het algemeene ligt de eigenschap, die onder den naam typisme door Lamprecht als de bij uitstek kenmerkende van den middeleeuwschen geest is gesteld. Zij is echter veeleer een gevolg van die onderschikkende behoefte van den geest, welke voortspruit uit het ingewortelde Idealisme. Het is niet zoozeer een onvermogen om het bijzondere aan de dingen te zien, als de bewuste wil om overal den zin der dingen aan te duiden in hun betrekking tot het hoogste, hun zedelijke idealiteit, hun algemeene beteekenis.' VW III: *Herfsttij der Middeleeuwen* (1919): 262.

to the very same need modern minds had, which is the need to establish meaning in life.[74]

At this point, I turn to a concern that might have struck the critical reader. The previous chapter understood *Autumntide* in the light of an 'ethics of preservation' by emphasizing its investment in historical 'reconstruction'. This section, on the other hand, has presented a part of *Autumntide* as a response to Huizinga's experience of an 'irreparable loss' and historical inquiry's ensuing need to come to terms with the past having been irretrievably lost. Regarding this tension, I wish to make two points. First, I think the tension is telling of the fact that *Autumntide* came about in a period that spanned over a decade and thus bore the marks of several episodes and experiences. Second, the tension opens an interesting window on Huizinga's post-war historiographic works. Huizinga wrote at least another three historical books after the war, and for all of these, he of course continued to read source material critically and diligently. He had not given up on the ideal of an 'objective' retrieval of historical facts in this sense. However, unlike in his pre-war works, and much in line with what has been said about *Autumntide* in this section, Huizinga repeatedly emphasized a new, therapeutic value of this retrieval, and this reorientation reappeared in his later (and increasingly technical) critiques of Lamprecht's *Methode*.

Huizinga's opposition to Lamprecht's *Methode* after 1919

Huizinga's growing appreciation of the self-cultivating potential of historical inquiry stuck with him, and by the late 1920s, Huizinga started describing the value of historical inquiry in terms of its ability to answer to the 'needs of life'. Historical inquiry, he then started arguing more systematically, was a way of organizing the mind, taming the passions and achieving serenity.[75] In these later historiographic works, Huizinga again presented his historical theory as a critique of Lamprecht's scientific understanding of history. Yet unlike his 1905 inaugural lecture, Huizinga did not continue to fault Lamprecht for his insensitivity to empirical and 'objective truth' – at least

74 Against this background, Huizinga's appreciation in 1919 of Burckhardt's *Weltgeschichtliche Betrachtungen* (1905) can be properly understood. Huizinga's recognized and appreciated the self-fashioning ability Burckhardt ascribed to historical inquiry in these lectures. As the previous chapter discussed, Huizinga had a different understanding and apprecation of that 'other' Jacob Burckhardt, the one who had authored *The Civilization of the Renaissance in Italy* (1860).

75 VW VII: *De taak der cultuurgeschiedenis* (1929): 41.

not primarily.[76] From the 1920s on, and in line with what he had already put into practice in *Autumntide*, Huizinga faulted Lamprecht's *Systemzwang* for its alleged inability to aid and appreciate the needs of the soul and the means by which a virtuous mind may be cultivated. The fact that Huizinga continued to fault Lamprecht for 'subjective' rather than 'objective' reasons after the Great War is itself indicative of the reoccurrence of a conviction developed during this war. In fact, Huizinga's two later historiographic writings – *The Task of Cultural History* (1929) and *The Science of History* (1937) – repeated arguments and convictions against Lamprecht that had first surfaced in Huizinga's 1917–18 correspondence. This last section examines the sense in which the Great War reappeared in the arguments these books launched against Lamprecht.

Juxtaposed with his 1905 inaugural lecture, an interesting commonality between these two books is revealed: Huizinga became and remained, if not less interested in the objective recovery of past facts, at least increasingly occupied with the value of historical research to the 'needs of life'.[77] In line with his opening lecture of 1919/20, previously discussed, Huizinga captured the purpose of historical inquiry in terms of the 'peace' and 'love' that historical research could instil in the human mind.[78] Of course, by 1929 and 1937, Huizinga was occupied with concerns quite distinct from the most immediate consequences of the Great War. By then, images of mass unemployment, the increased mechanization of labour, populist politics, a normalization of political violence and another, looming world war had taken a hold of Dutch media, and these episodes shall be discussed at length in upcoming chapters.[79] For now – and this point is of primary importance to the present chapter – what matters is that the historiographic works written by Huizinga in later decades reiterated the mark left by those experiences examined at this chapter's outset.[80] Huizinga's interest in the stoic virtues

76 'objectieve waarheid'. VW VII: *Het aesthetische bestanddeel van geschiedkundige voorstellingen* (1905): 27.

77 VW VII: *De taak der cultuurgeschiedenis* (1929): 41; VW VII: *De wetenschap der geschiedenis* (1937): 152. Judging from these publications, it is hard to say to which extent Huizinga may have been inspired by Nietzsche's *Vom Nutzen und Nachteil der Historie für das Leben* (1874). He mentioned this essay once in *De wetenschap der geschiedenis* but did not expand on his understanding and appreciation of it. VW VII: *De wetenschap der geschiedenis* (1937): 152.

78 HA: 9.I.2 (1919).

79 These later developments and their effect on Huizinga's historical outlook are discussed especially in relation to Huizinga's critiques of Spengler and Schmitt.

80 Huizinga himself, too, stated publicly that he expected the war to resonate in his work long after its conclusion when he said in 1919: 'We, who believed peace would come when war ended, experience once again that we always look to childishly simplify this confused world.' The original

of historical inquiry took shape in the period 1915–19, and this interest was reproduced and remediated in different contexts in the next decades.

First, a look at Huizinga's *The Task of Cultural History* (1929) is in order. The book was Huizinga's first proper historiographic publication, apart from an essay that had appeared in 1926.[81] As in his inaugural lecture of 1905, Lamprecht figured as one of Huizinga's main partners in conversation. Other than in 1905, Lamprecht was now faulted not for his aesthetics (or the lack thereof) and sociology but for his biological concept of *Entwicklung*. Lamprecht was one of those 'unshaken evolutionists' who imagined themselves to have the 'key' to all ages and times. For them, the universe had no more 'riddles', Huizinga remarked scathingly,[82] and in their works, a certain 'catholicity of knowledge' had been lost 'by no small degree'.[83] The gravity and importance of this 'loss' to Huizinga was not a theological concern. The 'loss of catholicity' was, and also in the case of Lamprecht, a virtuous concern: a loss of interest in particularities and 'antiquarian interest'.[84] By presupposing a biological relation of *Entwicklung* between occurrences, Lamprecht had robbed historical events of their character of 'riddles', making them mere 'shackles' in a chain. Virtues of humbleness and curiosity had dissipated.[85] Thus, Huizinga proposed an alternative: 'history' was not an 'evolution' of life but was itself a 'need of life' or a way of life.[86]

> Every culture is conditional upon being steeped in the past to a certain extent. In every civilization, certain images of a past reality exist that are dear to the community carrying this civilization. These images come in essentially different forms, but this difference does not absolve their communal character, because to the culture from which they spring, they

reads: 'Wij, die meenden, dat het vrede werd, als er een oorlog eindigde, ervaren opnieuw, dat wij altijd bezig zijn, deze verwarde wereld in onzen geest kinderlijk te vereenvoudigen.' VW VIII: *Toespraak tot de Maatschappij der Nederlandsche Letterkunde te Leiden* (1919): 477. My attention was brought to this talk by Du Pree, *Johan Huizinga en de bezeten wereld*, 110.

81 The book consisted of five independent essays addressing related themes: (1) a discussion of the anatomy of historical research questions, (2) the improper usage of the term 'development' in historical analyses, (3) the negative consequences of history's popularization, (4) the morphological nature of historical understanding and, lastly, (5) the proper role of historical periodization in historical inquiry.

82 'Ongeschokte evolutionisten'. VW VII: *De taak der cultuurgeschiedens* (1929): 50.

83 'Katholiciteit der kennis'. VW VII: *De taak der cultuurgeschiedens* (1929): 40.

84 'Antiquarische belangstelling'. Here Huizinga referred explicitly to Nietzsche's *Vom Nutzen und Nachteil der Historie für das Leben* (1872). VW VII: *De taak der cultuurgeschiedens* (1929): 41.

85 VW VII: *De taak der cultuurgeschiedens* (1929): 56.

86 VW VII: *De taak der cultuurgeschiedens* (1929): 56.

> are 'history'. [These images] fulfil not only a need of life, but also a need of truth. When the faith in the narrative dies, the form's productive time comes to an [end]. [The] less developed a culture is, within which forms realize themselves, the more evidently their [those forms'] relation to the culture is given away.[87]

Huizinga found that Lamprecht's interest in historical causality and the trans-historical evolution of culture did not amount to a 'narrative' through which a given culture could meaningfully relate to the past. Similar to his objections to Lamprecht in *Autumntide*, Huizinga faulted Lamprecht for not being interested in the 'needs' that historical actors tried to meet through their actions. Lamprecht's causality robbed past events of their symbolic content and, in effect, of the value these historical events may have had to readers and researchers in the present. In other words, causality of this kind fed into a retrospective arrogance through which the past could not be truly 'dear' to the historian. Causality of this kind turns history into a passed station unable to actually address the contemporary, spiritual needs of history's students. By giving content to the idea of an overarching totality (e.g. causal *Entwicklung*), historical inquiry loses its ability to address and cultivate those parts of the human soul where ethical virtues and moral longings reside. Historical inquiry, Huizinga held, requires a respect for the unknown:

> This not entirely transparent contact with the past concerns the entrance of a sphere – it is one of many forms given to the *mensch* whereby one treads outside oneself, and experiences a truth. It is neither an aesthetic appreciation [*kunstgenot*], nor a religious affection [*religieuze aandoening*], nor a trembling before nature [*natuurhuivering*], nor a metaphysical perception [*metaphysisch erkennen*], but it belongs in this chorus.[88]

87 'Elke cultuur van haar kant heeft als levensvoorwaarde een zekeren graad van gedrenkt zijn in verleden. In iedere beschaving leven zekere beelden van vroegere werkelijkheid, die de gemeenschap, welke die beschaving draagt, aangaan, ter harte gaan. Die beelden nemen essentieel verschillende vormen aan, zonder dat dit hun gemeenschappelijk karakter opheft, van voor de cultuur, die ze voortbrengt 'historie' te zijn. [...] Zij vervullen niet alleen een levensbehoefte, maar ook een waarheidsbehoefte. Sterft het geloof in de waarheid van het verhaalde af, dan is de productieve tijd van den vorm voorbij[.] [...] Hoe minder ontplooid de cultuur is, waarin die vormen zich realiseren, hoe duidelijker zij hun verband met den cultuur verraden.' VW VII: *De taak der cultuurgeschiedens* (1929): 58–59.

88 'Dit niet geheel herleidbare contact met het verleden is een ingaan in een sfeer, het is een der vele vormen van buiten zich zelf treden, van het beleven van waarheid, die den mensch gegeven zijn. Het is geen kunstgenot, geen religieuze aandoening, geen natuurhuivering, geen

Historical inquiry had the potential to negatively suggest a totality, not despite but precisely because of a continuous inconclusiveness. On this point, Huizinga was still in agreement with his 1905 inaugural lecture. Huizinga likened this alluding experience to 'treading outside oneself', similar to a 'metaphysical recognition', and he called this experience a 'historical sensation'.[89] Unlike the case in 1905, and in line with the views he had started developing in the late 1910s, Huizinga no longer took history to be an objective retrieval of past particulars. Rather, he now understood historical inquiry as an irreducible way of life, a form of thought in an overwhelming world of change and transition. Similar to his letters to Pirenne from 1917, and similar to his opening address of 1919, a proper dealing with images of the past seemed to be of the utmost importance to leading a good life. Whether one likes it or not, Huizinga argued, the world around us suggests a temporal dimension that is 'the past', and it simply must be accounted for.[90] It is in this vein that he introduced the term 'necessity' or 'need' of 'life' to describe the role of historical reflection.[91] These negotiations between present and past were, in turn, what Huizinga understood to be 'culture', and these negotiations could not be subjected to fashions and whims: 'history may become democratic, she must remain stoic.'[92]

In 1937 Huizinga's second historiographic monograph was published: *The Science of History* (*De wetenschap der geschiedenis*).[93] In this book, Huizinga again took up the project of objecting to the role the natural sciences had

metaphysisch erkennen, en toch een figuur uit deze rei.' VW VII: *De taak der cultuurgeschiedens* (1929): 71.

89 The value of Huizinga's 'historical sensation' to contemporary debates has been discussed most recently in e.g. M. Jay, 'Historical Explanation and the Event: Reflections on the Limits of Contextualization', *New Literary History* 42, no. 4 (2011): 557–71; H. Mah, 'The Predicament of Experience', *Modern Intellectual History* 5, no. 1 (2008): 97–119; F. Ankersmit, 'Huizinga and the Experience of Past', in *Sublime Historical Experience* (Palo Alto: Stanford University Press, 2005), 569.

90 'rekenschap geven van'. VW VII: *Over een definitie van het begrip geschiedenis* (1929): 100; VW VII: *De wetenschap der geschiedenis* (1937): 163.

91 'Life' had been already used by both Dilthey and Windelband as an analytical category in historiographic thought. Huizinga did not credit any author with having inspired the usage of the term, but his understanding of it was very similar to that of Dilthey, and Huizinga would have encountered it in his reading of Dilthey, e.g. W. Dilthey, *Der Aufbau der geschichtlichen Welt in den Geisteswissenschaften* (Leipzig/Berlin: B. G. Teubner, 1927), 130–38.

92 VW VII: *De taak der cultuurgeschiedenis* (1929): 65.

93 The book drew from a lecture series Huizinga had given at the University of Santander in 1934 and was followed by a number of smaller but similar historiographical essays: *Kort begrip van den aard der geschiedenis als cultuurverschijnsel en als wetenschap* (1937), *Historische grootheid* (1940) and *Over vormverandering der geschiedenis* (1941).

gained in history, and again Lamprecht figured as a central antagonist in conversation. This time, Huizinga objected to a more particular feature of Lamprecht's and others' conception of historical *Entwicklung*: the support it might lend to absolute politics. From this political angle, the book developed anew the idea of treating historical inquiry as a 'need of life' (*levensbehoefte*), as Huizinga had suggested already in *The Task of Cultural History*.[94] In order for historical inquiry to meet the needs of life, Huizinga now held, historical inquiry had to be freed from modern conceptions of political necessity and identity:

> To immerse oneself in history, I would say, is a way of finding gratification in the world and of blending into her consideration. History, as much as natural science, cures us of egocentrism, and cures us of overestimating the importance or validity of what immediately surrounds us. What is better for the *mensch* than seeing the borders of one's narrow personality disappear in time and space, rather than finding oneself bound to what has been and to what shall come? What is more healing than observing the eternal perfection and the eternal aspiration, the limitation of human ability, the dependence of even genius and heroism on a higher power?[95]

Historical inquiry in the proper sense of the term, Huizinga held, meant understanding phenomena in terms of their times. The ensuing awareness of the historical relativity of events, in turn, deflates 'egocentrism', complacency as well as images of 'genius and heroism.' Such images of genius, Huizinga held, had become the leitmotif of European nationalism and needed to be defused. Both political and personal vice, Huizinga argued, followed from conceptions of universal validity and judgment, not from moral modesty. The scientific historians and nationalist politicians caught up in schemes and supposed historical necessities had forgotten the moral potential of 'not knowing' (*niet-weten*) and 'not saying' (*ongezegd-laten*).[96] Modesty and

94 VW VII: *De taak der cultuurgeschiedenis* (1929): 59.

95 'Zich verdiepen in de historie, zou ik zeggen, is een vorm van behagen aan de wereld en van opgaan in haar beschouwing. Historie, evengoed als natuurwetenchap, geneest van egocentrisme, geneest van overschatting der belangrijkheid of geldigheid van het onmiddellijk omringende. Wat is beter voor den mensch dan de grenzen te zien wijken, in den tijd en in de ruimte, van zijn eigen enge persoonlijkheid, dan zich gebonden te zien aan wat voorafging en wat volgen zal? Wat is heilzamer dan de eeuwige onvolmaaktheid en de eeuwige aspiratie te zien, de beperktheid van alle menschelijk vermogen, de afhankelijkheid, ook van genie en heldendom, van hooger macht?' VW VII: *De wetenschap der geschiedenis* (1937): 154.

96 VW VII: *De wetenschap der geschiedenis* (1937): 159.

curiosity were not weak relativist vices; they were the powerful and moral virtues. Precisely because historical relativism fostered them, historical investigations had the potential to therapeutically transcend the 'narrow boundaries of personality' and address the 'needs of life', which, Huizinga held in both a stoic and Anabaptist vein, was to achieve a peaceful frame of mind and to accept that the world runs its course – and that any attempt to control and artificially dominate it only leads to more rather than less suffering.

This kind of historical modesty, Huizinga acknowledged, could seem, on the one hand, to lead to a kind of moral relativism, but this was not necessarily the case, he held. On the contrary, when properly understood, historical relativism required and supported rather than debunked solid moral virtues. Only if modesty and curiosity are properly in place can a certain kind of historical relativism take shape, and in this capacity, it draws from rather than against moral convictions. In 1937 Huizinga used the word 'historicism' (*historisme*) to refer to the historical particularism of contextual judgment, but given his own moral conceptions – both conscious and not – this 'historicism' did not appear to him as a moral threat the way it did to other, and especially German, historians of his time.[97] Huizinga's Anabaptist, stoic and pacifist moral convictions could be reconciled with a historicist interest in historical particulars and modest judgment, for this academic interest in particular facts and its aversion to absolutist claims fitted well into a moral culture of dialogue, compromise and cooperation. Huizinga wrote:

> The accusation that historicism produces relativism, that is, that historicism reduces everything to a merely relative value, is not entirely unfounded, but here, too, applies: the historical orientation brings forth the dangers of relativism only when she rests on the preceding loss of intellectual and virtuous benchmark outside of history.[98]

In this passage, it becomes apparent that to Huizinga, moral norms – 'intellectual and virtuous benchmarks' – are, and should be, distinct from the

97 H. J. Paul, 'A Collapse of Trust: Reconceptualizing the Crisis of Historicism', *Journal of the Philosophy of History* 2, no. 1 (2008): 63–82; A. Wittkau, *Historismus: Zur Geschichte des Begriffs und des Problems* (Göttingen: Vandenhoeck & Ruprecht, 1994), 11–22.

98 'Het verwijt, dat het historisme relativist maakt, d.w.z. alles tot een schaal van slechts betrekkelijke waarden herleidt, is niet geheel ongerechtvaardigd, maar ook hier geldt het zooeven gezegde. De historische oriënteering brengt slechts het gevaar van relativisme mede, indien zij berust op een voorafgaand verlies van intellecteele en zedelijke maatstaven buiten de historie gelegen.' VW VII: *Kort begrip van den aard der geschiedenis als cultuurverschijnsel en als wetenschap* (1937): 190.

results of intellectual inquiry. That is, 'historicism' becomes a problem when other norms erode.[99] But if history continued to answer those stoic Christian 'needs of life' identified by Huizinga, he held, then historicism could continue to produce and reproduce virtues valuable to a good human life. This observation, in turn, opens a new angle on the moral perspective of historicism's 'crisis' in the 1920s. Herman Paul has argued that this crisis should be understood not as an epistemic crisis, as had been advocated by, for example, Charles Bambach, but rather as a moral crisis, fostered by the fear among historians that historicism might be undermining a moral order in which they were, in fact, socio-culturally invested.[100] Regarding the particular case of Huizinga, Paul's thesis can be applied as follows: to Huizinga, not historical relativism (which he indeed associated with 'historicism') but the arrogance of Lamprechtian determinism seemed to pose a threat to the academic virtues and sound social norms. To Huizinga, not 'historicism' but rather a certain kind of academic ethics seemed to be in crisis.[101]

Huizinga's interest in virtues returned in his historical works, for instance in his *Patriotism and Nationalism in European History Until the End of the 19th Century* (*Patriotisme en nationalisme in de Europeesche geschiedenis tot het einde der 19e eeuw*) from 1940. In this book, he discussed the political tendencies that dominated his day and age. This history was not quite history. Towards the very end of the book, Huizinga took the liberty of addressing his public, and a tone reminiscent of that found in his 1919 lecture can be found here. Once again, a world had slipped away from beneath his feet; once again, the world had proven to be overwhelmingly powerful and larger than life; once again, 'history' was a way of ascetically undergoing an experience of loss. This time around, however, Huizinga did not express an unease about not understanding the whims of history. On the contrary,

99 The German 'crisis of historicism' was a topic of discussion in the Netherlands, too, but it did not spark nearly as much controversy as it did in Germany. One rather mundane but not unimportant reason for this was that the Dutch historical debates at the time were generally much less theoretical in nature, see Du Pree, *Johan Huizinga en de bezeten wereld*, 92; Krul, *Historicus tegen de tijd: Opstellen over leven en werk van J. Huizinga*, 236.

100 Paul, 'A Collapse of Trust: Reconceptualizing the Crisis of Historicism'; C. R. Bambach, *Heidegger, Dilthey, and the Crisis of Historicism* (Ithaca/London: Cornell University Press, 1995).

101 Interestingly, Huizinga thus had a positive appreciation of the term 'historicism' whilst the term emerged in various capacities first as a derogatory concept. L. Keedus, *The Crisis of German Historicism* (Cambridge: Cambridge University Press, 2015), 15. That being said, as has been shown, Huizinga, like many 'anti-historicists' at the time, emphasized time and again that history had to be of value of 'life'. Whilst speaking of Huizinga in relation to the 'crisis of historicism', semantics are thus of primary importance.

history had prepared him for this moment. At the age of sixty-eight, Huizinga commented on yet another European war in his lifetime. Unsure about what this war would release, he reflected:

> We return home as spectators who do not stick around to see the play through to the end. We close the curtain while the tragic complications become ever more tightly knotted, while wails of compassion and fright ring in the distance. We try to separate ourselves as unknowing, we try to be as those who still had to witness the turn of the century, our past selves from forty years ago, who imagined to see in the recently concluded peace conference the dawn of a glorious new era of progress and civilization.[102]

'We return home as spectators,' Huizinga held. Resignation, modesty and asceticism had become the heart of his historical response. By 1919 'aestheticism' had ceased to be the defining quality of historical research. By then history had become a way of relating to the chaos, loss and unpredictability of a present world, and in the 1920s, 1930s and 1940s, Huizinga put this conviction into practice – not only in his historiographic critiques of Lamprecht, but in his historical writings, too. To accept that one is a mere 'spectator' of the world, always and intrinsically unable to witness the end of history, feeds into both an epistemic virtue of modesty and spiritual state of resignation.

In sum, the following can be said: before the war, Huizinga understood the purpose of historical inquiry to reside in a truthful restoration of the past, and his first critiques of Lamprecht reflected this conviction. Historical truthfulness, Huizinga held, could only be approximated by recognizing the aesthetic factors involved in the historian's practice. After the war, Huizinga understood the purpose of historical inquiry to reside in the existential needs of life: a longing for tranquillity in a life of loss and upheaval. The virtues of 'self-disintegration' were at once *explanandum* and *explanans*. Not the objective recovery of the past but the cultivation of virtues towards this end lay at the heart of the historian's calling. Accordingly, Lamprecht was faulted after the war not for his aesthetic insensitivity but for the moral

102 'Wij gaan naar huis als toeschouwers, die het stuk niet uit zien. Wij trekken het gordijn toe, terwijl de tragische verwikkeling zich pas vaster knoopt, terwijl de jammerkreten van medelijden en schrik eerst in de verte hoorbaar worden. Wij trachten van ons onderwerp te scheiden als niet-wetenden, wij trachten te zijn als zij, die nog het eeuwcijfer gingen zien wisselen, ons eigen vroeger zelf van veertig jaar geleden, dat in de juist gesloten eerste vredesconferentie nog den dageraad kon wanen te zien van een glansrijk nieuwe aera van vooruitgang en beschaving.' VW IV: *Patriotisme en nationalisme in de Europeesche geschiedenis tot het einde der negentiende eeuw* (1941): 554.

impotence of his 'scientific' history. Lamprecht's conception of historical causality instilled arrogance rather than modesty, dogmatism rather than curiosity, and frustration rather than resignation. To Huizinga, the loss of his wife, the destruction of Ypres, and the horrors of the war more generally had revealed just how important a balanced, peaceful relationship with an irretrievable past was. Historical research could help achieve it.

Conclusion

This chapter has presented support for the following four claims:

1. During the course of 1914–19, Huizinga's experiences of loss shook and shocked his understanding of 'the past', history and the purpose of its investigation. The devastation of Ypres and loss of his wife symbolized the experience of a past that had suddenly, decisively and shockingly broken off and drifted away. In this period, Huizinga started writing of an 'irretrievably' lost past, and this sense of rupture influenced the role of historical inquiry in his present: Huizinga grew increasingly interested in the therapeutic capacities of historical research in times of uncertainty.
2. A comparison of Huizinga's historiographic publications from before and after the Great War reveals that these experiences resonated beyond his personal correspondence and even lectures. The ensuing transition is particularly pronounced in Huizinga's critiques of Karl Lamprecht. Lamprecht stood for a conception of history as a law-abiding, organic and systematic movement that could be subjected to diachronic study. Huizinga was especially suspicious of the understanding of 'historical causality', but for different reasons on either side of the war. Before the war, Huizinga accused Lamprecht's method of resulting in overly rationalized and empirically insensitive categories and historical periodization. After the war, on the other hand, Lamprecht was faulted for his method's alleged virtue-ethical bankruptcy.
3. *Autumntide* is arguably the first monograph by Huizinga attesting to this transition. In this book, Huizinga treated late medieval minds and their Gothic aesthetic as his contemporaries, equally invested in the trials and tribulations of frustration and fears, looking to achieve a peaceful frame of mind through universal considerations and self-cultivating, virtuous practices of play. When Huizinga described the medieval *horror vacui*, his contemporary images of cataclysmic catastrophe could not have been far away, and from this perspective of historical life, Lamprecht's diachronic

understanding of history as *Entwicklung* seemed to him to degrade rather than enhance the value that historical inquiry may have had to life. More specifically, Huizinga criticized Lamprecht in *Autumntide* for not having been sufficiently invested in 'the why': *why had medieval practices and images of universality become of primary importance to the medieval mind? Why did the medieval gaze turn towards the skies? From what need were these and such practices born?* Lamprecht's categories and taxonomies did not and could not address such historical needs of life. This observation, in turn, opens up a new angle on *Autumntide*. The previous chapter worked to show how the book drew from a pre-war culture; this chapter shows that other times travelled through this piece, too.

4. In his later, post-war historiographic output from the 1920s and '30s, the virtue-ethical convictions that first appeared in Huizinga's lectures, writings and publications from the period between 1915 and 1919 returned consistently. Of course, these publications drew from several post-war contexts, and these contexts are explored in the upcoming two chapters, but this does not challenge this chapter's central claim that after Huizinga's experiences of loss and shock during the war, his historiographical thought changed dramatically – literally. Huizinga started speaking consistently of the 'needs of life' that historical inquiry answered and the 'therapeutic' capacity of historical reflection, and his post-war retributions of Lamprecht followed suit.[103] Not unlike those medieval minds Huizinga described in *Autumntide*, which sought peace whilst their eyes travelled up and along Gothic pillars like those of the Thomaskirche, Huizinga's eyes peered out over the endless intricacies of 'history'. In the ensuing sense of an overwhelming totality much larger than any human could ever possibly stomach and comprehend, the historian may find not fear but acceptance, not anger but peace. Huizinga's relation to the image of Demosthenes had changed. It was by speaking and writing against the background of a loud sea that these virtues could be practised. The issue with Lamprecht's method was that it did not, and in fact could not, actually confront history's multitude.

103 Against the background of this therapeutic interest, some sense can now be made of a rather conspicuous feature of Huizinga's oeuvre: after the early 1920s, Huizinga's output became steadily less straightforwardly historical. Rather, he turned to more contemporary cultural concerns and anthropological reflections. Chapters 5 and 6 shall deal with this development in his output more concretely.

At this chapter's outset, two interventions into the secondary literature were announced. First, and following Strupp's book, this chapter looked to make a case for a virtue-centred reading of Huizinga's later historiography. Given the 1914–19 background against which his interest in stoic virtues emerged and his appreciation of historical inquiry first changed, such a reading appears not only warranted but urgent. Huizinga's post-war reflections on the nature of historical causality and moral-epistemic regulative ideals in history as well as the existential need for historical reflection repeated convictions and virtues that first appeared in his writings from this period. His later technical understanding of such historiographical themes thus continued to have a discursive proximity to his discussions during the war of how to deal with senses of uncertainty and vulnerability. Huizinga's experience of war and grief – the experience of sudden loss – informed these virtues, and these virtues informed his historiography. Huizinga's historical theory drew from these experiences, and, in turn, helped him narrate and give meaning to and deal with his experience of loss. Huizinga's history was a way of life.[104] Considered from this virtue-ethical and classical philosophical angle, Strupp is right in arguing for a philosophical understanding of Huizinga's cultural-historical works.

But this chapter has also challenged Strupp's analysis for all too quickly presupposing that war dominated and determined Huizinga's thought in the late 1910s. For now, enough has been said on how a description of Huizinga's 'experiences of loss' can be mobilized to bridge the gap between hunch and argument. Thus, I prefer to now turn to Strupp's methodological framework, Jörn Rüsen's invocation of Heidegger's *Daseinsorientierung*, by which Strupp invokes a tradition investigating 'the grounds for the fact that something like the discipline of "world history" is at all possible'.[105] This framework, I believe, allows for a certain empirical insensitivity. Philosophically, I am sympathetic to his angle; historically, I am less so, especially in relation to Huizinga. The development of Huizinga's historical thought, I believe I have shown, had less to do with 'orientation' than with the loss of it. Huizinga's historical thought was not a directedness but a way of dealing with a loss of direction. Through the usage of terms such as 'orientation', historians risk reifying historical characters, and especially in periods of crisis (such

104 M. Foucault, "Technologies of the Self', in *Technologies of the Self: A Seminar with Michel Foucault*, ed. L. H. Martin, H. Gutman, and P. H. Hutton (Amherst: The University of Massachussets Press, 1988), 16–49; P. Hadot, *Philosophy as a Way of Life: Spiritual Exercises from Socrates to Foucault*, ed. A. I. Davidson (Oxford: Blackwell, 1995).
105 M. Heidegger, *Being and Time*, ed. J. Stambaugh (New York: State University of New York Press, 1996), 20.

as Huizinga's), historical experiences of uncertainty, fear and loss are lost in translation. Against the background of Huizinga's sense of shock at the loss experienced in 1914–18, Polybius offers a more apt angle on Huizinga than Heidegger.[106]

Huizinga's experiences with the destruction brought about by 'mechanized' warfare and industrial means more generally, as well as his stoic response, returned in his discussion of another author during the 1910s. Ypres, Huizinga held, had not just been destroyed by humans. It was as much, if not more so, a 'revenge' of machines over their artificers, and according to Huizinga, the 'political scientists' of the nineteenth and early twentieth centuries had been too insensitive to the role played by the inanimate, material world in the political and social congregation of humans. So, by the late 1910s, one author in particular seemed to Huizinga to have been all too naive, if not arrogant, with regard to this mechanical revenge: Alexis de Tocqueville.

Bibliography

Anchor, R. 'History and Play: Johan Huizinga and His Critics'. *History and Theory* 17, no. 1 (1978): 63–93.

Ankersmit, F. 'Huizinga and the Experience of Past'. In *Sublime Historical Experience*, 109–39. Palo Alto: Stanford University Press, 2005.

Bambach, C. R. *Heidegger, Dilthey, and the Crisis of Historicism*. Ithaca/London: Cornell University Press, 1995.

Boterman, F. *Tussen utopie en crisis: Nederland in het interbellum 1918–1940*. Amsterdam: De Arbeiderspers, 2021.

Chickering, R. *Karl Lamprecht: A German Academic Life (1856–1915)*. Atlantic Highlands, NJ: Humanities Press, 1993.

Dilthey, W. *Der Afbau der geschichtlichen Welt in den Geisteswissenschaften*. Leipzig/Berlin: B. G. Teubner, 1927.

Dorsman, L. J. 'Periodisering als integrale benadering: Nederlandse historici in het Fin-de-Siècle'. *Theoretische geschiedenis* 16, no. 3 (1989): 277–95.

Foucault, M. 'Technologies of the Self'. In *Technologies of the Self: A Seminar with Michel Foucault*, edited by L. H. Martin, H. Gutman, and P. H. Hutton, 16–49. Amherst: The University of Massachussets Press, 1988.

106 '[The] surest and indeed the only method of learning how to bear bravely the vicissitudes of fortune, is to recall the calamities of others.' Polybius, *The Histories*, ed. W. R. Paton (Cambridge, MA/London: Harvard University Press, 1922), 3.

Gafijczuk, D. 'Dwelling Within: The Inhabited Ruins of History'. *History and Theory* 52, no. 2 (2013): 149–70.

Hadot, P. *Philosophy as a Way of Life: Spiritual Exercises from Socrates to Foucault.* Edited by A. I. Davidson. Oxford: Blackwell, 1995.

Hanssen, L. *Huizinga en de troost van de geschiedenis: Verbeelding en rede.* Amsterdam: Uitgeverij Balans, 1996.

Heidegger, M. *Being and Time.* Edited by J. Stambaugh. New York: State University of New York Press, 1996.

Huistra, P. *Bouwmeesters, zedenmeesters: Geschiedbeoefening in Nederland tussen 1830 en 1870.* Nijmegen: Uitgeverij Vantilt, 2019.

Jay, M. 'Historical Explanation and the Event: Reflections on the Limits of Contextualization'. *New Literary History* 42, no. 4 (2011): 557–71.

Keedus, L. *The Crisis of German Historicism.* Cambridge: Cambridge University Press, 2015.

Krul, W. E. *Historicus tegen de tijd: Opstellen over het leven en werk van J. Huizinga.* Groningen: Historische Uitgeverij Groningen, 1990.

Lamprecht, K. *Die kulturhistorische Methode.* Berlin: R. Gaertners, 1900.

Lem, A. van der. *Johan Huizinga: Leven en werk in beelden en documenten.* Amsterdam: Wereldbibliotheek, 1993.

Lem, A. van der. *Het eeuwige verbeeld in een afgehaald bed: Huizinga en de Nederlandse beschaving.* Amsterdam: Wereldbibliotheek, 1997.

Mah, H. 'The Predicament of Experience'. *Modern Intellectual History* 5, no. 1 (2008): 97–119.

Mann, T. *Betrachtungen eines Unpolitischen.* Frankfurt: Fischer, 1974.

Mann, T. *Reflections of a Nonpolitical Man.* Edited by Morris W. D. New York: New York Review of Books, 2021.

Moeyes, P. *Buiten schot: Nederland tijdens de Eerste Wereldoorlog.* Amsterdam/Antwerp: Arbeiderspers, 2001.

Paul, H. 'A Collapse of Trust: Reconceptualizing the Crisis of Historicism'. *Journal of the Philosophy of History* 2, no. 1 (2008): 63–82.

Paul, H. 'Performing History: How Historical Scholarship Is Shaped by Epistemic Virtues'. *History and Theory*, no. 50 (2011): 1–19.

Paul, H. 'The Scholarly Self: Ideals of Intellectual Virtue in Nineteenth-Century Leiden'. In *The Making of the Humanities Volume II: From Early Modern to Modern Disciplines*, edited by R. Bod, J. Maat, and T. Weststeijn, 397–412. Amsterdam: Amsterdam University Press, 2012.

Polybius. *The Histories.* Edited by W. R. Paton. Cambridge, MA/London: Harvard University Press, 1922.

Pree, C. du. *Johan Huizinga en de bezeten wereld.* Leusden: ISVW Uitgevers, 2016.

Raedts, P. *De ontdekking van de Middeleeuwen: Geschiedenis van een illusie*. Amsterdam: Wereldbibliotheek, 2014.

Rüsen, J. *Historische Vernunft. Grundzüge einer Historik I: Die Grundlagen der Geschichtswissenschaft*. Göttingen: Vandenhoeck & Ruprecht, 1983.

Schnädelbach, H. *Philosophy in Germany 1831–1933*. Edited by E. Matthews. Cambridge: Cambridge University Press, 2009.

Schulin, E. 'Weltkriegserfahrung und Historikerreaktion'. Frankfurt: Fischer Taschenbuch Verlag, 1997.

Shaw, D. G. 'Huizinga's Timeliness'. *History and Theory* 37, no. 2 (1998): 245–58.

Sicking, J. M. J. 'Woede en verontwaardiging: De Eerste Wereldoorlog in de Nederlandse literatuur'. *Armada* 14, no. 52 (2008): 44–52.

Strupp, C. *Johan Huizinga: Geschichtswissenschaft als Kulturgeschichte*. Göttingen: Vandenhoeck & Ruprecht, 2000.

Tollebeek, J. 'Wetenschap en waardering: Nederlandse historici in verzet tegen het positivisme (1890–1910)'. *Groniek*, no. 124 (1994): 69–96.

Wittkau, A. *Historismus: Zur Geschichte des Begriffs und des Problems*. Göttingen: Vandenhoeck & Ruprecht, 1994.

Woud, A. van der. *Het landschap en de mensen: Nederland 1850–1940*. Amsterdam: Prometheus, 2020.

4 The Future, a Machine

Abstract
This chapter examines how Huizinga's experiences of modern America and its alleged disregard for silence and spirituality informed his critique of Alexis de Tocqueville (1805–1859) and Tocqueville's history of American democracy.

Keywords: Johan Huizinga; American democracy; Alexis de Tocqueville; American history

Towards the end of 1925, Johan Huizinga received an invitation from the Laura Spelman Rockefeller Foundation to travel to the United States of America.[1] The invitation was part of a larger project meant to introduce European academics to the American university system and, in this way, inspire and expand the intercontinental exchange of students. Huizinga had been asked to represent Dutch universities, and he had happily accepted the offer.[2] The trip would be his first to America.[3] The next year, from 14 April to 19 June 1926 and in the company of a number of other invitees, Huizinga travelled across the United States from one university or factory to the next. Amongst his fellow travellers were the French sociologist Marcel Mauss (1872–1950), the Polish anthropologist Bronisław Malinowski (1884–1942) and the Italian economist Luigi Einaudi (1874–1961).[4] Huizinga soon befriended them and experienced the United States not only through his own eyes but through theirs, too. What struck him, however, was how similar their experiences were. Upon his return, Huizinga wrote the following about himself and his fellow travellers:

1 BW I: Huizinga–Van Vollenhoven (1925): 621. See p. 11 for further details on the references to Huizinga's collected works and letters as well as to the Huizinga archives.
2 BW I: Huizinga–Brummel (1926): 629.
3 The trip would also be his last to America.
4 Luigi Einaudi later became president of Italy (1948–55).

Rydin, T., *The Works and Times of Johan Huizinga (1872–1945): Writing History in the Age of Collapse.* Taylor & Francis Group 2023
DOI: 10.5117/9789463724593_CH04

> Strange: among us European travellers continuously arose, and with a striking solidarity among Romans, Germans and Slavs, that pharisaic feeling: we have something you do not; we admire your power, but we do not envy you. Your instrument of civilization and progress, your 'big cities', and perfect organization only give us *Heimweh* for what is old and silent, and at times your life seems to us barely worth living, not to speak of your future.[5]

'*Heimweh* for what is old and silent' – that, above all, was what the American cities, industries and universities had evoked in Huizinga and his fellow travellers. Yet this experience was not exactly new or unexpected. In a sense, this experience had been taking shape for over a decade, before Huizinga had ever set foot in America.[6] In his first book on American history from 1918, *Mensch and Crowd in America*, Huizinga had likened American life to the 'overwhelming', 'dissonant' and 'confusing' effects of atonal music.[7] Already then, the United States had become synonymous with a world of Taylorism, popular voting, attention-grabbing headlines, sensual movie scenes and the commodification of art. Already then, the US had seemed to Huizinga anything but 'old and silent' and appeared ever so different from those 'silently fishing, lost dreamers' called the Dutch. Now, there was nothing particularly 'dreamy' and 'silent' about Dutch colonialism and industry at the time, but Huizinga's descriptions are revealing of a certain fear, a sense of impending loss, which could be projected onto images of American culture.[8] In 1917, whilst the American intervention in the Great War was celebrated in Dutch newspapers, Huizinga wrote in an ominous

5 'Vreemd: bij ons Europeesche reisgenooten, in een treffende solidariteit van Romanen, Germanen en Slaven, rees in Amerika telkens dat farizeesch gevoel: wij allen hebben iets wat gij mist; wij bewonderen uw kracht, maar wij benijden u niet. Uw toestel van beschaving en vooruitgang, uw 'big cities' en uw volmaakte organisatie, geven ons slechts heimwee naar wat oud en stil is, en uw leven schijnt ons somtijds nu reeds nauwelijks meer waard om geleefd te worden, om van uw toekomst niet te spreken.' VW V: *Amerika levend en denkend* (1927): 479.

6 W. E. Krul, *Historicus tegen de tijd: Opstellen over leven en werk van J. Huizinga* (Groningen: Historische Uitgeverij Groningen, 1990), 194.

7 VW V: *Mensch en menigte* (1918): 251. From here on, the book will be referred by the abbreviated title: *Mensch and Crowd*.

8 A learned overview of the symbolic power of the United States in Europe throughout the 1900s and 1910s can be found in D. W. Ellwood, 'How the American Century Started', in *The Shock of America: Europe and the Challenge of the Century* (Oxford: Oxford University Press, 2012), 22–71. This image can, in turn, be contrasted to that of Boterman, who argues that the United States had a rather marginal cultural status, at least in the Netherlands, at the time. F. Boterman, *Tussen utopie en crisis* (Amsterdam: De Arbeiderspers, 2021), 73–105.

tone: '[n]ow nothing is more important than America. [...] A great influence awaits. And what do we know about it?'[9]

By 1918 Huizinga had found a name for this 'influence': 'mechanization'.[10] At the time, this word was uncommon in Dutch discourse. If used at all, the term mostly referred to technical developments in agriculture and industry.[11] Yet, in *Mensch and Crowd*, Huizinga argued that America had fallen prey to a 'mechanization *of culture*', which, in turn, had been nothing less than 'a fatal moment in the history of civilization'.[12] In the footsteps of 'culture', the term 'mechanization' was on its way to migrating from crop to cranium. In order to understand the term's semantic transposition, meaning and ensuing importance to Huizinga's analysis of American culture, Huizinga's relation to Alexis de Tocqueville (1805–1859) offers a helpful proxy. In *Democracy in America* (*De la démocratie en Amérique*, 1835–40), Tocqueville had described the collapse of aristocratic structures and the emergence in America of a new equality of 'social conditions' (*états sociaux*). Though equality had the potential to serve justice, Tocqueville argued, it could equally well enable new forms of 'tyranny'.[13] Huizinga shared Tocqueville's apprehension about democracy, and he admired Tocqueville's clarity and empirical scope. But, Huizinga wondered, 'how could such a clear mind' have been 'so deceived' so as to focus solely on the dangers of social groups and not on the 'revenge' of 'machines' in America?[14]

This chapter asks the following questions: (1) what role did experiences of loss play in Huizinga's understanding of 'mechanization'? (2) How did

9 'Juist nu niets belangrijker dan Amerika. [...] In ieder geval geweldige invloed te wachten. En wat weten wij ervan?' HA: 28 III 7. The importance of the Great War to Huizinga's interest in American history is discussed in C. Strupp, *Johan Huizinga: Geschichtswissenschaft als Kulturgeschichte* (Göttingen: Vandenhoeck & Ruprecht, 2000), 151–52; W. E. Krul, *Historicus tegen de tijd: Opstellen over leven en werk van J. Huizinga* (Groningen: Historische Uitgeverij Groningen, 1990), 177.

10 VW V: *Mensch en menigte* (1918): 290. An earlier but also more tentative usage of 'mechanic' in a similar capacity can be found in BW I: Huizinga–Colenbrander (1916): 168.

11 E.g. *Algemeen Handelsblad* 04-04-1912: 1; *Arnhemsche Courant* 06-04-1912: 6. Exceptions to this rule exist, too, see e.g. *Het Nieuws van den Dag voor Nederlandsch-Indië* 15-07-1912: 9.

12 In terms of its rhetorical effect and narratological function, Huizinga's 'mechanization' resembled Tönnies's *Vergesellschaftung*, Weber's *Entzauberung*, Heidegger's *Beherrschung*, Benjamin's *Reproduktion* and Spengler's *Zivilisation*, but Huizinga's understanding of 'mechanization' developed independently from these authors and vocabularies. By 1917–18 Huizinga had become at best aware of Tönnies and Weber; Spengler, Heidegger and Benjamin were still unpublished. Huizinga's relation to such concepts is discussed in the next chapter on Huizinga's relation to Spengler.

13 A. de Tocqueville, *Democracy in America*, trans. G. E. Bevan (London: Penguin Books, 2003), 292.

14 VW V: *Mensch en menigte* (1918): 292.

these experiences transpire in the arguments he launched in *Mensch and Crowd* against Tocqueville's understanding of American democracy? To the first question, this chapter argues that Huizinga's understanding and first invocation of 'mechanization' drew from two sources: (A) Huizinga's attendance in 1917 of a lecture series on Marxist theory by the Dutch astronomer and Marxist Anton Pannekoek, and (B) Huizinga's rejection of his father's 'mechanical' worldview. In both contexts, Huizinga related 'mechanization' to experiences of a loss of sentimental life. In response to the second question, this chapter argues that Huizinga's critique of Tocqueville reveals how these experiences tied into an entire way of imagining and viewing 'the past'. Huizinga's emphasis on 'economic factors' and 'mechanization' – which, according to him, Tocqueville had utterly failed to appreciate – was mobilized to (A) defuse sociological approaches to history and (B) to narrate the world that had been lost to machines, managerialism and men. Huizinga's history of the United States became a stage for his negotiation of experiences of loss, images of the 'maternal' and a longing for affection in a 'mechanical world'.

By making these two arguments, this chapter looks to support and develop a particular branch of literature on the relation between Tocqueville and Huizinga. Roughly, two distinct approaches on their relation can be discerned in this literature. First, Colie, Ankersmit, White and Otterspeer have offered historiographic analyses of the relation between Huizinga and Tocqueville and have underlined a number of commonalities in Huizinga's and Tocqueville's dialectic imageries and eschatological narratives.[15] On the other hand, more historical approaches by Krul, and most recently Torpey, have argued that Huizinga's emphasis on an ominous 'mechanization' fed into a narrative fundamentally different from Tocqueville's tale of a providential state of social equality.[16] By exploring the experiences of loss that informed Huizinga's usage and understanding of the term 'mechanization', this chapter builds on and contributes to this latter line of argumentation. By

15 W. Otterspeer, 'Huizinga and Tocqueville' (paper presented at *Huizinga Heute. Hundert Jahre* Herbst des Mittelalters, Vienna, 15–17 May 2019); R. L. Colie, "Johan Huizinga and the Task of Cultural History', *The American Historical Review* 69, no. 3 (1964): 620. This supposed affinity is linked to the fact that Huizinga and Tocqueville are commonly mentioned in the same breath by eminent historical theoreticians. F. R. Ankersmit, 'Historiography and Postmodernism', *History and Theory* 28, no. 2 (1989): 141; H. White, 'Historical Pluralism', *Critical Inquiry* 12, no. 3 (1986): 490. Note that Otterspeer's piece was a conference paper. I want to thank Otterspeer for sharing this paper with me and for giving me permission to cite it here.

16 J. Torpey, 'Huizinga on America', *Journal of Classical Sociology* 15, no. 3 (2015): 294; W. E. Krul, *Historicus tegen de tijd* (Groningen: Historische Uitgeverij Groningen, 1990), 184.

exploring how certain experiences of mechanical culture and generational friction from the 1910s informed Huizinga's usage of the term 'mechanization', this chapter looks to further explore and contextualize Huizinga's differences with Tocqueville.

For these purposes, the main body of the present chapter has been divided into four sections. The first section explores how the term 'mechanization' became important to Huizinga and unpacks the nature of Huizinga's investment in this concept. The second and third sections introduce Tocqueville's *Democracy in America* and Huizinga's rejection thereof, respectively. To Huizinga, 'mechanization' not only described a state of affairs in the world he inhabited but dialectically symbolized a once silent and timeless world of past ages, an idealized history, from which his contemporary culture had become estranged and for which he felt *Heimweh*. A fourth section draws from these observations and sets out to show how Huizinga's 'mechanical' experiences informed not only his general and historiographic apprehension about Tocqueville but amounted to a different reading of specific historical particulars. For this purpose, Huizinga's cultural-historical understanding of the nineteenth-century invention of barbed wire is used as an example.

A past turned silent

The term 'mechanization' (*mechaniseering*) became central to Huizinga's writing in the late 1910s. Both in *Mensch and Crowd* (1918) and *Autumntide of the Middle Ages* (1919), the term figured in a decisive analytical role in his argumentation, and it would reappear time and again in his later works in the 1930s.[17] When Huizinga started using the term in the 1910s, it did not yet belong to general Dutch discourse. In Dutch newspapers, the word was only infrequently and loosely used to describe trends away from manual labour in agriculture and warfare. To approximate Huizinga's early interest in this term, two figures from the 1910s are helpful signposts: the Dutch astronomer and Marxist thinker Anton Pannekoek (1873–1960) and the Dutch psychiatrist and novelist Frederik van Eeden (1860–1932).[18] Through an

17 E.g. HA 71 I: 1.M.2 (1917–18); VW V: *Mensch en Menigte* (1918): 290; VW III: *Herfsttij der Middeleeuwen* (1919): 122; VW VII: *In de schaduwen van morgen* (1935): 416.

18 Huizinga might also have encountered the term in a number of other books he read by other authors in the field of (political) philosophy: German neo-Kantians such as Wilhelm Dilthey and Wilhelm Windelband had popularized the term. Huizinga was familiar with these authors, but there is no evidence suggesting that he straightforwardly took inspiration from their technical usage of 'mechanization' – or even noticed their usage of the word for the matter. For a first

exploration of Huizinga's relation to these figures, whom he knew personally, two points can be made regarding Huizinga's usage of 'mechanization': (1) despite Huizinga's overall anti-Marxist stance, his usage of 'mechanization' was likely informed by Marxist discourse on alienation;[19] (2) Huizinga's understanding of 'mechanization' also referred to his father Dirk Huizinga (1840–1903), who as a professor of physiology had been a public advocate of a 'mechanical view on life'. In the next two subsections, these relations are unpacked.

Anton Pannekoek and Huizinga's historical materialism in 1917–18

In the autumn of 1917, Huizinga attended a lecture series on Marxist theory by the Dutch astronomer Anton Pannekoek (1873–1960) at Leiden University.[20] In prior years, Pannekoek had worked in Germany at astronomical observatories, labour unions and the school of the *Sozialdemokratische Partei Deutschlands*, where he had taught together with Rosa Luxemburg (1871–1919). During one of his visits to the Netherlands, in 1914, the Great War broke out, and he found himself unable to return to Germany. That same year, he successfully applied for a position at Leiden University. Since then, Pannekoek taught both astronomy and an introductory course to Marxist theory, the latter of which Huizinga took.[21] No records remain of Huizinga's impressions of Pannekoek's course, but a number of significant signs of Huizinga's ongoing introduction to Marxist theory at the time do appear in his *Mensch and Crowd*. Before he took Pannekoek's course and started working on his book on America, Huizinga barely mentioned Marx in his writings. Hence, even though Huizinga became more critical of Marxism in his later works, good reason exists to view Huizinga's 1917–18 materialist sympathies in relation to Pannekoek's course.

Around 1917–18 Pannekoek was invested in at least two different convictions: (1) Marxist historiography drew from economic determinants but was itself not deterministic, and (2) an acceleration of economic

insight into how the term emerged elsewhere in Europe in relation to American culture, see Ellwood, 'How the American Century Started'.

19 BW I: Henriette and Richard Roland Holst–Huizinga (1918): 225.

20 C. du Pree, *Johan Huizinga en de bezeten wereld* (Leusden: ISVW Uitgevers, 2016), 110; Krul, *Historicus tegen de tijd: Opstellen over leven en werk van J. Huizinga*, 181.

21 Proof of Huizinga's attendance is only to be found in notes by Pannekoek himself. W. E. Krul, *Historicus tegen de tijd* (Groningen: Historische Uitgeverij Groningen, 1990), 181.

developments not infrequently lead to a weakening of 'thought-life'.[22] Pannekoek's respective arguments appeared in print in his 'Historical Materialism' ('Het historisch materialisme') from 1919. Whether or not these ideas informed his introductory course to Marxism is impossible to tell. What does follow from this publication, however, is that Pannekoek nevertheless had a structural understanding of Marxism, one that was commonplace at the time.[23] Consider, for example, the following passage from his 1919 publication:

> Technology is the deepest foundation; that is why she is the most important production force, whilst the legal framework belongs to thereon-resting, thereof-dependent *Überbau*. Because the Law determines economics, people have strained themselves to organize justice and laws in a way that serves the economic construction of society. This adjustment of justice to the needs of technology, of the realization of a certain economic system, does not operate on its own and suddenly, but is a troubled process of the struggle of classes. [...] The foundations of society, of production forces, are not predominantly formed by technology. [...] They [technologies] grow to increasingly greater perfection, because the practice of labour directs the spirit of humans to the means to improve this labour, and to meet new needs.[24]

For present purposes, what matters most in the passage above is the following general observation: technology is described as 'the deepest foundation' (*de diepste grondslag*) of law and society. Coming from a Marxist theorist, this statement is not particularly surprising in itself. What is surprising is how similar Huizinga's basic historiographic conviction became in *Mensch and Crowd*. In later works, Huizinga mostly let go of this tone and

22 A. Pannekoek, 'Het historisch materialisme', *De Nieuwe Tijd* 24, no. 2 (1919): 15–22.

23 Still, Pannekoek often clashed with his peers, G. Voerman, 'Anton Pannekoek: A "Principled Theorist"', in *Anton Pannekoek: Ways of Viewing Science and Society*, 2019, 51–74.

24 'De techniek is dus de diepste grondslag; daarom is ze de belangrijkste productiekracht, terwijl het recht tot de daarop rustende, daarvan afhankelijke, bovenbouw behoort. Juist omdat het Recht de economie bepaalt, daarom spannen de mensen zich in om recht en wet zo te regelen, als voor deze bepaalde economische bouw der maatschappij nodig is. Deze aanpassing van het recht aan de behoeften der techniek, ter verwezenlijking van een bepaald economisch stelsel, loopt dus niet vanzelf en ineens, maar is een moeitevol proces van strijd der klassen. [...] De grondslagen van de maatschappij, de productiekrachten, worden dus nu hoofdzakelijk door de techniek gevormd. [...] Zij groeien tot steeds hoger volkomenheid op, doordat de arbeidspraktijk zelf de geest der mensen richt op de middelen, om deze arbeid te verbeteren, om aan nieuwe behoeften, te voldoen.' Pannekoek, 'Het Historisch Materialisme', 18.

perspective, but in this particular book, his exploration of the 'coercion of capital' and the 'unobstructed implementation of the motives of large capital' in America seemed so obviously Marxist that even his friends and close correspondents at the time took notice of Huizinga's sudden turn towards a materialist perspective.[25] Moreover, it was in this materialist vein that Huizinga introduced the term 'mechanization', which, as stated, was to remain an important concept to him throughout his later works. Consider first the following passage from *Mensch and Crowd*:

> Each tool, every organization is charged by its own activity, forcing the *mensch* to make the tool work, to make the organization run. Without this mechanization there is no civilization. [...] The *mensch* hopes to reach new and greater freedom through each perfection of the tool, of each effort-reducing aggregation of the will and the ability of many. [...] Epochs exist wherein the 'power to bind' belonging to the mechanism of culture appears greater than the 'power to liberate'. Similar to when life threatened to be petrified through perfection and elaboration of the Church in the later Middle Ages, this century, in which we now live, brings the enslavement of humanity by her own perfected means of material and social technology.[26]

'Mechanization' is the activity by which 'each tool and each organization' force humans to serve the function of the tool and organization in question rather than their original end. At first, tools work to channel and organize the forces belonging to the human being to creative effect, yet soon the 'power to bind' overrules the 'power to liberate'. At such moments (and Huizinga considered his own times to be one), humanity is 'enslaved' by material and social technologies.[27] Huizinga's suspicion towards 'mechanization' has been associated with the nineteenth-century clichés of Frankensteinian

25 VW V: *Mensch en Menigte* (1918): 290.

26 'Elk werktuig en elke organisatie is geladen met een eigen activiteit, die den mensch dwingt, het werktuig te doen werken, de organisatie te doen fungeeren. Zonder deze mechaniseering is er geen beschaving. [...] De mensch hoopt nieuwe grooter vrijheid om te leven van elke volmaking van het werktuig, van elke krachtbesparende samenvoeging van den wil en het kunnen der velen. [...] Er zijn tijdperken, waarin de macht-tot-binden, die aan het cultuurmechanisme eigen is, grooter schijnt dan de macht-tot-bevrijden. Zoo wanneer in de latere Middeleeuwen het leven dreigt te versteenen in de volmaaktheid en uitvoerigheid van den alomvattenden bouw der Kerk. Zoo in de eeuw, waarin wij leven, nu de menschheid wel de hulpelooze slaaf schijnt te worden van haar eigen volmaakte middelen van materieele en sociale techniek.' VW V: *Mensch en menigte* (1918): 335.

27 VW V: *Mensch en menigte* (1918): 335.

technologies taking over their creators.[28] On a metaphorical and literary level, this judgement seems fair. Huizinga's understanding of the relation between human creativity and technology was not terribly technical and rests on what were then, too, tropes of technological suspicion: petrification, submission. But in terms of its ideological heritage, Huizinga's technological thought in *Mensch and Crowd* was more than a Frankensteinian cliché. Like Pannekoek, Huizinga saw the submission of human culture to technology not only as the outcome of a Promethean catastrophe but as a dialectic tragedy: as technology transforms from tool to end, culture collapses to later reinvent itself, but only to then, and again, collapse.

This dialectic understanding of 'mechanization' – means turned to ends – helps to make sense of the variety of examples Huizinga catalogued under its entry. In *Mensch and Crowd*, Huizinga mentioned at least ten distinct examples of 'mechanization': (1) industrial labour and Taylorism, (2) 'scientific management',[29] (3) rail transport, (4) democratization and the professionalization of politics, (5) modern print technology and the commercialization of newspapers, (6) motion picture and modern entertainment culture as a whole, (7) the commercialization of exercise and sports, (8) the concept of and culture surrounding the word 'personality', (9) eugenics and the culture of 'hygiene' and (10) the discipline of criminology.[30] Based on how 'mechanization' was wielded at the time in popular discourse, it would be hard to understand how Huizinga stretched this term to include such a variety of cases. In light of the Marxist theory with which he had recently become acquainted, this, however, becomes understandable. In each case, Huizinga found, 'culture' and 'life' become 'mechanized': organizations, technologies, bodily perfection and exercise had become ends in themselves. The ideals they once served had vanished, and in this metaphysically void land, 'man' becomes a machine. Mechanization meant a loss of ideals and of 'spiritual life'.

Huizinga's depiction of mechanization and its consequences was cast in a vocabulary similar to the Marxist terminology he was likely to have encountered through Pannekoek. This is, of course, not to say that a

28 D. van Lente, 'Huizinga's Children: Play and Technology in Twentieth Century Dutch Cultural Criticism (from the 1930s to the 1960s)', *Icon*, no. 19 (2013): 57.

29 Huizinga likely had the bureaucracy of his own university in mind, too, when he spoke of such new managerialism. W. Otterspeer, *Het horzelnest: De Leidse Universiteit in oorlogstijd* (Amsterdam: Prometheus, 2019), 27.

30 These examples are mentioned throughout the work, but most can be found in the book's second chapter *The Instrumentalization of Communal Life*. VW V: *Mensch en menigte* (1918): 292–335.

Figure 4.1. In the 1910s and '20s, cinematographic culture was booming in the Netherlands as it was all over Europe. (A) Cinema Rembrandt in Amsterdam on Rembrandtplein (1927). (B) Interior of Cinema Tuschinski in Amsterdam (1921). (C) A film poster by Elias Ott (1926).

one-to-one relation exists between Pannekoek's Marxism and Huizinga's materialism. What is important for present purposes is that the technological developments identified by Huizinga in Dutch and American society as 'mechanization' had achieved this qualification because they exemplified a loss of culture. 'Mechanical' ways of improving one's character were urgent to Huizinga not because of the technical advancements themselves but because these advancements symbolized a loss of something else: the loss of ideals, silent images, the reason why one wishes to become a good person

Figure 4.2. A new kind of public sports such as cycling, gymnastics and football entered the public arena around 1900 in the Netherlands. (A) Bike race in Amsterdam around 1900. (B) Public display by the General Gymnastics Association in Amsterdam in 1906.

in the first place. So Huizinga's understanding of 'mechanization' was, in the sense discussed in the introductory chapter of this piece, a 'historical experience' of his 'inhabited ruins'.[31] To Huizinga, 'mechanization' meant the death of culture, and thus its discussion highlighted not only what there was but also what there had been, the retrospective experience of a

31 Gafijczuk, 'Dwelling Within: The Inhabited Ruins of History', 151.

Figure 4.3. (A) Employees in an Amsterdam sweatshop around 1900. (B) Employees in the Philips lightbulb factory in Eindhoven 1910–25.

culture destined to become a stranger. Mechanization was estrangement of the present from the past in the present.

Against this background, a new reading of the *Heimweh* expressed by Huizinga in this chapter's first citation is possible. His longing did not concern the home he had left behind in Leiden in order to embark on his journey nor did it draw merely from an aristocratic sense of European superiority. Huizinga felt Europe's future would be similar to America's.[32] His *Heimweh* thus mainly concerned a more general and abstract sense

32 Krul, *Historicus tegen de tijd: Opstellen over leven en werk van J. Huizinga*, 180.

of a loss of once timeless ideals. The 'mechanized' culture he found in the United States not only made present a modern world, but it also made present past ideals and an idealized past. Mechanization was not just a technical term but an experience of the world as history, and in this capacity 'mechanization' marked, to Huizinga, a loss of ideals of culture. This spiritual estrangement vis-à-vis mechanization was, however, laced with another generational dynamic, too, and it is to this dimension that the next subsection turns.

Frederik van Eeden and Huizinga's experience of generations

Throughout his life, Johan Huizinga thought highly of his father Dirk Huizinga (1840–1903). As a student in theology in the 1860s, Dirk Huizinga had come to reject his Christian faith altogether, and against the wishes of his own father, had turned to medicine instead.[33] Later, as a professor in anatomy, Dirk Huizinga had maintained a broad academic interest, and whilst Johan was still in secondary school, Dirk directed him through the most recent anthropological literature. His personal library was filled with books on both the physical and biological sciences as well as history and literature.[34] Nevertheless, Dirk Huizinga represented to Johan the world of 'Haeckel, Büchner, Lorentz and Maxwell', and Johan experienced his inability to arouse interest and competence in his father's world a 'shortcoming', the result of his 'hereditary half-blindness'.[35] By the end of the 1880s, when Huizinga was about to turn eighteen and complete secondary school, he was initially adamant about his wish to study Arabic. His father, however, turned him to Germanic philology. His reasoning was that if his son was to embark on a philological education, this branch would be more likely to prove professionally advantageous.[36] Johan Huizinga was not upset with his father's directions, but soon their disciplinary divergence transformed into a generational conflict.

33 Dirk Huizinga did, however, continue to receive support from his father, even if it came only reluctantly. A. van der Lem, *Johan Huizinga: Leven en werk in beelden en documenten* (Amsterdam: Wereldbibliotheek, 1993), 3–4; Krul, *Historicus tegen de tijd: Opstellen over leven en werk van J. Huizinga*, 32.

34 VW I: *Mijn weg tot de historie* (1947): 16–17.

35 His father, that is, had been 'half-blind' the other way around. VW I: *Mijn weg tot de historie* (1947): 17–18.

36 Van der Lem, *Johan Huizinga: Leven en werk in beelden en documenten*, 28; Krul, *Historicus tegen de tijd: Opstellen over leven en werk van J. Huizinga*, 54.

In the 1890s, whilst the student Johan Huizinga came to take a devoted interest in the passions and mysticism of *fin-de-siècle* literature, Dirk Huizinga became a public spokesman for a 'mechanical view on life'.[37] In 1893 Huizinga senior spoke at the Fourth Congress for Natural and Medical Studies (Vierde Nederlandsch Natuur- en Geneeskundig Congres) at Groningen University. His paper was titled 'On Vitalism and Mechanism' (Over vitalisme en mechanisme) and dealt with the vitalistic debates of the early 1890s, which he vehemently and resolutely denounced. Huizinga senior showcased the liberal and natural scientific ideals of his scholarly generation – devotion, diligence and discipline – and dismissed anything that seemed remotely speculative or metaphysical.[38] The world of particulars was the one that interested him, and this world was a mechanical one. Only every now and then, he stated, does the researcher deserve to cast an overarching glance at the world as a whole, but only to then find that:

> The country of science is a mountainous one. When we are in the valleys, each pass appears as a world in itself, secluded and inaccessible. The stonemason in one valley, the blacksmith in another, the lumberjack in the third – they do not hear each other. They do not exist for one another. But the person who rises in a tethered balloon, like us now, sees the mountains and dividing walls appear below, and a better understanding of the land's unity is achieved. [To] be reminded of the unity, the mutual belonging, the equal right and the equal laws, can sometimes be useful.
>
> And now we want the balloon to descend and store it away, and occupy ourselves with the more substantial results of the miners and the smiths and the other labourers from the mountains.[39]

37 For a discussion of both Dirk Huizinga's self-reflexive and public identity as a natural scientist, see Van der Lem, *Johan Huizinga: Leven en werk in beelden en documenten*, 8–11; Krul, *Historicus tegen de tijd: Opstellen over leven en werk van J. Huizinga*, 48–50.

38 H. Paul, 'Weber, Wöhler, and Waitz: Virtue Language in Late Nineteenth-Century Physics, Chemistry, and History', in *Epistemic Virtues in the Sciences and the Humanities*, ed. H. Paul and J. van Dongen (Amsterdam: Amsterdam University Press, 2017), 397–412.

39 'Het land der wetenschap is een bergland. Zijn wij in de dalen, dan schijnt ieder dal een wereld op zich zelf, afgesloten ontoegankelijk. De steenhouwer in het eene dal, de ijzergieter in het andere, de houthakker in het derde, zij hooren niets van elkander. Zij bestaan voor elkander niet. Maar wie, zooals wij ditmaal, met een ballon captif omhoog gaat, hem schijnen de bergen en scheidswanden lager, en beter begrip krijgt hij van de eenheid van het land. (Het) herinnerd te worden aan de eenheid, aan het bij elkander behooren, aan het gelijk recht en gelijke wetten voor allen, kan ook soms zijn nut hebben.
En nu willen wij den ballon omlaag halen en wegbergen, en verder ons bezighouden met de meer substantiëele resultaten van de mijnwerkers en de smeden en de andere arbeiders uit

Dirk Huizinga's message was: though the practice of natural science requires scientists to find the world fractured and diverse, their results belong to a universal structure. That same year, the Dutch psychiatrist and novelist Frederik van Eeden (1860–1932) gave a paper at the Society for Natural Sciences (Het Natuurkundig Genootschap) in Groningen wherein he directly countered Dirk Huizinga and his hot balloon metaphor in particular. According to Van Eeden, because scientists such as Dirk Huizinga spent so much time labouring in their 'mines', even their hot air balloon perspectives resulted in only more of the same: a 'unity of the landscape'. The concept of unity, Van Eeden held, was not the result of critical reflection and observation; it was a further and improper extension and extrapolation of the mine towards the rest of the world. Van Eeden objected to the carelessness with which materialist theory was transposed from its original, experimentally controlled environments to all other domains of life. For this reason, Van Eeden argued, vitalism was the less speculative of the two; it respected the boundaries of research rather than the supposition of ontological unity. This boundary was of tremendous importance. In his novel *Little Johannes* (*De kleine Johannes*) from 1884, he had shown the miserable world what its blurring had brought:

> A deafening noise reigned supreme – shaking, rattling, pounding, thumping. Large wheels turned, long belts slid and shook. Walls and floors were black, the windows dirty or broken. High above the black building, tremendous chimneys rose, vomiting thick, coiling columns of smoke. In that bustling of cogs and machines, Johannes saw a great number of people with white faces, black hands and black clothes working silently and restlessly. 'Who are they?', Johannes inquired. 'Cogs, more cogs,' Pluizer laughed, 'or people, if you wish.'[40]

In his novel, Van Eeden used a character named Pluizer to explore the ominous 'mechanical world' that had arisen in Amsterdam, where working-class

het bergland.' D. Huizinga, 'Over vitalisme en mechanisme', in *Handelingen van het vierde Nederlandsch Natuur- en Geneeskundig Congres* (The Hague: H. L. Smits, 1893), 30.

40 'Er heerste een oorverdovend lawaai, overal rammelde, ratelde, stampte en dreunde het, grote wielen draaiden en lange drijfarmen schoven slingerend voort. Muren en grond zagen zwart, de ramen waren vuil of gebroken. Hoog staken de geweldige schoorstenen boven het zwarte gebouw uit en braakten dikke, kronkelende rookzuilen uit. In dat gewoel van raderen en machines zag Johannes veel mensen met een bleek gezicht, met zwarte handen en kleren, zwijgend en rusteloos werken. "Wie zijn dat?" vroeg [Johannes]. "Radertjes, ook radertjes," lachte Pluizer, "of mensen, als je wilt."' F. van Eeden, *De kleine Johannes*, ed. D. Mok (Amsterdam: Abraxas, 2009), 99.

quarters and industry had grown significantly since the 1870s. Yet the 'deafening noise' and 'coiling columns of smoke' could just as well have figured in the cities of Rotterdam, Eindhoven, Tilburg or Enschede at the time.[41] All over the country, cities had become noisier, larger and busier. Life in them had become material and instrumental to such an extent that it was no longer obvious whether cogs had been replaced by workers or workers by cogs. Though this social dimension of mechanization did not appear explicitly in Van Eeden's technical critiques of Dirk Huizinga, it loomed between the lines: Van Eeden had already used 'mechanization' to describe the human degradation brought about by industrialization when he later used it to denounce the scientific worldview of Dirk Huizinga. To Van Eeden, Dirk Huizinga represented the liberal, positivist and industrialist worldview of his day, and when Huizinga junior took up 'mechanization' to launch cultural critiques similar to Van Eeden's, Johan Huizinga must have been aware that he was siding with a party opposing his father. Huizinga was most familiar with Van Eeden's work. Each time Huizinga junior invoked the term 'mechanization' to describe the ills of his time, columns of a generational conflict coiled in the background.[42]

This generational conflict signified not only an opposition between two irreconcilable worldviews – one mechanical, the other vital – but also a transition. Dirk Huizinga's positivism, liberalism and mechanical worldview marked their dialectic opposite: a static world of communal life and spiritual ideals that had been lost. To Johan Huizinga, 'mechanization' meant the 'suspension of all emotions', and here it appears warranted to conjecture that this 'suspension' of feelings might well have referred (consciously or not) to the upbringing of Johan by his father. Johan's mother had died when he was only two years old, and though he did not write much about the absence of his mother explicitly, he wrote about 'motherhood' as the domain of 'fantasy', 'courage' and 'ideals.' If 'mechanization' meant the demise of culture and emotional sentiment, his father might have been associated with the death of his mother, and

41 A. van der Woud, *Een nieuwe wereld: Het ontstaan van het moderne Nederland* (Amsterdam: Prometheus, 2021), 25–112.

42 As has been explored in the introductory chapter, Huizinga also grew aware of a generational conflict by other means: through the opposition between the *Tachtigers* and *Negentigers* consciously and explicitly fostered by the Dutch authors of the 1880s and '90s. In this chapter, however, I take Huizinga's opposition to his father to be, if not altogether independent from, at least distinct from this other and more general generational conflict. Perhaps the main reason for this distinction is that there is no direct evidence of these conflicts being married; at no point does Huizinga refer to his father as a *Tachtiger*.

Figure 4.4. Two murals by Jan Toorop from 1902. (A) *The Past*. (B) *The Future*. The former shows submission by workers and women to an unjust system; the latter reveals the just equality supposedly brought by industry and mechanical labour. A third mural, *The Present*, was completed the same year but is not included here. All three murals are to be found at the Amsterdam Stock Exchange.

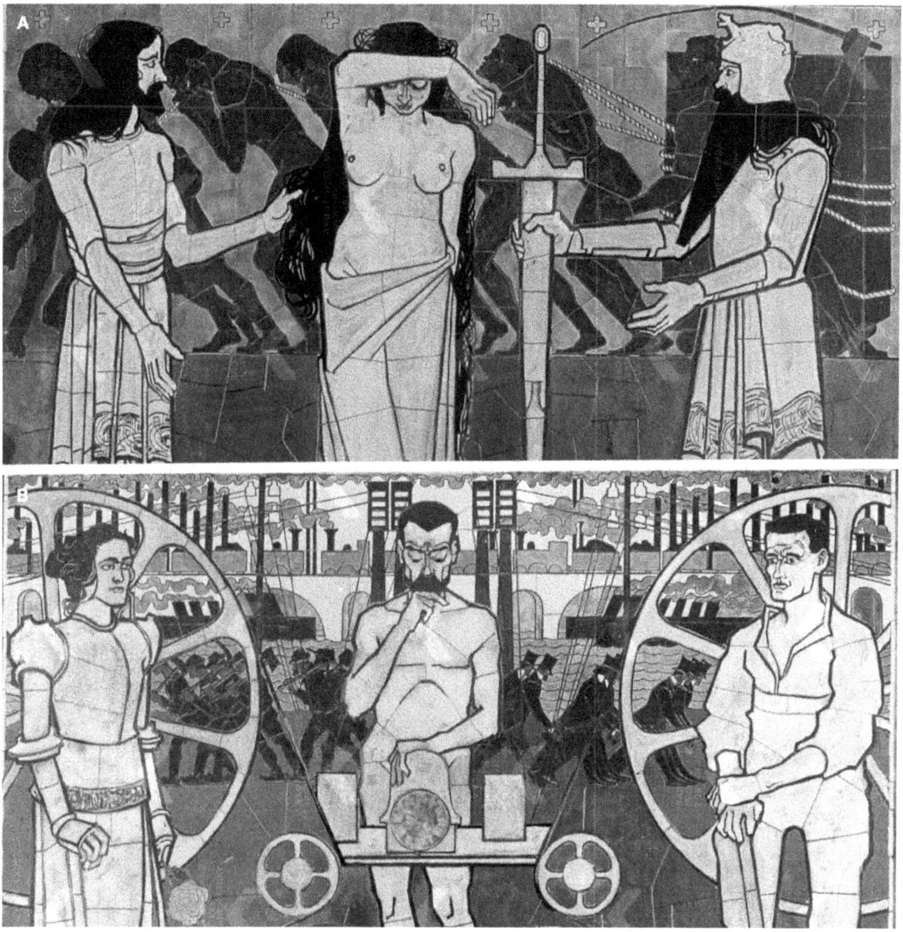

his mother might have become the screen on which he could project the emotions his father had lacked. In this vein, one could conjecture and speculate, Johan Huizinga's understanding of 'mechanization' symbolized the loss of his mother, and in this sense, too, 'mechanization' was the historical experience of a spiritual world's demise, a monument of a maternal past in a paternal world.

So far, two relations have been established regarding Huizinga's usage of the term 'mechanization'. First, and against the background of Huizinga's acquaintance with Marxist theory in 1917 through Pannekoek, Huizinga used the term 'mechanization' to signify loss and estrangement. The loss concerned the withering away of spiritual ideals whilst the means once used to pursue them became ends in themselves. The estrangement concerned what Huizinga later referred to as *Heimweh*: a longing for this lost world of supposedly timeless ideals, fantasy and silent reflection. Second came Huizinga's generational conflict with his father. Here 'mechanization', it was speculated, might well have represented a paternal figure, a mechanical figure dialectically making urgent a lost maternal world, one upon which Huizinga could project the affectionate ideals of a once spiritual life as well as a sense of estrangement vis-à-vis his father's world. In both cases – Marx and mother – 'mechanization' would have signified both a present and a 'broken off' world of past ideals. 'Mechanization' was an expression of having become a stranger in one's own home, an expression of estrangement. For this reason, both cases have been understood as 'historical experiences of the past' in Gafijczuk's sense of the term – that is, as monuments of an inaccessible past in the present.

Tocqueville's America: a social phenomenon

Huizinga read Tocqueville's book on America for the first time in either 1917 or 1918, and he soon discovered that he was reading a kindred author.[43] Both Huizinga and Tocqueville were weary of democracy, valued stoic virtues and universal ideals, emphasized the communal aspect of cultural life, rejected pantheism, were suspicious of narratives of progress, feared individualism, cherished aristocratic culture yet deemed it doomed, attributed primary importance to the role of media in the history of cultures and, last but not least, both reported on the historical seizure that was 'modernity'. 'The past no longer casts light upon the future; our minds advance in darkness,' Tocqueville wrote. 'What now?' Huizinga wondered.[44] Still, despite these commonalities in perspective and tone, Huizinga rejected Tocqueville's *Democracy in America*. Huizinga accused Tocqueville of 'playing the role of prophet' and of having 'dangerously large self-confidence'.[45] In fact,

43 HA: 71 I: 1.T.3 (1917–18).
44 De Tocqueville, *Democracy in America*, 819. HA: 9.I.2 (1919).
45 'speelt profeet'. HA: 42 I: 2.16. 'gevaarlijk groote zelfverzekerheid'. HA: 71 I: 1.T.3.

Huizinga held, Tocqueville had misunderstood the very medium, logic and nature of historical phenomena and developments: Tocqueville had improperly reduced history to sociology *avant la lettre*. Huizinga's critiques of Tocqueville can be understood by placing them against the background of the 'mechanization' discussed in the previous section. But first, a look at Tocqueville's *Democracy in America* is in order.

Democracy in America covers an overwhelming range of topics and domains. Tocqueville strung together arguments and observations concerning ethics, law, natural science, industry, philosophy, arts, politics, religion, social custom, military conduct, the education system and the horrors and historical development of slavery. Huizinga's critique of Tocqueville, which shall be discussed in greater detail in the next section, did not concern the particularities of Tocqueville's numerous examples. Instead, Huizinga objected to the string with which Tocqueville tied them all together. This 'string' was 'too much Montesquieu' and 'too little economic'.[46] But what, then, was this string? Tocqueville addressed his unifying logic at the very outset of the first volume of *Democracy in America*. For present purposes, it is helpful to cite from the opening in some length:

> Of all the novel things which attracted my attention during my stay in the United States, none struck me more forcibly than the equality of social conditions. I had no difficulty in discovering the extraordinary influence this fundamental fact exerts upon the progress of society; it sets up a particular direction to public attitudes, a certain style to the laws, fresh guidelines to governing authorities, and individual habits to those governed.
>
> Soon I came to recognize that this very fact extends its influence well beyond political customs and laws; it exercises no less power over civil society than it does over the government. It forms opinion, creates feelings, proposes ways of acting, and transforms anything it does not directly instigate itself.
>
> Consequently, as I studied American society, I increasingly viewed this equality of social conditions as the fact of which generated all the others and I discovered that it represented a central focus in which all my observations constantly ended.[47]

Tocqueville opened his book by stating several examples of the influence exerted by the 'equality of social conditions' in America and continued by

46 VW V: *Mensch en menigte* (1918): 292.
47 De Tocqueville, *Democracy in America*, 11.

extending its influence to civil society as a whole: opinions, feelings, ways of acting and 'anything it does not directly instigate itself'. He concluded only by stretching his previous statement even further: 'I increasingly viewed this equality of social conditions as the fact which generated all the others.' Throughout the two volumes of this book, Tocqueville gave one example after another of how the equality of social conditions was, if not determining, at least decisive to the particular fact under consideration. To Tocqueville, America was not just an object of inquiry; it was itself a *method* of inquiry, or at least the legitimization thereof. After all, he argued, 'America is the only country in which we have been able to watch the natural and peaceful development of a society and define the influence exerted by the origins upon the future of the states.[48] These origins, he found, lay in the Puritan and Methodist settler communities, and these communities formed the hereditary blueprint for the democratic society that ensued:

> Step back in time; look closely at the child in the very arms of his mother [...]. The entire man, so to speak, comes fully formed in the wrappings of his cradle.
>
> Something similar happens in the case of nations; they always carry the marks of their beginnings. The circumstances which accompanied their birth and contributed to their development affect the remainder of their existence.[49]

Tocqueville held the child of history in the wrappings of social constellations. The directions of history were those of the social developments, of the 'association' and 'assembly' of individuals. A few times in *Democracy in America*, Tocqueville emphasized that he was not attempting to offer a socially deterministic image of American culture.[50] On the contrary, he held: the human mind is not in a position to expose such universal tendencies. At other and more frequent moments, however, Tocqueville could not help attributing to these social conditions and their development a trans-historical lawfulness. In history, he then held, a tendency towards equality existed.[51] It had come to fruition in the United States, and it was on its way to Europe.

48 De Tocqueville, *Democracy in America*, 11.

49 De Tocqueville, *Democracy in America*, 37.

50 '[The reader] might come to the conclusion, on seeing me attribute so many different effects to equality, that I view equality as the only cause of everything which happens at the present time. That would be to suppose that I had a very narrow view.' De Tocqueville, *Democracy in America*, 489.

51 De Tocqueville, *Democracy in America*, 484–85.

Though wiggle room existed for politicians and citizens alike in dealing with this development, the development at large was unstoppable, and not just for any reason. According to Tocqueville, it belonged to a providential will concerning the totality of human history:

> The gradual unfurling of equality in social conditions is, therefore, a providential fact which reflects its principal characteristics; it is universal, it is lasting as it constantly eludes human interference; its development is served equally by every event and every human being.[52]

This providential direction of history was 'universal' and '[eluded] human interference'. One of its many implications, according to Tocqueville, was analytically similar to Huizinga's 'mechanization'. A culture of equality among individuals fed into a new conception of social mobility and intensified the pursuit of sensual satisfaction among common people. The consequences, according to Tocqueville, were unfathomable in their breadth: it affected the arts, historiography, philosophy and the sciences, as well as family structures and religious beliefs. The arts become more practical and bodily, history becomes more concerned with mass phenomena, philosophy becomes pragmatic, the sciences become more applied, generational bonds loosen and religion becomes more fundamentalist. For this reason, Tocqueville held that equality enabled the pursuit of 'material pleasures' to become an end in itself; it elevated 'machines' and 'means' to be considered 'the most magnificent work of the human intellect'.[53] This reversal between means and ends inspired in America a 'ceaseless nervousness', an unravelling of the social fabric that tied together not only families and communities but even generations.[54] Sons and daughters forgot their parents and drifted off in the 'isolation of their own hearts' as they chased the pleasures to which everyone was now equally entitled.[55] Tocqueville described the consequence as follows:

> In the United States, a man will carefully construct a home in which to spend his old age and sell it before the roof is on; he will plant a garden and will rent it out just as he was about to enjoy its fruit; he will clear a

52 De Tocqueville, *Democracy in America*, 15.

53 De Tocqueville, *Democracy in America*, 534.

54 De Tocqueville, *Democracy in America*, 624.

55 'My complaint against equality is not that it leads men to pursue forbidden pleasures but that it absorbs them completely in the search for those which are allowed.' De Tocqueville, *Democracy in America*, 620.

> field and leave others to reap the harvest. He will take up a profession and then give it up. He will settle in one place only to go off elsewhere shortly afterwards with a new set of desires. If his private business gives him some time for leisure he will immediately plunge into the whirlwind of politics. And, if towards the end of a year of unremitting work he has some time to spare, he will trail his restless curiosity up and down the endless territories of the United States. He will [cover] 1500 miles in a few days as a diversion from his happiness.[56]

Social equality led to the means of life – homes, travel, commerce, community and politics – becoming ends in themselves, and this, in turn, inspired a restlessness and loneliness without precedent. Nowhere had sensuous satisfaction become the dominant incentive for action in the way it had in America. This observation, consequently, led straight to Tocqueville's greatest fear of what he called a 'frightening spectacle', one that was, according to him, as inevitable as it was daunting: the commercialization of politics known as 'democracy'.[57] If professional politicians were in a position to dangle myopic promises of material satisfaction before the eyes of the majority of voters, what kind of oppression of minorities might the majority be willing to grant the government? The 'tyranny of the majority'(a coinage for which Tocqueville later achieved such fame) that may ensue could be avoided if politicians and citizens alike were educated to become a responsible *animale politico*, and Tocqueville's book was meant to be part of this education.[58] Nevertheless, the success of such education was not promised, and considering a rejection of democracy would be 'a struggle against God himself'. Tocqueville wrote: 'Positioned as we are in the middle of a rapid stream, we stare fixedly at a few ruins we can still see on the shore as the current drags us away backwards towards the abyss.'[59]

Tocqueville predicted seizure. Aristocratic times were over, and an entire way of being human was to collapse with it. In its place, democracy would grow. The consequences would be uprooting, fearful and exciting. Democracy also had at least the possibility of opening up new pathways to justice. For present purposes, what is important is that, according to Tocqueville, 'democracy' denoted more than legal, political, philosophical and economic

56 De Tocqueville, *Democracy in America*, 623.

57 De Tocqueville, *Democracy in America*, 16.

58 For this reason, Tocqueville presented his political thought as a 'new political science'. De Tocqueville, *Democracy in America*, 16.

59 De Tocqueville, *Democracy in America*, 16.

configurations of society. In fact, it was something both larger and more tangible: democracy was a relation among individuals, a way of meeting and speaking to one another, and though its consequences left nothing untouched, it also had a certain comforting ring to it. 'Democracy', however providentially willed, was *here* and *among people*. Democracy was a proto-sociological phenomenon, a human exercise. It was against the background of this observation that Huizinga found Tocqueville 'too political' and 'too Montesquieu'. Huizinga admired human activity as much as Tocqueville, but he was more suspicious of the role also played by extra-human and extra-conscious factors, whether in life, culture or democracy. The previous section explored how Huizinga came to call these factors 'economic' and 'mechanizing'. The next section explores how he mobilized these factors against Tocqueville.

Huizinga's America: a mechanical phenomenon

In April 1917, nearly all major Dutch newspapers published US President Wilson's plea before Congress to take military measures against Imperial Germany. Congress had complied, and ever since, Dutch papers were eagerly describing, if not blatantly praising, 'the unfathomable resources of the American military' and the 'new hope' these inspired for a rapid conclusion of the war.[60] In the Netherlands, a cultural interest in American history had existed for several decades, but the American intervention in the Great War sparked a different and more probing kind of inquiry: what would America bring to Europe? In his capacity as the Professor of General History, Huizinga felt called upon to pause the project on the late Middle Ages that had occupied him since 1906 and turn to this question instead.[61] In a brief period of only a few months, he developed from scratch a lecture series on American history and developed it into a book, which appeared before the war's conclusion.[62] This pace was unlike his usual research practice, typically slow and diligent, and perhaps for this reason he devoted a large part of his book to discussing what other authors had written about America's coming of age. Three authors

60 'ontzaglijke machtsmiddelen', 'nieuwe hoop'. *Algemeen Handelsblad* 05-04-1914.
61 As was discussed in the previous chapter, *Autumntide of the Middle Ages* was published in 1919. Though Huizinga paused this project to study American culture in 1917–18, the Middle Ages were still prominent in his mind. An interesting piece could be written on how his understanding of the United States and medieval Burgundy influenced each other.
62 C. du Pree, *Johan Huizinga en de bezeten wereld* (Leusden: ISVW Uitgevers, 2016), 111; Krul, *Historicus tegen de tijd: Opstellen over leven en werk van J. Huizinga*, 177.

became particularly important to him: Frederick J. Turner (1861–1932), Charles Beard (1874–1948) and Alexis de Tocqueville (1805–1859).[63]

At the time when Huizinga started his projects on American history, Tocqueville was not widely if at all cited in Dutch academic output. In fact, and unlike elsewhere in Europe, the publication of Tocqueville's first book on America had seemingly gone unnoticed in the Netherlands in the 1830s. Although Tocqueville's name had figured in Dutch newspapers during Europe's revolutionary period of 1848, it was only in relation to his political posts.[64] It is not quite clear how Huizinga arrived at Tocqueville's book on America between 1917 and 1918, but he did.[65] In fact, the very title of his ensuing book, *Mensch and Crowd in America*, appears to be a reference to Tocqueville's central discussion of 'individuals' and 'associations'. In turn, the work's subtitle, *Four Essays on Modern Cultural History*, might have been chosen to allude from the very start to an antagonism with Tocqueville's 'political science'.[66] These reflections are, however, mere speculation. One needs to turn to the content of Huizinga's book to find the antagonism with Tocqueville up close:

> Anyone reading these words must wonder how it is possible that such a clear mind as Tocqueville's, with the keen view of real relations which he had, could have been so deceived so as to ignore the great river that passed him. The rich: few and powerless! The spirits: lethargic! The relations of race, class and fatherland: weakened! On nearly all sides, the opposite appears to have come true. Could it have been that Tocqueville was too political a thinker, too deeply steeped in the school of Montesquieu, too invested in the reasonable explanations of Athenian political life, and without a sufficient awareness of economic factors?[67]

63 Frederick J. Turner (1861–1932) had become known for his 'frontier thesis', and Charles A. Beard (1874–1948) had at the time written three well-received books on American history dealing with its 'economic interpretations'. For Huizinga's relation to these authors, see e.g. Krul, *Historicus tegen de tijd: Opstellen over leven en werk van J. Huizinga*, 178. Accounts by Dutch historians and journalists of the United States had appeared since the 1830s, but Huizinga seems to not have given them any considerable attention. A. Lammers, *Uncle Sam en Jan Salie: Hoe Nederland Amerika ontdekte* (Amsterdam: Uitgeverij Balans, 1989), 11–28.

64 Only two of many examples are *Algemeen Handelsblad* 31-01-1848: 1; *Rotterdamsche Courant* 26-02-1848: 1. Later, Tocqueville was discussed more commonly in his capacity of historian and as a member of the French Academy, see e.g. *Algemeen Handelsblad* 14-09-1857: 4.

65 At any rate, not through Turner or Beard, who, to my knowledge, did not discuss Tocqueville.

66 De Tocqueville, *Democracy in America*, 16.

67 'Wie in dezen tijd die woorden leest, moet zich afvragen, hoe het mogelijk is, dat zulk een klare geest als Tocqueville, met den scherpen blik voor reëele verhoudingen, dien hij bezat,

Too political and not sensitive enough to economic factors: that is what Tocqueville appeared to be in Huizinga's eyes. Huizinga's critique did not concern particular readings of certain phenomena by Tocqueville; it engaged with Tocqueville's way of looking itself, which Huizinga found to be economically lacking. On one hand, this judgment was informed by Beard's work *An Economic Interpretation of the Constitution of the United States*, which appeared in 1913, and *Economic Origins of Jeffersonian Democracy*, published in 1915. In both books, Beard argued that individual economic incentives had been decisive in American politics. These observations might well have made the inclusion of 'economic factors' attractive to Huizinga, yet when he called Tocqueville 'too political', it was not only Beard's work he had in mind. In fact, Huizinga's critique was also, if not more strongly, propelled by perspectives similar to Pannekoek's: individual economic incentives are important, but they themselves are not self-sufficient, as they rest on other extra-individual and extra-conscious factors. Beard had analysed American culture in terms of new and old elites competing over economic advantage. Yet, according to Huizinga, and in line with Pannekoek's view, that could not be the ultimate explanation:

> Each technical finding, intended to liberate spiritual energy and disclose natural riches, captures human independence in the forms of the increasingly instrumental and efficient societal organizations that she enables. The organization, equipped with all those means that reduce human labour to management and regulation, becomes itself a machine, which is no longer completely controlled by humans, because management and regulation, seemingly flowing from free human deliberation, become [...] a submission to the machine, whose servants we all are. As the economic apparatus becomes more complicated and more technically perfect, the meaning of free individual reason in enterprise dwindles, whilst the mechanical element increases.[68]

zich zoo heeft kunnen bedriegen in de richting van den grooten stroom, die aan zijn oog was voorbijgevloeid. De rijken dun gezaaid en machteloos! de geesten niet energiek! de banden van ras, klasse en vaderland verslapt! Van bijna alles schijnt het tegendeel uitgekomen. – Zou het niet zijn, dat Tocqueville te uitsluitend een politisch denker is geweest, te zeer nog opgegroeid in de school van Montesquieu, zijns ondanks nog te zwaar beladen met de redelijke verklaring van het Atheensche staatsleven, en zonder genoegzaam besef voor de economische factoren?' VW V: *Mensch en menigte* (1918): 292.

68 'Elke technische vinding, bedoeld om geestelijke energie vrij te maken en natuurlijke rijkdommen te ontsluiten, legt tevens menschelijke zelfstandigheid vast in de vormen van de fijner bewerktuigde en doeltreffender maatschappelijke organisaties, die zij mogelijk maakt. De organisatie, toegerust met al de middelen, die den menschelijken arbeid reduceeren tot

As the organization of labour and finances increases beyond a certain critical point, Huizinga held, it extracts itself from human comprehension and control. Huizinga was not putting forth an intricate system of technical vocabulary to describe the relation between humans and technology but was most probably giving expression to a general perspective and attitude he had found as a professor-turned-student in the lecture halls of Leiden. Importantly, these explanations suited his suspicion towards phenomena of mass culture. His suspicions are explored in greater detail in the next chapter. For now, what matters is that Huizinga's divergence from Beard drew not only implicitly from materialist thought. Huizinga was prepared to make his affiliation explicit:

> The economic factors in America lie so clearly visible on the surface that the process, which has been postulated as all-encompassing by Marx, takes place repeatedly before our eyes. The powers of production – ultimately, the technical and natural resources of economic progress – determine a totality of historical development together with its societal, political and cultural formations.[69]

In America means had become ends, and technical advancements and economic progress determined cultural ideals, social constellations and political relations. Huizinga affiliated his observations with a name: Marx. What seems to have mattered to Huizinga, however, was not a critical engagement with Marx or Marxism. Huizinga mobilized this observation not in support of Marxist historiography but to make yet another point: to argue that American history should not be understood in narratives of 'individuals versus groups' (Tocqueville) or 'old versus new' (Beard) or as a struggle between classes (Marx). Rather, it should be understood in terms of another opposition, namely that of 'nature versus humans', which lay closer

een leiden en regelen, wordt zelf een machine, die de mensch niet meer volkomen in zijn hand heeft. Want het leiden en regelen, schijnbaar voortvloeiend uit het vrije menschelijk overleg, wordt, verdeeld als het is over velen, een gehoorzamen aan de machine, wier dienaars allen zijn. Naarmate de geheele economische toestel ingewikkelder en technisch volmaakter wordt, daalt de beteekenis van het vrije individueele vernuft in de daad van het bedrijf, en neemt het mechanisch element toe.' VW V: *Mensch en menigte* (1918): 300.

69 'De economische factoren liggen in de Amerikaansche geschiedenis zoo aan de oppervlakte, dat men het proces, door Marx als albeheerschend gepostuleerd, zich herhaaldelijk voor ogen ziet afspelen, namelijk, hoe de productiekrachten, – dat zijn in laatste instantie de technische en natuurlijke middelen van economische voortbrenging, – een geheele geschiedkundige ontwikkeling met al haar maatschappelijke, staatkundige en cultureele formaties regelrecht kunnen bepalen.' VW V: *Mensch en menigte* (1918): 255.

to Turner's frontier thesis.[70] This thesis held that the pioneer experience of the American western frontier was the central and decisive condition of American culture and democracy. At its heart, this experience referred to the opposition between tools and technology, on one side, and a pristine nature that was both beautiful and to be destroyed, on the other. This perspective resonated with Huizinga. To him, the technological prowess and organization of America ultimately not only symbolized the subjection of nature itself, but it pointed towards an absence.

In America, Huizinga found, something had been lost, and in its place had arisen what could only be considered a monument for what had disappeared. The taller the skyscrapers, the faster the trains and the busier the crowds, the more conspicuous the absence of what had once been: an idealized past of unconditional relations and non-instrumental bonds to one's kin. This perspective on time, loss and technology was deeply steeped in references to works by Ralph Waldo Emerson (1803–1882), Henry David Thoreau (1817–1862) and Walt Whitman (1819–1892). Huizinga devoted disproportionate attention to these American authors in *Mensch and Crowd*, and soon his interest spilled over into admiration for them, and this admiration, in turn, is telling of what America symbolized to Huizinga.[71] What united Emerson, Thoreau and Whitman was a longing to escape urban modernity and to live in pre-industrial times, when attention to natural detail and silence had not been disturbed by the whir of machinery. A discussion of their works is beyond the scope of this chapter, but what matters at the moment is less what they wrote and more what Huizinga selected from their writings and included in *Mensch and Crowd*, especially in the concluding section. As the book drew to a close, Huizinga included the following from Whitman's poem 'Song of Myself' (1892):

> I am he that walks with the tender and growing night,
> I call to the earth and sea half-held by night.

70 Upon the publication of *Mensch and Crowd*, Huizinga and Turner had a brief and friendly correspondence. Huizinga, interestingly, wrote: 'The notions I borrowed from your "Significance of the Frontier" of 1893 have helped me more than most of my other reading to form a clear understanding of the main substance of American history. It has been a great surprise to me to see how much we Europeans could learn from American history, not only as to the subject itself, but also with regard to historical interpretation in general.' BW I: Huizinga–Turner (1919): 248. This last statement supports this chapter's contention that Huizinga was less interested in particular historical events than he was in narrative and structure.

71 In *Mensch and Crowd*, Huizinga referred to Tocqueville fourteen times, all of which in a more or less critical fashion. He mentions Whitman no less than fifty times and Emerson forty-three times, all in a more or less approving manner. For a discussion of Huizinga's relation to American authors, see W. Otterspeer, *Orde en trouw: Over Johan Huizinga* (Amsterdam: De Bezige Bij, 2006), 192–93.

Press close bare-bosom'd night – press close magnetic nourishing night!
Night of south winds – night of the large few stars!
Still nodding night – mad naked summer night.
Smile O voluptuous cool'breath'd earth![72]

Whitman's descriptions read as a stereotypical example of nineteenth-century conceptions of gender and nature. His nature is one of maternal affection; she is 'bare-bosom'd', 'nourishing' and 'voluptuous.' Though Huizinga was often sceptical of lyrical and passionate prose, he appreciated Whitman's poem as a window on American modernity – that is, as a window on the natural life American modernity had forsaken. Whitman's poem brought together the estranging effect of industrial modernity (by exposing what had been lost through it) and a male longing for maternal affection (symbolized by 'nature'). Onto Whitman's poem, Huizinga could project (1) images of the alienation brought about by modern labour division and as taught by Pannekoek, as well as (2) a longing for maternal affections, as was demonstrated in relation to Van Eeden. Huizinga's America represented a world of paternal domination, the 'task of mastering the gigantic continent through labour'[73] and, consequently, a space representing the disappearance of a mother – mother nature, mother Huizinga. If one were to consider only the dynamics of social congregation, such as Tocqueville, one would remain blind to what had been lost: not just a world of small communities but a world that had not yet succumbed to a father figure.

In *Autumntide of the Middle Ages*, which appeared a year after *Mensch and Crowd*, Huizinga elaborated further on the gendered dimension of this 'yearning' for unconditional, non-instrumental, 'authentic' modes of life. In his analysis of late-medieval Burgundian culture, Huizinga argued that this 'yearning' had lain at the heart of its ascetic culture.[74] This ascetism, he held, was 'closely entangled with the erotic nature of its life attitude, and might be only the ethical processing of an unsatisfied desire'.[75] This

72 VW V: *Mensch en menigte* (1918): 415.

73 'In de plaats van de tegenstelling tusschen oude en nieuwe levensvormen in de maatschappij zelf, treedt hier de nog geweldiger tegenstelling tusschen de natuur en den mensch. De macht, wier inertie hier overwonnen moet worden, is die van de materie zelf. De taak, die het Amerikaansche volk van den beginne af gebiedend voor ogen ziet, is de bedwinging van het reusachtig continent door den arbeid.' VW V: *Mensch en menigte* (1918): 253.

74 'zucht'. VW III: *Herfsttij der Middeleeuwen* (1919): 34.

75 The role of *eros* in Huizinga's historiography has not been explored in the literature and deserves to be taken seriously in future research on Huizinga. The original reads: 'De diepe trek van ascese, van moedige zelfopoffering, die het ridderideaal eigen is, hangt met den erotischen grond van die levenshouding ten nauwste samen, is misschien slechts de ethische verwerking van onbevredigd verlangen.' See VW III: *Herfsttij der Middeleeuwen* (1919): 88.

Figure 4.5. **Huizinga's image of American culture and its cultural degeneration is for several reasons typical of the male perspective of his times. The Dutch women's suffrage movement typically cultivated a much brighter image of American culture. (A) A poster from the Dutch women's suffrage movement is displayed. Democracy and universal suffrage symbolized the inclusion and benefit of 'female insight'. (B) The countries and American states marked in black had not yet granted suffrage to women by 1913; Europe was lagging behind the developments in the United States.**

medieval desire, in turn, consisted of a longing for 'the indirectly erotic', 'promise', 'loyalty' and a 'noble meekness'.[76] This observation, Huizinga argued, revealed just how male-oriented medieval sources had been, for these qualities of yearning had been typical of the male conception of female

76 'indirecte erotiek', 'belofte', 'trouw', 'edele zachtmoedigheid'. VW III: *Herfsttij der Middeleeuwen* (1919): 135.

love. Hence, when Huizinga not only lauded but even adopted Whitman's narrative of a 'lost, motherly past of nature', he was consciously expressing *a* male affection towards a pre-paternal, pre-mechanized world. Huizinga's America was part of a male experience of female affection. After all, to most members of the Dutch female intelligentsia at the time, words such as 'democracy', 'industry', 'science', 'newspapers' and the professional opportunities of 'bureaucracy' – Huizinga's 'mechanization' – symbolized a new future, not a lost past.

To conclude, two points from this section deserve to be highlighted and repeated. First, Huizinga's first and most general critique of Tocqueville's proto-sociological account of America and its 'ceaseless nervousness' drew from an explicitly Marxist perspective. Tocqueville, he found, had been too insensitive to the role of tools and technology to human thought life. This insensitivity is what Huizinga had in mind when he called Tocqueville 'too political'. Second, Huizinga's critique of Tocqueville was fortified by another dynamic that remained mostly implicit. The world from which modern society had become alienated through the dominance of technology was a maternal and affectionate one. This understanding shines through Huizinga's choice of sources for explaining America: Emerson, Whitman and Thoreau were not mere historical objects to Huizinga. They actually revealed the dialectic loss of modernity; they had made the absence of unconditional maternal love present. Their ways were fundamentally part of a male gaze projecting images of the natural and authentic onto the female figure. Both these points tie into what was above called Huizinga's experience of loss, and the next section reveals just how concretely this experience manifested itself in Huizinga's particular cultural-historical analyses.

Man's land and no man's land

So far, the distinction between Tocqueville's and Huizinga's historiography of the colonized North America has been understood in terms of a different appreciation of beginnings and ends. Tocqueville – or rather, Huizinga's version of Tocqueville – had written American history in terms of a 'beginning': modern America *was* the emergence of a new sociological fact, a new dynamic between individuals and associations. On the other hand, Huizinga's America was drafted in terms of a loss, of an end. Modern America consisted of the erosion of reflection by means of 'mechanical' ways of life. This distinction is one of narratological form, and though it has been understood against the background of Huizinga's more or less concrete experiences of loss, the distinction remains a formal one. This section sets out to articulate the formalistic distinction between Tocqueville's America and Huizinga's America with an empirical case in point to show how Huizinga's experiences not only affected his general apprehension about Tocqueville but also informed the differences in their analyses of particular historical events. For this purpose, this section compares Tocqueville's and Huizinga's accounts of changing conceptions of property in post-revolutionary America.

With each mobilizing a different perspective, Tocqueville and Huizinga offered quite different accounts of what 'property' meant in the United States, which is hardly surprising given the eight decades separating their analyses. A comparison is nonetheless revealing of what set their perspectives apart. Regarding the study of post-revolutionary times, Tocqueville wrote he had been 'astonished that commentators old and new have not attributed to the laws of inheritance a greater influence on the progress of human affairs'.[77] After the revolution, new laws had been passed granting equal inheritance rights for all the deceased's legal offspring, and by failing to appreciate the magnitude of this legal development, Tocqueville held, previous historians had missed entirely how 'property' and 'land' had been fundamentally re-defined. Previously, the eldest had taken over whatever estate and property had been accumulated, and when this changed, both family trees and agricultural crop changed shape.[78] Tocqueville wrote about the new equal inheritance laws:

> When the law of inheritance institutes equal divisions, it destroys the close relationship between family feeling and the preservation of the land which ceases to represent the family. For the land must gradually diminish and ends up disappearing entirely since it cannot avoid being parcelled up after one or two generations.
>
> The instant you remove from a landowner that interest in the preservation of his land which is fuelled by his family feeling, memories, pride, ambition, you can take it as certain that sooner or later he will sell up. He has a great incentive to sell up, for moveable assets produce greater returns than other assets and more readily satisfy the passions of the moment.
>
> Once shared out, great estates never come together again; for the small landowner earns proportionally a better return from his land than the large landowner does from his and sells it for a higher sum.[79]

77 De Tocqueville, *Democracy in America*, 60.
78 De Tocqueville, *Democracy in America*, 61.
79 De Tocqueville, *Democracy in America*, 62.

A social institute – inheritance law – contributed to a new conception of family and property. The equalization of inheritance rights effectuated a continuous erosion of those bonds and objects around which peoples had previously congregated into 'families'. 'Democracy makes men forget their ancestors,' Tocqueville concluded, and against this background and others, 'anyone living in the United States learns from birth that he must rely upon himself to combat the ills and obstacles of life.'[80] Yet the longing for community does not diminish, and thus a new family is sought, and this longing sets in motion the meaning-creating dynamics of the social associations in America. Released from family bonds, bonds of faith, belief and a common cause take centre stage, altering the ways science, art, rhetoric and politics take shape. Of course, Tocqueville was not proposing a one-directional causal relation between inheritance law and the whole of social fabrics in post-revolutionary America. He was showing a relation of mutual exchange and influence, revealing how culture could be and in fact was mediated by the social dynamics that his 'new political science' meant to study.[81]

In the first part of *Mensch and Crowd*, Huizinga, too, reflected on the developments that had taken place in the conceptions of 'property' in nineteenth-century America. Unlike Tocqueville, however, Huizinga did not commence from law or other social institutes, but from the 'forces of production'. He highlighted a number of such technologies from the long nineteenth century: Eli Whitney's 'cotton-gin' (short for 'cotton engine'), Joseph Glidden's barbed wire, the creation of train lines, and the development of refrigeration technology. All these 'forces of production' played a 'peculiarly direct role' in how law, commerce and American culture at large took shape in nineteenth-century conceptions of property in America.[82] To be sure, Huizinga's analyses were not logically incommensurable with Tocqueville's, but they showcased different sensitivities and interests – different understandings of the relation between, on the one side, human agency and social relations and, on the other, the material world and technology. According to Huizinga, the invention of barbed wire, for one, had meant nothing less than a revolution in how land and property appeared. A new, modern industrial-agricultural property and cattle 'volume' had become not only conceivable but possible.[83]

80 De Tocqueville, *Democracy in America*, 589, 220.
81 De Tocqueville, *Democracy in America*, 16.
82 VW V: *Mensch en menigte* (1918): 256.
83 VW V: *Mensch en menigte* (1918): 255–56.

Figure 4.6. (A) The barbed wire's 'revenge' at the Dutch-Belgian border as depicted by the Dutch cartoonist Albert Hahn (1877–1918) in 'Deathwire' in *De Notenkraker*, 24 July 1915. (B) The mural *The Homestead and the Building of Barbed Wire Fences*, by John Steuart Curry (1897–1946). (C) Barbed wire incentivized the development of new military technology. Displayed here is a *Boirault machine*, a French landship designed to flatten and cross barbed wire on both regular and irregular terrains. The machine was later deemed impractical, but the first French tank models grew from its design.

Barbed wire had, Huizinga argued, enabled a new kind of world: unfathomable surfaces of land could now be distinguished and defended at low cost, and consequently, the concept of land itself had been transformed.

Now that less manual labour was demanded for cotton production, a new moral-legal attitude towards slavery was enabled. New train connections enabled a 'distribution' of human 'resources'. The possibility of refrigerating foods renegotiated centre-periphery relations. Hence, whilst Tocqueville argued 'how democracy fosters industrial development and multiplies without limit the number of industrialists' through the new inheritance laws and other social transformations, Huizinga had a greater interest in the reverse relation: what had the effects of 'capital concentration' been on property?[84] After Huizinga's own trip to America in 1926, he repeated in more urgent terms what he had already come to formulate in 1917–18: industry had transcended its role as 'means to an end'. It had become a way of thought, a purpose itself, as 'factories and standardization were not merely a more intensive and cheaper form of production but a higher form in the full sense of the word.'[85]

As previously stated, the discrepancy between Tocqueville's and Huizinga's approach to the shifts in conceptions of property in modern America is hardly surprising. Tocqueville wrote before industrialization had caught proper momentum; the examples Huizinga cited were all invented later. The anachronism of their comparison on this point, is, however, still productive. By the time Huizinga decided to write on modern American culture, the inventions he deemed decisive to its development had all executed their 'revenge' on the inventor. These technologies had acquired applications well beyond and independently of the initial human intention; they had revealed an agency of their own. In the Great War, tens of millions had been transported from all corners of Europe, like cattle, to Flanders. The millions who did not return alive were consumed by a mechanically ploughed soil – that is, if they had not been fed into the budding refrigerated funeral industry or been left dangling in the barbed wire that belonged to neither side but to 'no man's land'. At the sight of this destruction and loss, mechanical forces roared loader than any human, and Huizinga's mechanical estrangement, together with that of many others, grew stronger and stronger.

Conclusion

This chapter set out to establish two points. The first concerned Huizinga's understanding of the term 'mechanization' in his *Mensch and Crowd* (1918)

84 De Tocqueville, *Democracy in America*, 645. VW V: *Mensch en menigte* (1918): 255.
85 VW V: *Amerika levend en denkend* (1927): 467

and how it tied into an 'experience of loss'. The other concerned how Huizinga mobilized this term in his critique of Tocqueville's *Democracy in America* (1835–1840) to deflate a sociological approach to American history, if not history at large. The former point, to begin, rested on a double argument: it has been argued that Huizinga's usage of the term was most likely influenced by (1) the lecture series he attended on Marxist theory by Anton Pannekoek the same year he started working on his book and (2) the paternal image Huizinga projected onto the term. In more than one way, Huizinga's father was a symbol for the 'mechanical world' Huizinga sought to denounce, and in effect his prematurely deceased mother became its dialectic counterpart. Before the paternal image of a mechanical world grew dominant, another affectionate and unconditional perspective had existed. Huizinga's image of modern America was also the image of his father, who was a public advocate of mechanical views on life and the technical improvement of society. In 'mechanization', Huizinga captured not only a worldly state of affairs but also (and to an equal extent) the demise of a different, affectionate world that had been eclipsed and exhausted.

'Mechanization' would remain part of Huizinga's vocabulary throughout his oeuvre, but he first used it in *Mensch and Crowd*, and then again to debunk an argument made by Tocqueville in *Democracy in America*, namely that the modern world – with its commerce, politics, arts, sciences, music, literature and historiography – was ultimately conditioned by a particular and new 'social state' among individuals. That Huizinga responded so strongly against Tocqueville is, at first sight, quite surprising given their moral-ideological similarities. This antagonism perceived by Huizinga makes sense, however, when considered against the background of what 'mechanization' meant to him, what his experience thereof symbolized. To Huizinga, Tocqueville's alleged omission of technical and industrial advancements in his account of America represented an omission of the tragic silencing of an affectionate life that had taken place, whether in America, the Netherlands or his own life. Huizinga's call to be more sensitive to 'economic factors' in American history was thus not so much a plea for economic history as it was an argument against social history. Huizinga's alternative meant the inclusion of what had come under pressure by a paternal world of mechanics, namely the human longing for an unconditional maternal figure. The name of this alternative history was in his book's subtitle: 'cultural history'.

Huizinga's cultural-historical project of 1918 was a response to the experience of loss as described above. But this response says as much about the experience as it does about the subject that underwent it. This cultural history of America by Huizinga was part of a male perspective on maternal love; it was part of a life invested in a world that appeared to be under pressure by what was called modernity. American history as depicted by Huizinga in 1918 thus always belonged to that of a male author yearning for maternal affection in a world dominated by fathers, either directly or symbolically. In effect, history became the substitute for the maternal figure missed and, moreover, a canvas onto which the needs unsatisfied by a paternal figure could be projected. Huizinga's pre-mechanical world was his mother image: it was pristine, undemanding, loving, unchanging, a 'bare-bosom'd night', a 'nourishing night' – that is, until 'modernity' had dawned. Huizinga's present was a modern one, and in this capacity, it reminded him precisely of an imagined opposite. America figured as the ruins of a lost maternal past. Huizinga's second book on America appeared in 1926, was less technical in tone and made no reference to Tocqueville at all. Still, it opened with a reference to that same, estranging 'fatal moment in the history of civilization':[86]

> Progress is a terrible thing.[87]

Huizinga's suspicion towards the 'mechanical' world and the narratives of progress that came with it was not unique. By the time Huizinga wrote about Tocqueville, an entire body of critiques of the modern and mechanical world had taken shape in most European languages. As the 1910s drew to a close, one of the most famous authors on this front in the Netherlands was Oswald Spengler, whose *Der Untergang des Abendlandes* (1918) could capitalize on the prophetic timing of its publication. Spengler's name continued to make waves throughout the 1920s, and Huizinga became acquainted with his work no later than 1921. Spengler and Huizinga shared a variety of antipathies and concerns. Still, Spengler's critique of modernity and mechanization was too morose, too romantic and too 'irrational' for Huizinga's taste. In fact, to Huizinga, Spengler's work soon seemed to pose a threat to humanistic thought as much as the phenomena of mechanization itself.[88] Unlike Spengler, Huizinga's response to the loss of silence and tranquillity to machines and industry was entangled with his response to another loss, too: an experienced loss of an internationalist and 'civilized' world. In Huizinga's eyes, Spengler's work symbolized an equally unwelcome alternative to the 'mechanized' world.

86 VW V: *Mensch en menigte* (1918): 291.
87 Huizinga cited from the American psychologist-philosopher William James (1842–1910). VW V: *Amerika levend en denkend* (1927): 419.
88 VW VII: *Taak en termen der beschavingsgeschiedenis* (1926): 33.

Bibliography

Ankersmit, F. R. 'Historiography and Postmodernism'. *History and Theory* 28, no. 2 (1989): 137–53.

Boterman, F. *Tussen utopie en crisis: Nederland in het interbellum 1918–1940*. Amsterdam: De Arbeiderspers, 2021.

Colie, R. L. 'Johan Huizinga and the Task of Cultural History'. *The American Historical Review* 69, no. 3 (1964): 607–30.

Eeden, F. van. *De kleine Johannes*. Edited by D. Mok. Amsterdam: Abraxas, 2009.

Ellwood, D. W. 'How the American Century Started'. In *The Shock of America: Europe and the Challenge of the Century*, 22–71. Oxford: Oxford University Press, 2012.

Gafijczuk, D. 'Dwelling Within: The Inhabited Ruins of History'. *History and Theory* 52, no. 2 (2013): 149–70.

Huizinga, D. 'Over vitalisme en mechanisme'. In *Handelingen van het vierde Nederlandsch Natuur- en Geneeskundig Congres*, 18–36. The Hague: H. L. Smits, 1893.

Krul, W. E. *Historicus tegen de tijd: Opstellen over het leven en werk van J. Huizinga*. Groningen: Historische Uitgeverij Groningen, 1990.

Lammers, A. *Uncle Sam en Jan Salie: Hoe Nederland Amerika ontdekte*. Amsterdam: Uitgeverij Balans, 1989.

Lem, A. van der. *Johan Huizinga: Leven en werk in beelden en documenten*. Amsterdam: Wereldbibliotheek, 1993.

Lente, D. van. 'Huizinga's Children: Play and Technology in Twentieth Century Dutch Cultural Criticism (from the 1930s to the 1960s)'. *Icon*, no. 19 (2013): 52–74.

Otterspeer, W. *Orde en trouw: Over Johan Huizinga*. Amsterdam: De Bezige Bij, 2006.

Otterspeer, W. *Het horzelnest: De Leidse Universiteit in oorlogstijd*. Amsterdam: Prometheus, 2019.

Otterspeer, W. 'Huizinga and Tocqueville'. Paper presented at conference *Huizinga heute. Hundert Jahre* Herbst des Mittelalters, Vienna, 15–17 May 2019.

Pannekoek, A. 'Het historisch materialisme'. *De Nieuwe Tijd* 24, no. 2 (1919): 15–22.

Paul, H. 'Weber, Wöhler, and Waitz: Virtue Language in Late Nineteenth-Century Physics, Chemistry, and History'. In *Epistemic Virtues in the Sciences and the Humanities*, edited by H. Paul and J. van Dongen, 397–412. Amsterdam: Amsterdam University Press, 2017.

Pree, C. du. *Johan Huizinga en de bezeten wereld*. Leusden: ISVW Uitgevers, 2016.

Strupp, C. *Johan Huizinga: Geschichtswissenschaft als Kulturgeschichte*. Göttingen: Vandenhoeck & Ruprecht, 2000.

Tocqueville, A. de. *Democracy in America*. Edited by G. E. Bevan. London: Penguin Books, 2003.

Torpey, J. 'Huizinga on America'. *Journal of Classical Sociology* 15, no. 3 (2015): 293–301.

Voerman, G. 'Anton Pannekoek: A "Principled Theorist"'. In *Anton Pannekoek: Ways of Viewing Science and Society*, 51–74, 2019.

White, H. 'Historical Pluralism'. *Critical Inquiry* 12, no. 3 (1986): 480–93.

Woud, A. van der. *Een nieuwe wereld: Het ontstaan van het moderne Nederland*. Amsterdam: Prometheus, 2021.

5 The Delay of the 'Grotian Hour'

Abstract
This chapter examines how the dissolution of internationalist communities during the 1910s and '30s informed and altered Huizinga's critiques of Oswald Spengler (1880–1936) and Spengler's *Der Untergang des Abendlandes* in particular.

Keywords: Johan Huizinga; internationalism; Oswald Spengler; historical determinism

On 6 March 1935, Oswald Spengler visited Leiden University on Huizinga's invitation to give a lecture on the history of maritime law.[1] The topic of maritime law and the codification of naval warfare in particular had become an urgent issue in recent decades, and to Dutch ears, its history had two immediate implications. Historically, it concerned the intellectual heritage and canonization of the Dutch legal theorist Hugo de Groot (1583–1645), better known as Grotius. Second, and perhaps more importantly, the topic also had direct bearing on what had already been dubbed in 1918 the 'Grotian hour' of Europe: the founding of the Permanent Court of International Justice in The Hague in 1922.[2] Spengler had acquired tremendous fame and notoriety upon the publication of the first volume of his *The Decline of the West* (*Der Untergang des Abendlandes*, published first in 1918), and most of the audience of his lecture in Leiden would have known what to expect before Spengler even uttered a single word: Spengler interpreted the legal codification of state policy as the coffin of culture, the ruination of the

1 The title of his paper was 'The Emergence of Nautical Travel and its Influence on World History' ('Die Entstehung der Seefahrt und ihr Einfluß auf die Weltgeschichte'). C. Krumm, *Johan Huizinga, Deutschland und die Deutschen* (Münster/New York/Munich/Berlin: Waxmann, 2011), 154. I have not been able to recover how, or even why, the invitation came about.
2 C. van Vollenhoven, *De drie treden van het volkenrecht* (The Hague: Martinus Nijhoff, 1918), 54; J. Vervliet, *The Peace Palace Library Centennial: The Collection as a Mirror of the Historical Development of International Law, 1904–2004* (The Hague: Martinus Nijhoff, 2004), 17–26.

Rydin, T., *The Works and Times of Johan Huizinga (1872–1945): Writing History in the Age of Collapse.* Taylor & Francis Group 2023
DOI: 10.5117/9789463724593_CH05

Raubtier that was humanity.[3] Huizinga, at any rate, knew what Spengler thought was at stake in humanitarian law, and though Huizinga had invited Spengler, Spengler's views appalled him. An attending journalist from the liberal-leftist newspaper *Het Vaderland* reported the next day:

> I saw Prof. Huizinga smile uncomfortably several times as he gracefully thanked Spengler and monopolized him for Leiden as a matter of course. The evening must have been torturous for him [Huizinga]; it became highly doubtful whether Spengler cares about 'the sober parts of intelligible reality' (as Prof. Huizinga wrote in *De Gids* in February) as he swung on horse and by ship from one culture to the next, tying together with a '*wahrscheinlich*' statements about which the real historian would doubt, weigh and arrest himself for forty years.[4]

'Torturous' – that is what the reporter imagined Spengler's lecture to have been to Huizinga, and he had ample reason to make this inference in such explicit language. In previous publications, Huizinga had accused Spengler of showcasing 'naive arrogance' on 'nearly every page'; Spengler was the author of the 'most violent dichotomies and artificial concepts', utterly insensitive to facts and the virtues of *humana civilitas*.[5] Later Huizinga went as far as stating that Spengler was nothing less than the 'greatest enemy of humanistic thought'.[6] Inspired by Nietzsche's 'philosophy of life' and his critique of *Zivilisation* in *Zur Genealogie der Moral* (1887) in particular, Spengler had improperly reduced to biological impulse and violent reflex all human activity and thought, whose trajectory was now no different from a river's inevitable dissolution into lower waters. And just in case Huizinga's rejection had not been explicit enough in his published works, the reporter in question would have had abundant opportunities to learn about

3 'Denn der Mensch ist ein Raubtier'. O. Spengler, *Der Mensch und die Technik* (Munich: C. H. Beck, 1931), 14.

4 'Ik zag prof. Huizinga, die Spengler tot slot en besluit zoo hoffelijk dankte en hem daarbij als vanzelfsprekend voor Leiden monopoliseerde, meermalen smartelijk glimlachen. De avond moet voor hem een kwelling zijn geweest; want of het Spengler te doen is om "het sobere deel der kenbare waarheid" (zooals prof. Huizinga het in de "Gids" van februari genoemd heeft) mag ernstig ten twijfel worden getrokken, als men hem al doceerend per schip en te paard van de eene cultuur naar de andere ziet vliegen en met een "wahrscheinlich" alles van zich afschudden, waarvoor de echte historicus veertig jaar lang dubbend en wegend halt houdt.' *Het Vaderland*: 07-03-1935: 9.

5 VW IV: *Twee worstelaars met den engel* (1921): 443–44. See p. 11 for further details on the references to Huizinga's collected works and letters as well as to the Huizinga archives.

6 VW VII: *Taak en termen der beschavingsgeschiedenis* (1926): 33.

Huizinga's antipathy by other means, for the reporter was Menno ter Braak (1902–1940), Huizinga's cousin, and they had corresponded in some length about the issue of Spengler's popularity in the Netherlands and beyond.[7]

Given the vehemence, frequency and technical prose with which Huizinga wrote about Spengler's philosophy of history, it is surprising to see how little Huizinga's arguments against Spengler have been inventoried and structurally discussed in studies on Huizinga.[8] On the one hand, the crudeness of Huizinga's rejection has understandably invited fairly general and superficial characterizations of his antipathy: Spengler, a generalist with an unrelenting predisposition towards the morose and mythical; Huizinga, an admirer of the particular, intellectual modesty and ethical duties.[9] These crude characterizations, however, leave untouched the most substantial question, which deserves to be taken seriously: what exactly in Spengler's works horrified Huizinga so deeply? Spengler has now become synonymous with the early twentieth-century narrative of pessimism and decadence, but he was by no means exceptional in this respect at the time. Why, of all authors, was Spengler the 'greatest enemy of humanistic thought'? What exactly turned Spengler in particular into a figure of considerable danger? What in Spengler's writings was at stake for Huizinga?

The answers to these questions become even less obvious once one considers the commonalities between Spengler and Huizinga. Amidst their aforementioned differences, their historical writings shared not insignificant features. Both authors wrote tragic historical narratives of cultural decline and draped them in geological, seasonal and spiritual metaphors. Moreover, their narratives commonly identified similar antagonists: industrial society, natural scientific inquiry, instrumental morality, democracy, popular media, impressionism, modern architecture and the latest fashions in urban planning. Spengler and Huizinga identified and addressed similar trends and developments within Europe's modernization and found in them the sources of cultural decline. Spengler wrote of the 'frigid, petrifying' cosmopolitanism. Huizinga, in turn, wrote about the 'petrification' of the soul brought

7 For an overview of their relation, see e.g. C. du Pree, *Johan Huizinga en de bezeten wereld* (Leusden: ISVW Uitgevers, 2016), 231–38.

8 The most comprehensive studies on Huizinga's critique of Spengler have been offered in Krumm, *Johan Huizinga, Deutschland und die Deutschen*, 134–55; C. Strupp, *Johan Huizinga: Geschichtswissenschaft als Kulturgeschichte* (Göttingen: Vandenhoeck & Ruprecht, 2000), 63–65. However close these two accounts may be, they do little to tie together Huizinga's critiques (and the developments therein) to Huizinga's lived life and its context.

9 The article by Ter Braak cited above serves as a good example of how such images were distributed and perpetuated.

about by urban industriousness.[10] Against this background, the questions stated above return with increased salience: given these similarities in tone, vocabulary, perceived threats and spiritual inclinations, why did Huizinga single out Spengler the way he did?

This chapter makes three claims. To begin, it sets out to show that Huizinga drew from a system of ideals and virtues typical of the generation of Dutch artists, authors and academics born in the 1860s and 1870s. These authors, who as a group shall be referred to as the 'Peace Palace generation', entertained ideals of dialogue and internationalism and cherished virtues of compassion and love. The war had revealed just how vulnerable these ideals and virtues were, and the experience of their loss inspired among this generation a heightened identification with them. Secondly, this chapter argues that Huizinga's opposition to Spengler should be understood in relation to this ethical and optimistic background. Huizinga faulted Spengler for epistemic reasons, but the urgency of these faults was fuelled by moral investments that often remained implicit.[11] To Huizinga, Spengler represented not just a theoretical or historical perspective; Spengler's book was an assault on his ethics, identity and international way of life. Lastly, this chapter shows how this moral discrepancy had implications for how both authors interpreted cultural-historical objects. By examining their interpretations of Rembrandt van Rijn (1606–1669), this chapter shows how their moral-historiographic differences were reflected even in their most empirical cultural histories.

For the purpose of this argument, the remainder of the chapter has been divided into five sections. The first spells out a heuristic ideal type of the early twentieth-century culture of internationalism in the Netherlands, in which Huizinga took part. It is argued here that Huizinga's self-reflexive internationalism was induced by experiences of loss and exemplified by an ensuing sense of hope. The second and third sections then discuss, respectively, Spengler's distinction between *Kultur* and *Zivilisation* and Huizinga's morally invested rejection thereof. A fourth section shows how the moral discrepancy of Spengler and Huizinga helps explain not only their theoretical epistemic differences but also, on a more empirical level, how these authors selected and read particular historical material for their

10 'die steinerne, versteinernde Weltstadt'. O. Spengler, *Der Untergang des Abendlandes: Umrisse einer Morphologie der Weltgeschichte, Erster Band* (Munich: C. H. Beck, 1920), 44; 'verstenen'. VW III: *Herfsttij der Middeleeuwen* (1919): 252.

11 Generally, Huizinga's intellectual opposition to Spengler can be divided into two: before 1933 Huizinga took issue mainly with Spengler's historiography, while after 1933, Huizinga addressed Spengler's theory of *Zivilisation*. Both critiques were, however, informed by the values and virtues of the 'Peace Palace generation'.

cultural histories. A fifth and concluding part ties together all the previous sections to show how Huizinga's experience of hope and his culture of internationalist and civil virtues figured in his critiques of Spengler on historiographic, epistemic and empirical fronts.

Huizinga and the 'Peace Palace generation'

The life and works of the Dutch poet Albert Verwey (1865–1937) exemplify the international cultural infrastructure that appeared around the turn of the century in the Netherlands and Europe at large.[12] Verwey helped his German friend and fellow poet Stefan George (1868–1933) translate the Florentine dialect of Dante's *Divina Commedia* into German, corresponded with the Viennese author Hugo von Hofmannstahl (1874–1929), frequented Belgian and British literary societies and, together with his fellow poet and wife Kitty van Vloten (1867–1945), translated works from English, Finnish and Swedish into Dutch.[13] When war broke out in August 1914 and nationalist sentiments started to affect his correspondence networks, Verwey was horrified, and in 1916 he composed the following poem:

> He who lies in the trench, who has lost
> Hope for life and love and lust,
> He knows his death will contribute not,
> To the nation that chose him
> As warrior against his kin;
> Only hope he can,
> To hurry the arrival,
> Of the kingdom that unites us all.[14]

12 A. van der Woud, *Een nieuwe wereld: Het ontstaan van het moderne Nederland* (Amsterdam: Prometheus, 2021), 27–46.

13 M. de Keizer, *Als een meeuw op de golven: Albert Verwey en zijn tijd* (Amsterdam: Prometheus, 2017), 9–11.

14 The original reads:
'Wie in de loopgraaf ligt, wie hoop op leven,
En liefde en lust verloor
Hij wete dat zijn dood geen baat zal geven
Aan 't volk dat zich hem uitverkoor
Als strijder tegen zijn naasten;
Hopen kan hij alleen
De komst te verhaasten
Van het rijk dat ons allen Vereen.'
A. Verwey, *Het zwaardjaar* (Amsterdam: W. Versluys, 1916), 12.

In this piece, Verwey's 'hope' was not focused on, say, peace, pacifism, reason or mutual understanding, though he surely had all of these in mind. Instead, Verwey's hope went out most explicitly to the arrival of 'the kingdom' that would unite 'us all'. To Verwey, the war had meant the destruction of the international life he had led for decades: correspondence was halted, travel became impossible, conferences were cancelled and international publications were postponed or terminated. Consequently, in his eyes, the recovery from war would have to bring about the re-emergence, restoration and further solidification of the international, multilingual space he had known, as well as the legal and cultural codes enabling it. If any positive developments were to come from the unspeakable violence of war, the poem above reads, it could only be towards a further acceleration of such codes and laws required for the arrival of a 'kingdom for all'.

The ideals and hopes expressed in Verwey's poem are typical of those beliefs held among the educated members of his generation during and after the Great War.[15] For example, in 1916 the author and anarchist Frederik van Eeden (1860–1932), who was discussed in the previous chapter, recollected in his diary dream visions of an island where Americans and Germans had replaced war with a set of playful, rule-bound games 'pour rire' (for fun).[16] The painters Piet Mondriaan (1872–1944) and Theo van Doesburg (1883–1931), in turn, presented their art in 1918 as a new engagement with 'the universal', a new, collective thought enabled by the 'destruction of the old'. The legal scholar Cornelis van Vollenhoven (1874–1933) announced that same year how the war had enabled the awaited 'hour' of a new internationalism.[17] And in 1920 the poet and socialist Henriette Roland Holst (1869–1952) wrote that the 'treachery and cowardice' experienced in Europe could be answered only by 'enduring' one's opponent by 'allowing him into one's fortress'.[18] Before the war, these authors had been deeply steeped in pan-European networks, grassroots politics, art movements, journals and conferences, and to authors with this background and of this generation, the war had only made more urgent what it had destroyed: the arenas and communities of international exchange.

15 A similar analysis regarding the academic community at Leiden University during this period has been made in Otterspeer, *Het horzelnest: De Leidse universiteit in oorlogstijd*, 36–41.

16 F. van Eeden, *Dromenboek* (Amsterdam: Bert Bakker, 1979), 304.

17 The first manifesto for 'De Stijl' was published by Doesburg, Mondriaan and others in *Algemeen Handelsblad* 26-11-1918: 9. For the 'Grotian' hour, see Van Vollenhoven, *De drie treden van het volkenrecht*, 54.

18 H. Roland Holst, *Tusschen twee werelden* (Rotterdam: W. L. & J. Brusse, 1923), 70–71.

Perhaps the most impressive and megalomaniac manifestations of such internationalist dreams in the Netherlands are, however, to be found in the sketches and drawings of the architects Hendrik P. Berlage (1856–1934) and Karel de Bazel (1869–1923). Already before the war, in 1905, De Bazel had suggested furthering the international spirit through the construction of a 'capital city of the world', which, of course, was to be built in the Netherlands. For this purpose, De Bazel envisioned an octagonal city of symmetry right next to the Peace Palace, whose construction had already been commissioned in The Hague.[19] After the war, in 1919, Berlage went even further: to be built on a high hill surrounded by an empty plane and in the heart of the European continent, a 'Pantheon for Humanity' was to be erected both in commemoration of the lives lost in the war and in support of the international ideals of pacifism.[20] Amidst the towers of 'love and courage' and 'inspiration and discretion', a dome was to keep the 'international community' dry.[21] Berlage presented his drawings in a pamphlet alongside poems by Henriette Roland Holst, mentioned above. In these poems, she spelled out the world this pantheon announced: fathers, brothers, sons and husbands lay dead in the fields of Flanders, so women would build the socialist universe for their children in a mausoleum.[22]

From early on in his life, Huizinga had engaged with and been shaped by international traffic and the ideals of international cooperation: he had been a visiting student in Germany and later worked as a teacher in a secondary school of a kind (HBS) that had been devised to educate international scientists and scholars. Later Huizinga took part in international delegations, had numerous contacts and friends in both the international socialist and pacifist movements, corresponded in at least six different languages with authors and friends in Europe and North America and visited international conferences throughout Europe.[23] The ideals he found in these social spaces resonated with the ideals of pacifism and cooperation of his Anabaptist upbringing[24] and resurfaced time and again in his academic writing. 'A history that has lost its living contact with national and international culture cannot be healthy,' he reasoned, because if the historical discipline is to

19 *Arnhemsche Courant* 13-09-1905: 2; *De Courant* 21-11-1906: 2.

20 H. P. Berlage, *Het Pantheon der Menschheid* (Rotterdam: W. L. & J. Brusse, 1919).

21 Berlage, *Het Pantheon der Menschheid*, 13.

22 H. Roland Holst-van der Schalk. 'Ter gedachtenis', in *Het Pantheon der Menschheid* (Rotterdam: W. L. & J. Brusse, 1919), 22–23.

23 Du Pree, *Johan Huizinga en de bezeten wereld*, 181–87.

24 A. van der Lem, *Johan Huizinga: Leven en werk in beelden en documenten* (Amsterdam: Wereldbibliotheek, 1993), 20–24.

Figure 5.1. Drawings from Berlage's manifesto *The Pantheon of Humanity* (1919).

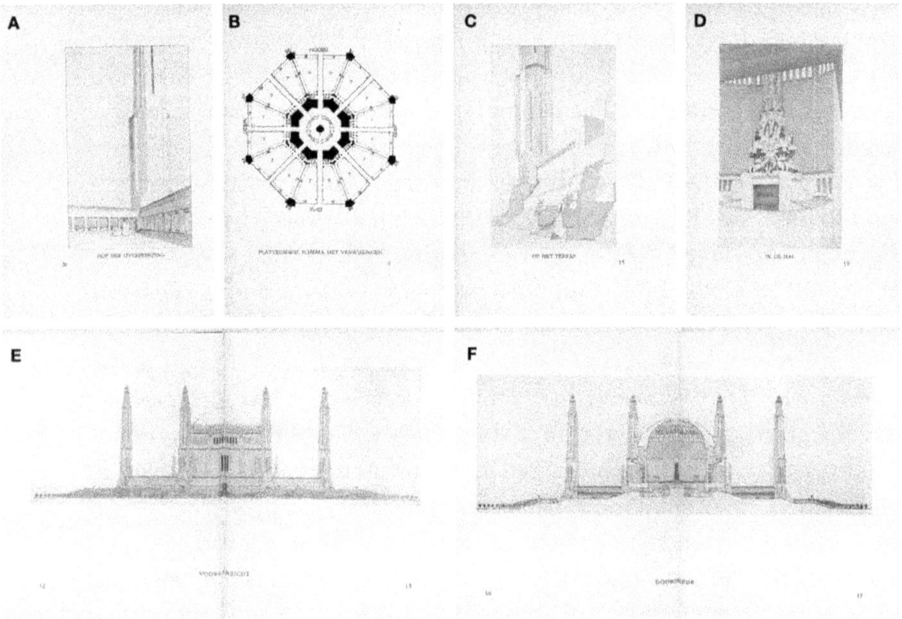

be of value to life, it must recognize and accept that 'the international interpenetration of peoples will continue'.[25] In fact, Huizinga thought, we must allow this tendency to 'develop freely on our soil and keep our Dutch heads cool'.[26]

To be sure, tremendous differences existed among the internationalist authors mentioned. Some of these authors belonged to the nineteenth-century traditions of classical liberalism (such as Van Vollenhoven), while others drew from revolutionary socialism (such as H. Roland Holst). Some wished to dispense with all historical authority (such as De Bazel) so as to liberate human creativity from the yoke of rules, while others found traditional norms to be the foundation of freedom and creativity (such as Huizinga). Such widely differing convictions led to equally diverse political positions. Nevertheless, these authors shared the basic ethical conviction that international exchange and cooperation were desirable. They celebrated ideals of debate, mutual understanding and the civil virtues required to this end. These ideals were often mediated by the contexts of their time, which are unpacked below: a colonialist claim to rational morality, a bourgeois

25 VW VIII: *De taak der cultuurgeschiedenis* (1929): 108.
26 VW VII: *Nederland's geestesmerk* (1935): 292.

Figure 5.2. Another example of Dutch internationalist culture at the beginning of the twentieth century: several board games celebrating peace and cooperation were brought onto the market in the 1900s and 1910s, both by commercial and public institutions. In the game pictured, the winner was the player who got their pawn to the Peace Palace's table of negotiation quickest.

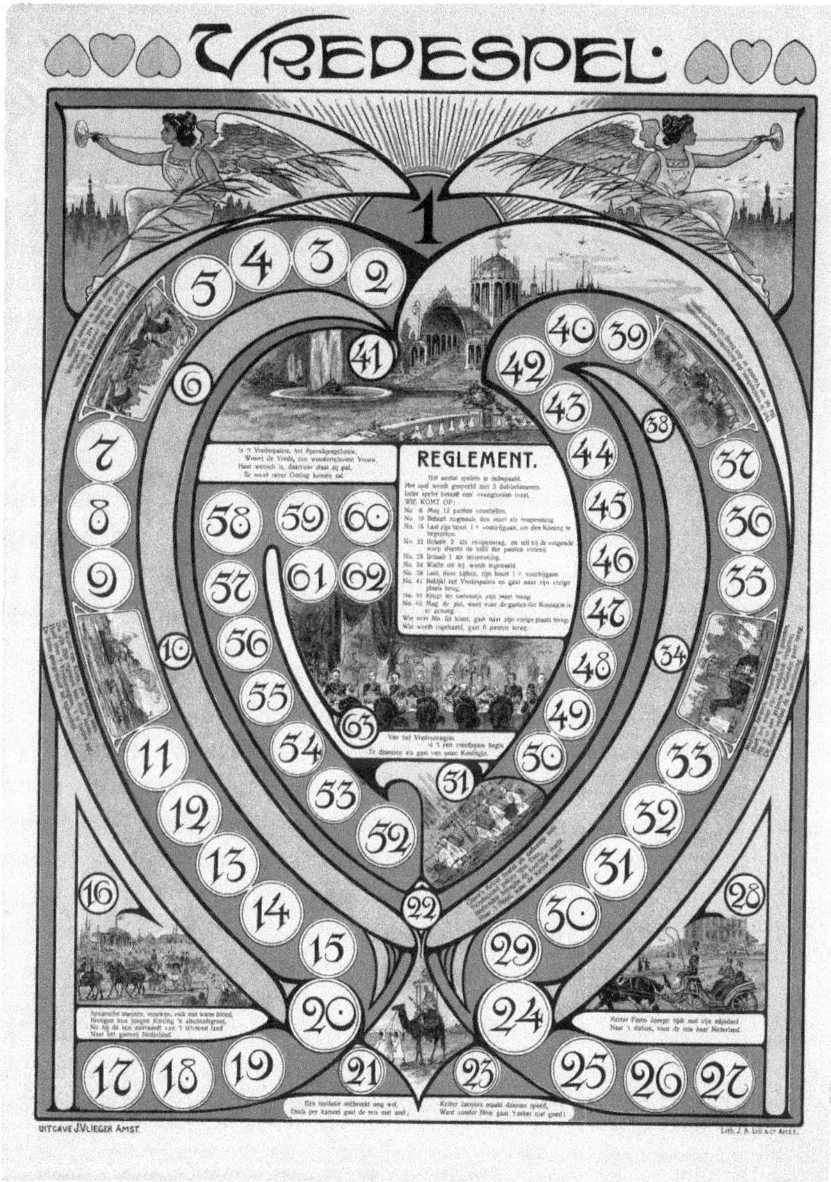

entitlement to travel and commerce and an appreciation of Christian values and eschatology. Hence, by emphasizing the shared internationalist perspectives of these authors, no claim is made regarding the origins or implications of their internationalism. The single claim made here is this: in the 1910s and '20s, a generation of Dutch academics celebrated international cooperation and virtues of love and empathy.

Huizinga's relation to these generational ideals of international cooperation can be traced and conceptualized particularly well through his appreciation and understanding of the Peace Palace in The Hague. Throughout his life, the Peace Palace and the international courts it accommodated remained a banister for his hopes and a reoccurring point of reference in his writings. For this reason, the community with which Huizinga shared his international ideals shall typically be dubbed the 'Peace Palace generation'. But what exactly did the Peace Palace symbolize to Huizinga and others? In the next section, the internationalist ideals of Huizinga's generation, and Huizinga's specific position therein, is discussed in terms of this 'Peace Palace generation'.

Huizinga and the Peace Palace

In 1913 the Peace Palace was completed in The Hague and presented in a grand style to both the Dutch public and its diplomatic elites. Newspapers had reported on its advancement for years – the architect's choice of carpets, chandeliers and even lightbulbs had been meticulously discussed – and once the inauguration came around, endless attention was devoted in newspapers to all its international guests. And for good reason. A Russian initiative supported by Andrew Carnegie's industrial dollars, the Permanent Court of Arbitration was performative of the internationalism it sought to perpetuate, and the Dutch, celebrating a century of independence from Napoleonic rule, were keen to see their geopolitical role secured among military powerhouses both new and old.[27] In this sense, the image of internationalism became an object of national pride. The Court was celebrated as the materialization of the centuries-old ideals of Hugo de Groot (1583–1645), or Grotius, and the minor military power of the Netherlands was perceived in terms of the country's supposed unique ability to view the world 'objectively' and

27 'De rol der kleine staten is niet uitgespeeld, integendeel zij begint pas. Het blijft nu eenmaal, ook bij de onbeperkte communicatiemogelijkheden van heden, een waarheid, dat goed staatsbestuur het best kan worden uitgeoefend in een betrekkelijk kleinen kring over een niet overgroot gebied.' VII: *Geschonden wereld* (1945): 601.

Figure 5.3. A committee headed by the Dutch Catholic architect Pierre Cuypers (1827–1921) was installed to judge the proposals for the Peace Palace. Above, submissions by (A) F. Wendt, (B) Greenley and Olin, (C) L. Cordonnier and (D) F. Schwechten have been included. Cordonnier was awarded first prize.

'impartially'.[28] One newspaper from 29 August 1913, the day of the palace's opening, wrote of a blissful display of what had proven possible 'after Napoleon': a 'tremendous turn towards international morality', a 'world historical moment'.[29]

Then war came. Exactly one year after the publication of said newspaper article, and only just outside the Dutch coast, the naval forces of Great Britain and Imperial Germany clashed for the first time in a war that would last years. In a matter of only a few autumn days, the Peace Palace was transformed in public perception from an announcement of the future to a monument of past naiveté and ignorance. The palace had become the living example of 'history's irony', as one newspaper article remarked, if

28 'objectief', 'onpartijdig', in e.g. *Arnhemsche Courant* 16-09-1905: 5; *Eindhovensch Dagblad* 11-09-1914: 1.
29 *Algemeen Handelsblad* 29-09-1913: 1.

not itself the 'image of a broken peace', as yet another piece declared.[30] And still, after the war's conclusion, the Peace Palace continued to exert sufficient symbolic force to accommodate the newer post-war institutions devoted to the propagation of humanitarian law. From 1922 onwards, the palace housed the Permanent Court of International Justice, for which it is now best known. In this capacity, the Palace transformed into a physical emblem of the perseverance and continuation of the internationalist peace movements before the war. In this temporal vein, the Leiden Professor of Law Cornelis van Vollenhoven published a piece announcing the final yet delayed arrival of the 'Grotian hour':

> Could 1920 achieve what had been impossible in 1625? Our times have a number of advantages over those of Grotius. First of all, an utterly controlled planet: not only politically distinguished and organized, but manageable as one by steam engines, telegraphs and technological force. Moreover, [it has] the ability to judge independently and impartially over and between nations through international councils. Finally, as a result of these two points, [it has] the possibility of [founding] international bodies and organizations, wherein the world as a whole selflessly recognizes her will and her power.[31]

Van Vollenhoven's account of the emergence and maturation of humanitarian law was imbued with a temporal understanding of internationalism: 1920 was understood as the substantiation of the promise of 1625, the practical fulfilment of a dormant principle of justice.[32] In the eyes of Van Vollenhoven, the technical advancements of his day – steam engines and telegraphs – enabled the completion of a historical hope for an impartial tribunal of international law. 1920 had become 'the hour of Grotius', whose legal thought in the seventeenth century had rested on the theoretical conviction that

30 *De Tijd* 28-08-1914: 2; *De Tijd* 16-12-1914: 1.

31 'Zou 1920 vermogen wat voor 1625 ondoenlijk was? Een paar zeer grote dingen in elk geval heeft onze tijd op dien van Grotius voor. Vooreerst, een volkomen beheerschte aardbol: niet alleen staatsrechtelijk verdeeld en geordend, [maar] door stoomwezen, telegraaf en technische kracht als eenheid te hanteren geheel. Voorts, de mogelijkheid van internationale colleges [die] onafhankelijk en onpartijdig tusschen en over natiën kunnen beslissen. Eindelijk, als resultante van die twee, de mogelijkheid van internationale organen en organisaties, waarin de wereld als geheel haar wil erkent en haar kracht onbaatzuchtig belichaamd ziet.' Van Vollenhoven, *De drie treden van het volkenrecht*, 64–65.

32 The 'promise' here described could be catalogued under what has been called an interwar 'memory of the future'. W. Linmans, *De oorlog van morgen: Nederlandse beeldvorming van een volgende oorlog 1918–1940* (Amsterdam: Prometheus, 2021), 17.

'war itself will lead us to peace, as to its proper end'.[33] Van Vollenhoven's references to Grotius were not mere rhetorical devices for the construction of authority and the celebration of a national heritage, but even if they had been, these devices would reveal not only the cultural status of Grotius at the time but also how the political temporality of the internationalist movement was constructed by Van Vollenhoven and others. In the wake of the Hague Conventions of 1899 and 1907 as well the conclusion of the subsequent war, Grotius's *De jure belli ac pacis* (1625) received a new kind of attention in the Netherlands as well as abroad. In 1913 the Peace Palace Library acquired fifty-five editions of Grotius' magnum opus, and in 1920 the Grotius Society was founded in London.[34]

Soon after Huizinga moved from Groningen to Leiden in 1915 for his professorship, he became a close friend and avid reader of Van Vollenhoven, and as a former professor of Dutch history, Huizinga was knowledgeable on Grotius and his times.[35] Grotius had been a legal scholar of unfathomable importance during the Dutch Eighty Years' War (1568–1648), when the Dutch needed central laws and rules of military conduct at sea. At the time, these statutes were necessary not only to supply their naval forces with general rules for unforeseeable circumstances on faraway seas but also to accumulate diplomatic authority among foes and allies to explain decisions and policies.[36] Authority of this kind had traditionally been accumulated through the genre of war manifestos, whereby heads of states and kingdoms bolstered their military actions with some kind of theoretical argument. The legal theory of Grotius now offered a systematic 'law of war and peace' through which such manifestos could be assessed and compared on ostensibly impartial theoretical grounds. For this reason, Huizinga held, the early twentieth-century endeavour to create a space for dialogue and exchange synchronized the efforts of Grotius, Van Vollenhoven and the institutes of the Peace Palace:

> The external events and conditions had further invigorated a hopeful internationalism. The Netherlands had accommodated the first and

33 H. Grotius, *The Rights of War and Peace; Including the Law of Nature and of Nations*, ed. A. C. Campbell (Washington/London: M. W. Dunne, 1901), 17.

34 Vervliet, *The Peace Palace Library Centennial: The Collection as a Mirror of the Historical Development of International Law, 1904–2004*, 11.

35 For a comparison between Huizinga's conception of Grotius and that of other Dutch historians at the time, see A. van der Lem, *Het eeuwige verbeeld in een afgehaald bed: Huizinga en de Nederlandse beschaving* (Amsterdam: Wereldbibliotheek, 1997), 121–24.

36 O. A. Hathaway and S. J. Shapiro, *The Internationalists and Their Plan to Outlaw War* (London: Penguin Books, 2018), 31–55.

> second peace conferences. First the Permanent Court of Arbitration, then the Peace Palace, and then the Permanent Court of International Justice were established in The Hague. The spirit of Grotius had blossomed anew in Asser and Van Vollenhoven. [...] And now comes the League of Nations to mould our most hopeful expectations in a sustainable and solid form. It was no wonder that the League of Nations was not able to find a supporter more loyal than the Netherlands.[37]

In Huizinga's eyes, the Peace Palace and its institutes radiated symbolic meaning on at least two levels. On a temporal level, they symbolized the continued potency and relevance of the 'most hopeful expectations' and thus the continued role of pre-war dreams in the post-war reality. In this capacity, the palace offered a physical manifestation of a moral continuity amidst unfathomable political transitions. On a second, national-political level, this emblem of internationalism became, at once, an emblem of national pride.[38] The emergence of the international court was narrated in terms of the maturation of Grotius's thought, and so the palace also became an attestation of the role the Netherlands could play in the violently disrupted geopolitical landscape. In fact, Huizinga argued, the virtue-ethical features supposedly typical of the Dutch nation were indispensable to the success of the humanitarian and internationalist movement: as 'impartial', only 'mildly Christian' and 'civilian', the Dutch had been destined to be international.[39] To Huizinga, the Peace Palace symbolized the continued role and importance of the Dutch historical heritage and civilian culture, as well as of the cultural heritage of his own generation. This neo-Gothic cathedral of Grotian peace launched the Dutch past into the international future: 'the light of [Grotius's *De jure belli ac pacis*] shines like a beacon over the tumultuous sea of our time.'[40]

37 'De uiterlijke gebeurtenissen en omstandigheden hadden dien trek van een hoopvol internationalisme nog versterkt. Nederland had de eerste en de tweede vredesconferentie mogen herbergen. In Den Haag was eerst het Hof van Arbitrage, dan het Vredespaleis, dan het Permanente Hof van Internationale Justitie gevestigd geworden. De geest van Grotius had er opnieuw gebloeid in Asser en Van Vollenhoven. [...] En thans kwam de Volkenbond al onze hoopvolste verwachtingen in een duurzamen, hechten vorm gieten. Het was geen wonder, dat de Volkenbond geen trouwer aanhanger vond en behield dan Nederland.' VW II: *Terugblik op Nederlands groei in de veertig jaren van het regeeringsjubileum* (1938): 553.

38 Van der Lem, *Het eeuwige verbeeld in een afgehaald bed: Huizinga en de Nederlandse beschaving*, 171–75.

39 VW II: *Hugo de Groot en zijn eeuw* (1925): 391; VW VII: *Nederland's geestesmerk* (1935): 288–292.

40 '[Het] licht van dat boek schijnt als een baken over de woelige zee van onzen tijd.' VW II: *Grotius' plaats in de geschiedenis van den menschelijken geest* (1925): 382.

The 'light of Grotius' reached Huizinga's eyes through at least two media: one colonial, the other religious. From the seventeenth century onwards, Grotius's legal theory appealed to states who sought to draw material benefit from military might. The theory allowed a 'right by might' principle in the absence of judicial means and was in this sense tailored to colonial bodies of power in distant lands. Hathaway and Shapiro have pointed out the irony of Grotius's popularity among the internationalists of the 1910s, but perhaps this irony is better conceived of as blatant opportunism. Grotius' legal theory was recognized and respected by colonial forces and cultures, and by re-appropriating it in an internationalist and peaceful spirit, colonial heritage could be whitewashed amidst political-ethical transitions.[41] Grotius, a towering figure of Dutch intellectual heritage, was dressed in twentieth-century fashion and re-invented in such a way as to prohibit newer military powers from 'seizing' the 'property' of nations whose colonial military prowess was waning. Huizinga's interest in and appreciation of Grotius was not a conscious decision for colonialist ends in any direct fashion, but it did draw from a longer and more structural culture of Dutch colonialist interests that had helped propagate and disseminate Grotius's work and name through the centuries.

A second contextual piece of Grotius's appeal to Huizinga was religious in nature: in Huizinga's Mennonite upbringing, the ideal of pacifism was one of the central pillars. The Mennonites were followers of the Frisian pastor Menno Simons (1496–1561), who, in turn, had been influenced by Anabaptists who fled from Switzerland to the northern provinces of the Dutch Republic in the sixteenth century.[42] In Switzerland these Anabaptists had been prosecuted for their refusal to enter the military apparatus, and their pacifism resonated with Simons after his loyalty had swung back and forth between Luther and Zwingli. The Anabaptists were tolerated in the Calvinist Netherlands after Menno had laboured towards their institutionalization and ever since their pacifism had been re-invented time and again by his followers. Johan Huizinga's grandfather, Jakob Huizinga, was a Mennonite pastor and taught his grandchildren: 'break what breaks peace, grow what grows love.'[43] During the 1920s several Dutch Mennonites ran into trouble with the law for refusing military service. Huizinga was not a

41 Hathaway and Shapiro, *The Internationalists and Their Plan to Outlaw War*, 69–72.

42 G. Mak et al., *Verleden van Nederland: Een nieuwe geschiedenis* (Amsterdam/Antwerp: Atlas Contact, 2021), 152–54; J. C. Kennedy, *Een beknopte geschiedenis van Nederland* (Amsterdam: Prometheus, 2018), 98.

43 As cited in Van der Lem, *Johan Huizinga: Leven en werk in beelden en documenten*, 22–23.

pacifist himself, but he never broke with his Anabaptist upbringing, and the re-interpretation of Grotius along internationalist lines would have struck him as being congenial with his personal religious upbringing and education.

Against these two backgrounds to Huizinga's internationalism – one colonialist, the other religious – an important point needs to be made: Huizinga's sympathy for the Peace Palace and the moral-political project it symbolized did not stand alone; it drew from an older and structural set of values and ideals. The term 'Peace Palace generation' is not meant to explain Huizinga's internationalist perspective but rather to characterize it in terms of its ideal type. As is explored below, Huizinga launched what was at its heart a moral attack on Spengler's historical thought. This moral perspective did not follow from his appreciation of the Peace Palace; his appreciation of the Peace Palace followed from and can be seen to symbolize his moral perspective. That is, Huizinga did not need to be a proponent of the Peace Palace's institution to dislike Spengler. But he was a proponent, and this sympathy is a helpful guide to exploring the nature and implications of his critiques of Spengler.

As was explored above, Huizinga found that the Peace Palace project 'had further invigorated a hopeful internationalism', and on this account, the palace belonged to an experience of both loss and hope shared by Huizinga and his peers.[44] To Huizinga, the palace symbolized not so much an eschatological hope as it mediated an entelechial one: a Grotian civilization that had been on the verge of slipping away through the cracks of nationalism and violence yet had nevertheless been solidified in stone and marble once again.[45] In this sense, the Palace and its Permanent Court of International Justice were doorways to history: they opened up and mobilized the past in the present. Huizinga's internationalist ethic became a part of his self-reflexive identity the moment its loss became a possibility. The palace represented not a 'broken-off' past in the way that other experiences did. Rather, it symbolized a past ideal that could not be taken for granted; it was vulnerable and had to be defended. Hence, this past was not ruinous in the way that Ypres or medieval architecture had been. It was ruinous not in actuality but in potentiality, and the window for subsistence was what Huizinga found hopeful. Its 'temporal hesitation' belonged not to an experience of decay but to history's hesitant perseverance.[46]

So far, what matters most is neither the actual content of Grotius's legal theory nor the question of Huizinga's (and Van Vollenhoven's) technical

44 VW II: *Terugblik op Nederlands groei in de veertig jaren van het regeeringsjubileum* (1938): 553.
45 I thank Herman Paul for suggesting this perspective on the nature of Huizinga's hope.
46 Gafijczuk, 'Dwelling Within: The Inhabited Ruins of History', 163.

interpretation thereof. For present purposes, what matters is that Huizinga's appreciation of Grotius reveals a temporal, moral and experiential orientation congruent with that of many other Dutch authors of his generation after the Great War. This generation had been shaken and horrified by the war, but the war had not completely shattered their pre-war moral ideals and internationalist experiences. Quite on the contrary, the war had made urgent what had previously remained only implicit: international, impartial communication had become salient precisely because of its temporary collapse, and now it had to be defended. To this end, the foundation of institutes such as the Permanent Court of International Justice served as the symbol and sense of hope that past ideals had a chance to serve in what was to come. The palace was not only a judicial body. The palace itself, its concrete existence, revealed a past that was seemingly in a position to persevere into the future. Huizinga and his peers' belief in the possible maturation of internationalist ideals was part and parcel of their experience of a loss of morals.

In this sense, authors such as Huizinga and Van Vollenhoven were the Dutch equivalents of those Weimar intellectuals born in the 1870s who, after 1918, defended the nineteenth-century ideals of cultural exchange, debate and political pluralism.[47] In the Netherlands, these antiquated idealists could have found no better material representative than the historicist architecture of the Peace Palace. Still, members of this generation were in their fifties during the 1920s, and whilst they identified as progressives, a younger generation of authors and cultural critics conceived of them as apolitical monuments of a poorly aged past.[48] These younger authors were less interested in abstract ideals of humanity and cooperation than they were in concerns of gender, nations or classes. Of these younger authors, a certain group was particularly vocal in their critiques of the ideals and virtues of the Peace Palace generation. These authors read not Spinoza but Spengler; they longed not for citizenship in the cosmopolis but for membership in a community.

Spengler's critique of *Kosmopolitismus*

Discontent with the ideals of internationalism had been looming long before the Peace Palace opened. By the time Spengler's *The Decline* appeared in

47 P. E. Gordon and J. P. McCormick, 'Introduction: Weimar Thought: Continuity and Crisis', in *Weimar Thought: A Contested Legacy*, ed. P. E. Gordon and J. P. McCormick (Princeton/London: Princeton University Press, 2013), 7–8.
48 Du Pree, *Johan Huizinga en de bezeten wereld*, 224–25.

1918, critiques of 'civil' virtues and cosmopolitan ideals had been widespread across European book markets for decades.[49] For different reasons but with a similar effect, members of the intellectual avant-garde of Germany, Austro-Hungary, France, England, Scandinavia and the Netherlands produced abundant critiques of those cosmopolitan values and ideals dear to Huizinga and his peers, and from the 1890s onwards, cultural pessimism of this kind could be found among anarchists, Catholics, proto-fascists, industrialists, aristocrats and socialists alike throughout Europe.[50] In terms of the attention it attracted, Spengler's book did have at least one advantage over similar and related books: its seemingly prophetic timing. In the Netherlands, too, the war had mobilized sentiments of both vulnerability and revolution, and these were carried far and wide by new media. For example, one widely read, moderate newspaper from January 1918 wondered: 'will the Netherlands be dragged along with the worldwide revolution?'[51] Against this background, Spengler's story of doom received tremendous attention, but before its Dutch reception can be analysed, I will illustrate Spengler's central point.

'To understand the world is to offer resistance to it,' Spengler wrote in 1922 in the foreword to the first volume's second edition, and this statement aptly captures the book's central mission and tone.[52] Spengler's book was not an academic reflection responding to earlier academic exposés and arguments. Its stated purpose was to respond to the world and its modern transformations and to prepare and equip its readers with those tools required for withstanding the alleged ills and confusion of their time. Spengler's perceived enemies were many and diverse. They ranged from the natural scientific metaphors in the social sciences to modernist architecture, atonal

49 F. C. Beiser, *Weltschmerz: Pessimism in German Philosophy 1860–1900* (Oxford: Oxford University Press, 2016), 1–12; J. F. Dienstag, *Pessimism* (Princeton: Princeton University Press, 2006), 3–48.

50 De Keizer, *Als een meeuw op de golven: Albert Verwey en zijn Tijd*, 19–21; B. Spaanstra-Polak, *Het symbolisme in de Nederlandse schilderkunst 1890–1900* (Bussum: Thoth, 2004), 1–28.

51 In the Netherlands, male suffrage had been pushed through parliament in 1917, and in 1918 revolutionary movements of different kinds had spread through the country with significant force and visibility. A year later, female suffrage followed, and all the while one central question grew increasingly pronounced in Dutch media: 'will the Netherlands be dragged along with the world-wide revolution?' Would the civilized ideals of a new democracy give way to a collapse of culture? *Algemeen Handelsblad* 28-01-1918.

52 'Die Welt verstehen nenne ich der Welt gewachsen sein.' O. Spengler, *Der Untergang des Abendlandes: Umrisse einer Morphologie der Weltgeschichte, Erster Band*, 2nd ed. (Munich: C. H. Beck, 1923), ix. Unless stated elsewise, other references to *Der Untergang* in this chapter concern the edition from 1920.

music, urbanization, industrial economy and, perhaps most pervasively, *Kosmopolitismus*, that cemetery of culture called *Zivilisation*.[53] In Huizinga's eyes, Spengler's rejection of internationalism and cosmopolitanism eclipsed the value of any antipathy they might have had in common. Spengler's rejection of cosmopolitanism was, however, more technical than Huizinga's support thereof. In order to get a proper grasp of both Spengler's rejection of cosmopolitanism and Huizinga's subsequent retribution, Spengler's criticism of Kant (1724–1804) – or rather, of 'the world of Kantians' he left behind with a belief in history's 'infinite progress towards perfection'[54] – must be examined.

If 'historians are the coroners of political-moral convictions,' as Stefan-Ludwig Hoffmann recently wrote, then Oswald Spengler was the coroner of any belief in historical progress.[55] Strapped to Spengler's table for dissection, the belief in a rationally organized history of societies appeared to him as the very corpse of what had once been vital 'culture'. Neatly defined structures of periods and parishes, reigns and revolutions merely symbolized the modern inability to come to terms with what would, in fact, always remain the destiny of humanity: the tragedy of a will unable to fulfil itself in the natural world. This was not because the will to create, love and lust is weak, but because the natural predicaments are inherently and tragically insensitive to the will's call.[56] In view of this once-deemed-rational history, Spengler set out to devise an alternative organization of history. This new history was to dispense with the chronological organization of events prescribed by the historical sciences of the nineteenth century and replace it with an organization that did justice to the 'tragedy' and 'destiny' of historical cultures.[57] But what exactly did it mean, according to Spengler, for a historian to do justice to 'tragedy'? And if not chronologically, how, then, was the coroner-historian to organize historical events? And for what purpose?

53 Spengler, *Der Untergang des Abendlandes: Umrisse einer Morphologie der Weltgeschichte, Erster Band*, 44–46. His distinction between 'culture' and 'civilization' was not uncommon at the time; not dissimilar invocations of these terms can be found in the works of Thomas Mann. T. Mann, *Reflections of a Nonpolitical Man*, ed. Morris W. D. (New York: New York Review of Books, 2021), 23.

54 Spengler, *Der Untergang des Abendlandes: Umrisse einer Morphologie der Weltgeschichte, Erster Band*, 168, 29.

55 'Die Historiker sind die Leichenbeschauer unserer politisch-moralischen Überzeugungen.' S. L. Hoffmann, 'Rückblick auf die Menschenrechte', *Merkur* 71, no. 812 (2017): 5.

56 Spengler, *Der Untergang des Abendlandes: Umrisse einer Morphologie der Weltgeschichte, Erster Band*, 218.

57 'Die Zivilisation is das unausweichliche Schicksal einer Kultur.' Spengler, *Der Untergang des Abendlandes: Umrisse einer Morphologie der Weltgeschichte, Erster Band*, 43.

In order to understand Spengler's rejection of a chronologically organized history, 'scientific history', and his suggested alternative, one of Spengler's central distinctions must be introduced first: his distinction between 'nature' and 'history'.[58] With this distinction, Spengler sought to oppose all those historians who had understood history as a natural science in at least one of three ways: (1) history as a collection of empirical particulars from the past, (2) history as a study of the natural causes of historical developments or (3) history as a natural lawfulness supposedly revealed by a chronological organization of these particulars.[59] In all three cases, Spengler held, historians had failed to recognize that the principle according to which they organized their material – chronology – was itself the outcome of a historical development. Had these historians been aware of the historical contingency of their historical principle, Spengler argued, they might have arrived at the question Spengler sought to address: *how does one write the history of history?* Or, in Spengler's terminology: how does one write the spiritual history of natural history? Which spiritual development underpinned and had led to a natural scientific understanding of historical inquiry? Modern historians, Spengler held, had failed to achieve this awareness and thus could not ask such questions:

> One may therefore perhaps say, and later one will, that a true historiography in the Faustian style has been lacking, that is, a historiography able to sufficiently distance itself in order to see in the image of world history the present [...] as something infinitely distant and alien, as an epoch that does not way heavier than others, without the standards of whatever ideal, without reference to itself, without wish, worry and personal interest, just as practical life demands; a distance that enables one to see the entirety of the phenomenon of human history with the eyes of a God, like the peaks of a mountain range.[60]

58 Spengler, *Der Untergang des Abendlandes: Umrisse einer Morphologie der Weltgeschichte, Erster Band*, 10.

59 Spengler, *Der Untergang des Abendlandes: Umrisse einer Morphologie der Weltgeschichte, Erster Band*, 6–10.

60 'Man darf deshalb vielleicht sagen, und man wird es später einmal tun, daß es an einer wirklichen Geschichtsschreibung faustischen Stils überhaupt gefehlt hat, einer solchen nämlich, die Distanz genug besitzt, um im Gesamtbilde der Weltgeschichte auch die Gegenwart – die es ja nur in Bezug auf eine einzige von unzähligen menschlichen Generationen ist – wie etwas unendlich Fernes und Fremdes zu betrachten, als eine Epoche, die nicht schwerer wiegt als alle andern, ohne den Maßstab irgendwelcher Ideale, ohne Bezug auf sich selbst, ohne Wunsch, Sorge und persönliche innere Beteiligung, wie sie das praktische Leben in Anspruch nimmt; eine Distanz also, die [...] es erlaubt, das ganze Phänomen der historischen Menschheit wie mit

The ideal of 'objectivity' formulated by modern historians (historians who deemed their inquiry to amount to 'scientific history') disabled them from experiencing the true historical difference of which Spengler held history to consist. The ideal of a disengaged objectivity was itself profoundly embedded in modern times and cast a shadow over the process whereby objectivity itself had emerged. Spengler was not specific about what these modern objective methods were and who exactly practised them. Rather, he addressed them by taking Kant as their main spokesperson, whose philosophy, in Spengler's understanding, epitomized the idea that all events could be causally related in a framework fixed by time and Euclidean space.[61] For present purposes, it is not necessary to delve into the quality of Spengler's reading of Kant (or the lack thereof). What does matter is the following question: how was the historian, according to Spengler, to reconstruct and do justice to historical differences if not through 'objectively' established events in a given space-time? The answer concerned a certain way of looking, or of 'believing':

> Here follows the dissolution of all older problems into the genetic. The conviction that everything that *is* has also *become*, that something historical underlies all that is natural and knowable, the world as it really is consists of a possible *Ich* that realized itself, the insight [of the fact that] all which has ever been must also be the *expression of a living thing*.[62]

To think properly in terms of history, Spengler wrote, the historian must approach history not as a collection of events which 'have become' and are now ready to be understood – that is, subsumed in categories, structures and series.[63] Rather, the historian should seek inspiration

dem Auge eines Gottes zu überblicken, wie die Gipfelreihe eines Gebirges.' Spengler, 135–36. In the next edition, Spengler prefaced a revised version of this passage with: 'Es gehört zum Stolz moderner Historiker, objektiv zu sein, aber sie verraten damit, wie wenig sie sich ihrer eigenen Vorurteile bewußt sind.' Spengler, *Der Untergang des Abendlandes: Umrisse einer Morphologie der Weltgeschichte, Erster Band* (1923), 127.

61 Spengler's reading of Kant was incredibly poor.

62 The full passage in the original German reads: 'Hier erfolgt die Auflösung aller älteren Probleme ins Genetische. Die Überzeugung, daß alles was *ist*, auch *geworden* ist, daß allem Naturhaften und Erkennbaren ein Historisches zugrunde liegt, der Welt als dem Wirklichen ein Ich als das Mögliche, das sich in ihr verwirklicht hat, die Einsicht, daß nicht nur im Was, sondern auch im Wann und Wie lange ein tiefes Geheimnis ruht, führt auf die Tatsache, daß alles, was immer es sonst sei, auch *Ausdruck eines Lebendigen* sein muß.' Spengler, *Der Untergang des Abendlandes: Umrisse einer Morphologie der Weltgeschichte, Erster Band*, 64.

63 Spengler, *Der Untergang des Abendlandes: Umrisse einer Morphologie der Weltgeschichte, Erster Band*, 64–65.

from the vocabulary of those 'morphological' interpretations of the world – that is, those interpretations seeking to understand the world in its 'becoming': biology and geology.[64] In particular, Spengler wished to adopt from the former the terms and methods of 'homology' and 'morphology'. Following Goethe's vitalistic understanding of biology as the study of *Urphänomenen*, Spengler distinguished biology, and in particular its morphological subdiscipline, from the natural sciences and placed biological investigations firmly in the camp of what he understood to be the historical sciences.[65]

In biology, 'morphology' had become the study of the common genetic ancestry of physiological structures among different species, whether or not these structures continued to evolve towards vastly different ergonomic functions.[66] This kind of analytical tool, Spengler found, allowed for an altogether different appreciation of history, a kind of appreciation supposedly required to narrate history properly. Homology namely allowed one to treat different anatomical structures in different species at different times as simultaneous events, even though they might not appear at the same point in measurable time. This way of examining structures across natural time as genetically simultaneous, Spengler found, opened up a new way not only of comparing structures – be they biological or cultural – but of identifying differences between structures. This 'between' was not a chronological unit of time but concerned a genetic difference among organisms, revealing a development quite independent of any particular configuration of the hands on a clock. Now, in order to achieve a more intuitive understanding of Spengler's distinction between diachronic natural time (the time of scientific history) and genetic time development (the time of Spengler's morphological history), consider the following passage:

> Because the living act of counting is somehow related to time, one has again and again confused number with time. But counting is not a number, just as little as drawing is an image. Counting and drawing are a 'becoming' [*ein Werden*]; numbers and images are 'have-becomes' [*Gewordenes*]. Kant and the others have had in view the living act (the counting) in terms of

64 Spengler, *Der Untergang des Abendlandes: Umrisse einer Morphologie der Weltgeschichte, Erster Band*, 64–65.

65 Spengler, *Der Untergang des Abendlandes: Umrisse einer Morphologie der Weltgeschichte, Erster Band*, 151–52.

66 Spengler, *Der Untergang des Abendlandes: Umrisse einer Morphologie der Weltgeschichte, Erster Band*, 150.

> its result (the formal relations among completed images). One belongs to the region of life and its direction; the other to extension and causality.[67]

The living activity of computing cannot be expressed along a line of computed intervals. The 'becoming' of computation is fundamentally different from the computation 'that has become'. The former belongs to the 'domain of life and time', the other to the domain of 'extension and causality'. This domain of life, Spengler held, was characterized by a certain directedness, a 'destiny'.[68] The structure of the lived experience of action is always 'longing' and 'wanting' without always being able to say what it is one longs for and wants, because the object of one's longing is not objectively in existence. Later, in the second edition of *The Decline*, Spengler would approvingly cite from Nietzsche's *On the Genealogy of Morality* (*Zur Genealogie der Moral*, published 1887): 'Definierbar is nur Das, was keine Geschichte hat.'[69] (Definable is only that which has no history.) The duty of historians, Spengler wrote, was to describe and compare the different phases of this spirit of 'longing' and the cultures they exemplified. Such a comparison would yield a perspective on history that did justice to the non-diachronic, non-scientific lived experience of human life. The result was a historical periodization radically different from any other classical periodization presented before.

Spengler's distinction between 'nature' and 'history' thus manifested itself in his understanding of historical temporality and the historical comparative method as well as its purpose. However, it manifested itself on at least one other level that would later become of primary importance to Huizinga's interpretation of Spengler's book: the 'corpse' mentioned at the outset of this section. Spengler argued that his historical comparisons revealed at least one fundamental feature of the 'destiny' of the lived experience of culture: culture's degenerative inclination to view itself as part of 'nature', that is, as part of the world determined by rules, laws and natural causality.

67 'Weil der lebendige Akt des Zählens mit der Zeit irgendwie in Berührung steht, hat man immer wieder – auf ein Schema versessen – Zahl und Zeit vermengt. Aber Zählen ist keine Zahl, so wenig Zeichnen eine Zeichnung ist. Zählen und Zeichnen sind ein Werden, Zahlen und Figuren sind Gewordenes. Kant und die andern haben dort den lebendigen Akt (da Zählen), hier dessen Resultat (die formalen Verhältnisse der fertigen Figur) ins Auge gefaßt. Das eine gehört in den Bereich des Lebens und der Richtung, das andre in den der Ausdehnung und Kausalität.' Spengler, *Der Untergang des Abendlandes: Umrisse einer Morphologie der Weltgeschichte, Erster Band*, 177.

68 Spengler, *Der Untergang des Abendlandes: Umrisse einer Morphologie der Weltgeschichte, Erster Band*, 172, 181.

69 Spengler, *Der Untergang des Abendlandes: Umrisse einer Morphologie der Weltgeschichte*, 207. See § 13 of the second part of *Zur Genealogie der Moral*.

As cultures unfold through their genetic history, they become prone to a certain 'forgetfulness'. They 'forget' their organic historical nature and subject themselves and their activities (e.g. 'counting') to the outcome of their activities (e.g. 'numbers'). The moment cultures start to view their activities as objects demanding respect and assimilation rather than as potentials for development, they petrify and die.[70] The coroner, who found himself already living in a petrified world, reserved a particular word for this death of culture: civilization.

Civilization is the inevitable destiny of a culture. A pinnacle is reached here, from where the last and most complicated questions of historical morphology become solvable. Civilizations are the most extreme and artificial conditions of which a higher kind of human being is capable. They are the end; they follow the becoming as a 'has-become', life as death, development as rigidity, the countryside and spiritual childhood, as evident in the Doric and the Gothic, as spiritual elderliness (*geistige Greisentum*) and as a stony, petrifying cosmopolitan city (*steinerne, versteinernde Weltstadt*). They are an end, irrevocable, but they are reached with an essential necessity time and again.[71]

Civilization: the 'destiny of culture', the 'most artificial state of being' in which it becomes clear what culture had actually been. Like how the petrified city highlights the splendour of undeveloped land, like how youth reveals itself most strongly only once a mature age has been reached, civilization reveals what culture had been and would become. At the end of its life, when culture has become its antithesis (the artificial subjection to rules, laws, customs and technology), it appears in its original historical splendour of which it had not, could not and needed not be aware. Only at dusk could culture become aware of its day, only in decline could it become aware of its fertility and only the *Abendland* could know the power of birth. Nowhere, Spengler argued, was this revelatory contrast as pronounced as it was in the

70 Spengler, *Der Untergang des Abendlandes: Umrisse einer Morphologie der Weltgeschichte, Erster Band*, 153.

71 'Die Zivilisation ist das unausweichliche *Schicksal* einer Kultur. Hier ist der Gipfel erreicht, von dem aus die letzten und schwersten Fragen der historischen Morphologie lösbar werden. Zivilisationen sind die *äußersten* und *künstlichsten* Zustände, deren eine höhere Art von Menschen fähig ist. Sie sind ein Abschluß; sie folgen dem Werden als das Gewordene, dem Leben als der Tod, der Entwicklung als die Starrheit, dem Lande und der seelischen Kindheit, wie sie Dorik und Gotik zeigen, als das geistige Greisentum und steinerne, versteinernde Weltstadt. Sie sind ein Ende, unwiderruflich, aber sie sind mit innerster Notwendigkeit immer wieder erreicht worden.' Spengler, *Der Untergang des Abendlandes: Umrisse einer Morphologie der Weltgeschichte, Erster Band*, 43–44.

modern industrial and international *Weltstadt*.[72] Here facts replaced passions, society trumped state and life was buttressed by financial transactions rather than convictions. The cemetery of culture that was the metropolis had one defining feature: *Kosmopolitismus*, the cold, disengaged relation to an abstract, intangible image of humanity.[73]

So far, two pivotal elements of Spengler's *The Decline of the West* have been spelled out, and each one was related to Spengler's central distinction between 'history' and 'nature': (1) Spengler's non-chronological, non-natural scientific understanding of historical time and the cross-cultural comparisons it enabled, and (2) the supposed death of culture once it fails to acknowledge the distinction between history and nature and comes to view its historical development as a part of rules, laws and nature – in this sense, its literal petrification. This death, according to Spengler, was the destiny of culture – civilization and cosmopolitanism – and it was precisely because Spengler understood himself to live at the height of civilization that he imagined himself to be in the position to have achieved, 'for the very first time', an understanding of cultural history and its destiny that only the coroner could appreciate.[74] From this, it is apparent that Spengler's historiography was deeply morally invested. What troubled Huizinga, however, were not necessarily Spengler's most explicit ethical convictions but the virtues and ideals that Spengler's account of the history of culture inspired implicitly. In order to see what Huizinga read between Spengler's lines, a close look at Huizinga's writings is finally in order.

Huizinga's hope

The first volume of Spengler's *The Decline* was met by considerable hostility in Dutch newspapers.[75] One reviewer from the Reformed newspaper *De Standaard* accused Spengler's book of preaching a moral 'relativism of the most despicable kind': the historical necessity of which the book spoke

72 Spengler, *Der Untergang des Abendlandes: Umrisse einer Morphologie der Weltgeschichte, Erster Band*, 44.

73 Spengler, *Der Untergang des Abendlandes: Umrisse einer Morphologie der Weltgeschichte, Erster Band*, 46.

74 Spengler, *Der Untergang des Abendlandes: Umrisse einer Morphologie der Weltgeschichte, Erster Band*, 3.

75 The reception of Spengler in the Netherlands has been discussed in A. Pos, 'Dichterlijk cultuurpessimisme: Spengler in het werk van A. Roland Holst en H. Marsman', in *De Pijn van Prometheus*, ed. R. Aerts and K. van Berkel (Groningen: Historische Uitgeverij, 1996), 119–21.

was a threat to its reader's very sense of moral responsibility and duty.[76] For identical reasons, another reviewer from the Catholic newspaper *De Maasbode* accused Spengler's book of an 'unsurpassable nihilism'.[77] A third reviewer in the liberal newspaper *Het Algemeen Handelsblad* devoted more attention to Spengler's supposed academic vices: his 'mysticism' and 'anti-intellectualism' would not have been tolerated in the academic community only a few years earlier, this author mournfully stated.[78] A fourth reviewer, in the liberal magazine *De Gids*, accused Spengler of all of the above at once: Spengler's book was the hallmark of the anti-intellectualism of its day and a morally reprehensible failure to conduct 'sincere research' whilst showcasing 'naïve arrogance' instead.[79] This fourth and most damning review came from Johan Huizinga, and together with the other reviews, it reveals a central feature of the Dutch Spengler reception. The rejection of Spengler's historiography was a moral, virtue-ethical one.

Throughout his publications between 1921 and 1945, Huizinga fairly consistently gave three main arguments against *The Decline*: (1) it was not sufficiently well (if at all) steeped in historical details;[80] (2) it was improperly neo-Platonic in its incorrect determinism, insensitive to the unpredictability of life and the fallibility of findings;[81] lastly, (3) it gave a morally reprehensible definition of 'civilization', 'culture' and their historical interrelation.[82] Of these three critiques, the third is of the greatest importance to understanding Huizinga's distrust towards Spengler, for it is the only argument Huizinga launched against Spengler specifically. Huizinga had accused other authors of not being sufficiently empirically minded (e.g. H. G. Wells) or historically deterministic (e.g. Ernst Bernheim, Karl Lamprecht).[83] The issue of culture and civilization, on the other hand, appeared mainly in Huizinga's discussion of Spengler.[84] For this reason, the rest of this section is devoted to understanding Huizinga's rejection of Spengler's definition of civilization,

76 *De Standaard* 05-10-1920.

77 *De Maasbode* 21-07-1923.

78 *Algemeen Handelsblad* 01-02-1921.

79 VW IV: *Twee worstelaars met den engel* (1921): 443–44.

80 E.g. VW IV: *Twee worstelaars met den engel* (1921): 464; VW VII: *Taak der cultuurgeschiedenis* (1929): 78;

81 E.g. VW IV: *Twee worstelaars met den engel* (1921): 469; VW VII: *De wetenschap der geschiedenis* (1937): 115.

82 E.g. VW IV: *Twee worstelaars met den engel* (1921): 450; VW VII: *Taak der cultuurgeschiedenis* (1929): 33.

83 E.g. VW VII: *De wetenschap der geschiedenis* (1937): 115.

84 The next chapter, however, shall show that Huizinga faulted the works of Carl Schmitt along similar lines.

which appeared in three publications spread over two phases: (1) in the aforementioned review from 1921, titled *Two Wrestlers with the Angel*, after 1933 in (2) *In the Shadow of Tomorrow*, published in 1935 and, lastly, in (3) *The World Violated*, posthumously published in 1945.

Huizinga's critique of Spengler in 1921

Huizinga's aforementioned review of Spengler's book appeared in *De Gids* in 1921. In this piece, Huizinga juxtaposed Spengler with an unlikely companion: H. G. Wells, whose *The Outline of History* appeared in 1920. On the one hand, Huizinga conceded, one could barely have found two more dissimilar authors. Wells wished to introduce the methods and purpose of positivist science to historiography, while Spengler laboured indefatigably to achieve the exact opposite.[85] Still, Huizinga found, these authors were similar in one rather profound respect: in their respective books, both sought to expose a grand historical plan behind the world of empirical details. As was discussed in Chapter 3, Huizinga fundamentally rejected such aims and therefore both books. Still, between Spengler and Wells, Huizinga found that Spengler's determinism was significantly more morally corrosive than Wells's was. Though Wells had shown himself to be utterly 'anachronistic' in his thought and insensitive to the 'tragic greatness of history', he had extracted 'wise lessons' from his studies in history: 'gentleness', 'endless confidence' and 'solid hope' regarding the human ability to create beauty.[86] Spengler's analysis, on the other hand, had nothing of the sort to offer. On the contrary, Huizinga argued, Spengler taught his readers how not to 'love' the world.

What does someone who does not 'love' the world do, according to Huizinga? In the case of *The Decline*, Huizinga was referring to Spengler's distinction between culture and nature, or culture and civilization. This distinction was wrong, he held, not because of a theoretical flaw in Spengler's reasoning or because of Spengler's omission of this or that historical fact. The critique launched by Huizinga against *The Decline* and its distinction between history and nature in this review was based almost entirely on what can ultimately only be described as a virtue-ethical commitment. Spengler's understanding of civilization was wrong, according to Huizinga, because it inspired reprehensible vices and dismantled the virtues central to the belief system of the Peace Palace generation. In those passages where

85 VW IV: *Twee worstelaars met den engel* (1921): 471.
86 VW IV: *Twee worstelaars met den engel* (1921): 489, 495–96.

the 1921 review took its most explicit stance against *The Decline*, it was in terms of ethical virtues and vices, not theoretical objections or historical counterexamples. Spengler's book showcased a 'boundless self-confidence' revealing a 'naïve self-righteousness'; its 'titanic seriousness' would have been utterly 'comic' had it not been the product of 'a lack of simple sincerity'.[87] In the review, Huizinga lauded Spengler's incredible rhetorical capacity to create a world and system but loathed its moral effect:

> Spengler looks down on facts, and rather introduces his mysterious living cultures. He does not care about the living people in history. He lacks love – and here might lie what I just called his fundamental mistake. Throughout his entire book, there is no sound of compassion, reference or devotion; there is no peace of heart, and no hope.[88]

'Love', 'compassion', 'admiration', 'devotion', 'peace of heart', 'hope' – Spengler's book, Huizinga held, did nothing to cultivate these virtues. In fact, according to Huizinga, Spengler's inability to display these abilities in writing was not a mere peripheral issue or outcome of Spengler's thought and prose. Spengler's inability to display these qualities, Huizinga argued, was inevitably tied into the narratological fabric of Spengler's historical 'romanticism'.[89] Spengler had defined and narrated 'civilization' and 'nature' as the sterile, rule-bound society of commerce that would inevitably succeed and replace the original, creative 'culture' from which it nevertheless sprung. The 'neo-Platonic' independence thus granted by Spengler to 'culture' (separating it from the inherent degeneration of culture called 'civilization') turned not only 'facts' but even the need for facts into the hallmarks of decline.[90] Huizinga understood this narration of history not only as a particular organization of historical consciousness but as a blow to moral human agency:

> By attempting to understand history with the means of mysticism, Spengler has created an absurd world of historical thought; a world wherein

87 VW IV: *Twee worstelaars met den engel* (1921): 443.

88 'Spengler [minacht] de feiten, en komt met zijn geheimzinnig levende culturen. Om de levende menschen in de historie geeft hij niet. Want wat hem ontbreekt, en daar zit wellicht weer de grond van wat ik zooeven zijn fundamenteele fout noemde, is de liefde. [...] En in zijn heele boek klinkt niets van medelijden, niets van eerbied of overgave, niets van den vrede des harten, noch van hoop.' VW IV: *Twee worstelaars met den engel* (1921): 468.

89 VW IV: *Twee worstelaars met den engel* (1921): 451, 466.

90 VW IV: *Twee worstelaars met den engel* (1921): 449.

> each event becomes meaningful only in terms of the realization of a purpose given to it [the event] as the original symbol of culture. [Events become] only the realization of, not the seed or ground for that which is to come. I do not live in such a world, of that I am sure.[91]

The temporality of Spengler's historiography inherently fostered a world of decline, a world wherein historical events were understood in terms of their distance from an imagined original state of culture rather than in terms of the world they enabled. Spengler's world was a waning fire awaiting its extinguishment. Huizinga's issue with this narration was not an empirical one. Huizinga did not, say, list historical counterexamples, for that would indeed be to miss Spengler's point. Rather, Huizinga's issue continued to be moral in nature. Immediately after the passage cited in full above, Huizinga complemented his rejection of Spengler's thought with a result:

> Spengler's book has healed me homoeopathically: it has freed me from my own dark despair over the future of our humanity, as his hopelessness certainty made me feel that I still had hope and 'not-knowing'.[92]

Spengler's history had been drafted from a temporality antithetical to what Huizinga understood to be 'hope'. Against the background of what has been called the Peace Palace generation and the future-oriented temporality of their values, the above passage by Huizinga can be read as attesting to more than a rhetorical nicety or moral platitude at Spengler's expense. Huizinga's claim that Spengler's book had 'homoeopathically' rekindled hope reveals a direct relationship in Huizinga's writing between temporality and his ethical conviction. Spengler mobilized his distinction between 'nature' and 'history' to create a temporality of cultural decline, whereas the Peace Palace generation had cultivated a temporal orientation wherein war and suffering were understood as the possibility of restoring and fortifying a

91 'Door de historie te willen verklaren met de middelen der mystiek heeft Spengler een absurde historische denkwereld geschapen. Een wereld, waarin elk gebeuren slechts de beteekenis heeft van een verwezenlijking van strekkingen, die als oersymbool de cultuur waren meegegeven, enkel van verwezenlijking, niet van een kiem of voedingsbodem voor het later komende. Een wereld derhalve, die afbreekt achter ieder oogenblik. In zulk een wereld leef ik niet, dat weet ik heel zeker.' VW IV: *Twee worstelaars met den engel* (1921): 469–70

92 'Spengler's boek heeft op mij homoeopathisch genezend gewerkt, mij een weinig bevrijd van eigen duistere vertwijfeling aan de toekomst onzer beschaving, doordat zijn hopelooze zekerheid mij deed voelen, dat ik de hoop nog bezat en het niet-weten.' VW IV: *Twee worstelaars met den engel* (1921): 470.

way of life, a life exemplified by traditional ideals of dialogue, exchange and travel. Hence, to Huizinga, the temporal orientation of *The Decline* amounted not only to a narration of the past or a mere theoretical account of human agency. To Huizinga, *The Decline* appeared as a threat to the ethics of an international life of cultural exchange.

Huizinga's critique of Spengler after 1935

After his 1921 review of *The Decline*, Huizinga only occasionally returned to Spengler in his writings. This changed in 1935.[93] Until that year, Huizinga had identified himself as a 'cultural historian', though he had made no systematic attempt at defining 'culture'.[94] The change of powers in Germany and the ensuing proliferation in the Netherlands of the National Socialist Movement (the *Nationaal-Socialistische Beweging*, or NSB), however, made this issue urgent to Huizinga almost overnight.[95] Throughout the 1920s and early 1930s, a steady stream of critiques of 'modern culture' had been published in the Netherlands, and they soon formed one of the intellectual backbones of the Dutch National Socialist Movement.[96] With the success of the NSB – at first through its social visibility, later through its electoral push – the term 'culture' had migrated from academic and literary contexts to the heart of the political vernacular. Given his political identity and social standing, Huizinga, the undisputed authority on 'cultural history' in the Netherlands, could not but make explicit the political position of his works. For this purpose, Huizinga returned to Spengler in his subsequent writing. This time around, Huizinga's critique of Spengler concerned not the latter's historiographical distinction between culture and civilization but his definition of 'culture' and 'civilization' more directly.

The Dutch National Socialist media outlets had written of culture, virility and national organic force with references to Spengler's narration and

93 This development has been mentioned, however briefly, in Krumm, *Johan Huizinga, Deutschland und die Deutschen*, 155.
94 Of course, Huizinga had written about 'culture' (*cultuur*) and 'cultural history' (*cultuurhistorie* or *cultuurgeschiedenis*) ever since the early 1900s – see e.g. VW I: *Over studie en waardeering van het Buddhisme* (1903): 148 – but until the 1930s, he had used 'culture' in a relatively loose capacity to distinguish his historical perspective from that of 'political' history.
95 The significance of this event to Huizinga's thought is explored at length in the next chapter.
96 R. te Slaa and E. Klijn, *De NSB: Ontstaan en opkomst van de Nationaal-Socialistische Beweging, 1931–1935* (Amsterdam: Boom, 2021), 37–38.

definition of such terms in opposition to a supposedly sterile 'civilization'.[97] From 1935 onwards, Huizinga consistently defined culture in explicit opposition to both Spengler and the National Socialist rhetoric. Culture thus concerned all those activities through which (1) the material world is reorganized according to ideals and dreams whose imagined purpose (2) transcends the pursuit of individual well-being and is rather experienced as the dutiful service of a community, and it (3) involves the subjection of and control over the natural world.[98] The moment human activity loses sight of either its metaphysical, non-material ideals or collective duty, or both, Huizinga held, it becomes an irrational endeavour subject to the whimsicalness of the human will, bodily impulses and the violent emotions of hate and jealousy. Yet at the very same time, the moment thought subjects itself solely to neo-Platonic ideals, it becomes ascetic and unable to take part in an erotic fashion with the world it inhabits. This suggested balance was a response to Spengler's all too strict bifurcation between civilization and culture.

Alongside this point, Huizinga additionally and more strongly emphasized a second new critique of Spengler, though like in his 1921 review, Huizinga's main opposition to Spengler's dichotomy was to remain a moral one. In his publications from the 1930s, Huizinga repeatedly stated that those virtues necessary to the proliferation of culture were those belonging to classic *civilitas*:

> Spengler has wrongly understood 'civilization' in opposition to 'culture' as the lesser in opposition to the higher. Civilization [rightly understood] speaks of the human as a citizen, a legal companion, [it speaks] of the human, who has become conscious of his full worth. It speaks of order, law, justice, and excludes barbarism. [...] *Civilitas* or *civiltà* expresses so pristinely one of the most essential elements of the concept of culture, that is, the consciousness of full-fledged citizenship, that it should serve as the most fortunate descriptions of the phenomenon of culture.[99]

97 NSB, *Nationaal Socialistische Beweging in Nederland: Toelichting van het programma* (Utrecht: s.n., 1932).

98 VW VII: *In de schaduwen van morgen* (1935): 328–31.

99 'Ten onrechte had Spengler in "civilisatie" tegenover "cultuur" het mindere tegenover het hoogere gehoord. Civilisatie spreekt van den mensch als staatsburger, als rechtsgenoot, van den mensch, die zich zijn volle waarde is bewust geworden. Het spreekt van orde, wet en recht, en sluit barbarie uit. [...] Civilitas of civiltà drukt een der meest essentieele elementen van het cultuurbegrip, d.w.z. het bewustzijn van volwaardig staatsburgerschap, zoo zuiver uit, dat het als de gelukkigste van al de algemeene termen voor het verschijnsel cultuur mag gelden.' VW VII: *In*

Had Spengler been more historically and etymologically minded and less mythologically inclined, Huizinga held, he might have realized that the relation between 'culture' and 'civilization' is symbiotic rather than antagonistic. The agricultural root of the term 'culture' was steeped in activities of 'caring' (*verzorgen*), 'breeding' (*kweeken*) and 'cultivating' (*verbouwen*), that is, activities entirely congruent with those ideals classically associated with *civilitas* (e.g. order, law, justice) as well as its virtues.[100] From this comparative etymological perspective, Huizinga argued, civilization was an integral element of culture, or at least by no means antithetical to it, and its virtues could not have been more different from those of Spengler's cultural *Raubtier*:

> The word *civilis* developed a number of meanings in classical Latin, each assuming 'that which belongs to citizenship.' *Civilis* referred to the political in general, and in contrast to the military; it meant popular, kind, polite, ordinary, courteous and temperate.[101]

The non-military virtues of the citizen enabled the exchange and presentation of different perspectives and ideas within and between societies. Courtesy and temperance were inherent to a climate of debate. As such, when Spengler fundamentally opposed culture to civilization, Huizinga understood Spengler quite literally to be the 'greatest enemy of humanistic thought'.[102] This research, as Huizinga had come to know it, abided by the codes of civilized, international exchange: the possibility of imagining different cultural perspectives, however different, and taking stock in a conversation facilitated by temperance, charity and politeness. When this international exchange and imagination disappeared, those virtues Huizinga associated with civilization and *civilitas* occurred to him as the cornerstones of the culture he had known. His understanding of culture was, in this sense, directly influenced by what had been lost. In other words, his culture appeared precisely when and because it had vanished. In the 1930s and '40s, what Huizinga had previously loosely understood to be

de schaduwen van morgen (1935): 488–89. Huizinga returns to this theme in VW IV: *Patriotisme en nationalisme in de Europeesche geschiedenis tot het einde der 19e Eeuw* (1941): 499–503.

100 VW VII: *Geschonden wereld* (1945): 489.

101 'Het woord *civilis* had in het klassiek Latijn een reeks van beteekenissen ontwikkeld, uitgaande van die van 'wat des staatsburgers is.' *Civilis* duidde het staatkundige in het algemeen aan, in tegenstelling tot het militaire; het beteekende populair, minzaam, beleefd, voorkomend, hoffelijk, gematigd.' VW VII: *Geschonden wereld* (1945): 483.

102 VW VII: *Taak en termen der beschavingsgeschiedenis* (1927): 33.

culture proved to be conditional on the 'recovery of international rule of law, peaceful exchange between states, political leadership by agreement and deliberation'.[103] But as the 1930s unfolded and spilled into the next decade, this recovery appeared to Huizinga to rely on virtue in particular:

> Still, the last words must be: we will not let go of the hope for better and the will to improve. Humanity cannot do without that inestimable heritage we call civilization.[104]

'That inestimable heritage we call civilization' is what needed to be mobilized and nurtured anew; the reactivation of past ideals figured as the object of hope, and so the moral temporality of Huizinga's generation after the Great War reappeared in the 1930s. Not hope for the dawn of a radically new day, not hope for a release from the past and all its ills, not hope for merely the end of violence. No, it was the hope required to dutifully carry on serving the traditional ideals of civilization as it tumbled into the future. Huizinga's tomorrow was an opportunity for yesterday's potential, and within this temporal organization took shape those virtues Huizinga cherished most: courage, perseverance, modesty, patience and, above all, the aforementioned virtue of hope. These virtues and their temporal investment belonged not to an author of some kind of exceptional moral standing but to someone who was part of a generation that had spent enough time in the traditional nineteenth-century societies to hold on to their moral investment and the cultural capital with which they had grown up. To have hope and courage, Huizinga wrote, was a moral duty, and by the time 1935 came around, the obligation to defuse Spengler's determinism had become one, too.[105]

In sum, Huizinga's most explicit rejection of Spengler's *The Decline* can be subdivided into two kinds, one on each side of 1933. In the 1920s, Huizinga took issue with the supposed intellectual vices of Spengler's historiography of civilization. By the mid-1930s, Huizinga targeted the ethical vices of Spengler's definition of civilization. Both critiques, however, concerned Spengler's distinction between 'culture' and 'civilization', and both critiques were explicitly virtue-ethical in nature. They often implicitly drew from a particular set of ideals and virtues that became of primary importance

103 VW VII: *Geschonden wereld* (1945): 599.

104 'Desondanks moet het laatste woord zijn: de hoop op beter en den wil tot beter laten wij niet varen. Het menschdom kan niet afzien van dat onwaardeerbaar erfgoed, dat wij beschaving noemen.' VW VII: *Geschonden wereld* (1945): 478.

105 'To work and have hope, to not despair humanity, society, and civilization.' HA 9.I.2 (1919). See Chapter 3 for a lengthier discussion of this lecture.

among the Peace Palace generation after the Great War. Without reference to this moral system and the experiences of loss from which it drew, it is hard to understand why Huizinga responded as fiercely as he did to Spengler. Huizinga read *The Decline* as an attack on precisely those virtues he thought were of greatest importance to European societies post-1918. Even worse, he found Spengler inspired selfishness and apathy and disabled sentiments of moral duty. Read in this moral vein, *The Decline* was an assault not only on, say, neo-Kantians but also on the very identity and experiences of the Peace Palace generation. In Spengler's eyes, the hope for civilian virtues that Huizinga and his generation applauded was itself a marker of a culture's progressing self-denial.

Spengler's Rembrandt versus Huizinga's Rembrandt

The moral discrepancy between Huizinga and (his image of) Spengler reappeared directly in concrete differences between their applied cultural-historical analyses. In fact, Huizinga's ethical and epistemic virtues played a decisive role in his narration of historical particulars and set his historical perspective apart from Spengler's. Possibly the clearest example of this empirical divergence can be found in Spengler's and Huizinga's respective art-historical analyses of Rembrandt van Rijn (1606–1669). In this last section, this example is explored in some detail to finish an otherwise incomplete circle: Huizinga's critique of Spengler was part and parcel of his Christian internationalist ethics and experience of moral loss on the way, and these had a direct bearing on his cultural-historical analyses. Spengler extracted from Rembrandt's proto-impressionist inclinations a supposedly novel representation of infinity, sensuousness and restless striving. Huizinga, on the other hand, saw reflected in Rembrandt's works not restless striving but a sense of modesty and balance. Huizinga's Rembrandt was the Great War's. In order to unpack these differences, consider first Rembrandt's *Syndics of the Drapers' Guild* (1662).

In this painting, Rembrandt depicted the syndics of the drapers' guild in Amsterdam.[106] Among its other responsibilities, this board served to carry out quality control of the fabrics that entered the market. During these controls, special attention was paid to the dyes that had been used to colour the fabric. The scene displayed by Rembrandt shows the guild syndics comparing colours to this end. For this purpose, they used

106 J. Bikker et al., *Late Rembrandt* (London: National Gallery Company, 2014), 124–31.

Figure 5.4. Rembrandt's *Syndics of the Drapers' Guild* (*De Staalmeesters*), painted in 1662.

metal plates with standardized colours (hence the name *staalmeesters*, 'steel masters'). The guild had commissioned a portrait with its members sitting at a table. Rembrandt took the liberty of placing a servant in the background and putting one of the guild's members in a more upright position. Four guild members have turned their gaze straight at the spectator. All six figures carry a probing, inquiring expression. Against the background of the walls and even in contrast to the men, the tablecloth exudes presence through its lavish colour. This painting is about colour in more than just one or two senses, and the readings of Spengler and Huizinga attest to this.

Spengler's appreciation of Rembrandt can hardly be overstated. Throughout both volumes of *The Decline*, Spengler lists Rembrandt with Plato, Goethe and Beethoven as the most powerful bearers of culture. Spengler's admiration of Rembrandt was multifaceted, yet he isolated and analysed one specific trait of Rembrandt's in particular: his usage of different shades of brown (*Rembrandtbraun*).[107] Rembrandt and his Dutch contemporaries, Spengler found, had introduced a new kind of 'bright brown' whose

107 Spengler, *Der Untergang des Abendlandes: Umrisse einer Morphologie der Weltgeschichte, Erster Band*, 845.

'atmosphere' expressed nothing less than 'Destiny, God, the meaning of life'.[108] In Spengler's understanding, brown had achieved this symbolic status through the 'deep lights of Gothic church-windows and the twilight of the high-vaulted Gothic nave', and through this architectural application, brown had become the 'colour of the soul'.[109] Rembrandt, Spengler held, had 'best understood this colour' and had used it to elicit an unprecedentedly dynamic effect on canvas: 'The transcendent brown of Rembrandt is the colour of the Protestant world-feeling.'[110] By layering shades of brown, Rembrandt created a spatial experience congruent with the movement of human longing and striving for destiny. In the second edition of *The Decline*, Spengler wrote:

> Finally, with Rembrandt, figures dissolve into mere impressions of colour [...]. Impressionism captures the brief moment that is once and never returns. The landscape is not a being and standing still but a fleeting moment in its history. Just as in a Rembrandt portrait it is not the anatomical relief of the head that is rendered, but the *second visage* in it that is confessed; just as the art of his brush-stroke captures not the eye but the look, not the brow but the experience, not the lips but the sensuousness; so also the impressionist picture in general present to the beholder not the nature of the foreground but again a *second visage*, the look and soul of the landscape.[111]

Rembrandt's shades of brown had become 'the symbol of spatial infinity' wherein each event was a unique occurrence and moment rather than a

108 Spengler, *Der Untergang des Abendlandes: Umrisse einer Morphologie der Weltgeschichte, Erster Band*, 845, 257.
109 Spengler, *Der Untergang des Abendlandes: Umrisse einer Morphologie der Weltgeschichte, Erster Band*, 845.
110 'das transzendente Braun Rembrandts als die Farbe des protestantischen Weltgefühls'. Spengler, *Der Untergang des Abendlandes: Umrisse einer Morphologie der Weltgeschichte, Erster Band*, 390.
111 'Endlich, bei Rembrandt, verfließen die Gegenstände zu bloßen farbigen Eindrücken [...]. Der Impressionismus fesselt den kurzen Augenblick, der einmal ist und nie wiederkehrt. Die Landschaft ist kein Sein und Verharren, sondern ein flüchtiger Moment ihrer *Geschichte*. Wie ein Bildnis Rembrandts nicht das anatomische Relief des Kopfes, sondern das *zweite Gesicht* in ihm anerkennt, wie es nicht das Auge, sondern den Blick, nicht die Stirn, sondern das Erlebnis, nicht die Lippen, sondern die Sinnlichkeit durch das Ornament der Pinselstriche bannt, so zeigt das impressionistische Gemälde überhaupt nicht die Natur des Vordergrundes, sondern auch da ein zweites Antlitz, den Blick, die Seele der Landschaft.' Spengler, *Der Untergang des Abendlandes: Umrisse einer Morphologie der Weltgeschichte, Erster Band*, 386–87.

variation.[112] With Spengler's aforementioned conception of organic history in mind, one can appreciate why Spengler thought of brown as 'a historical colour'.[113] In Spengler's view, these browns stretch, move and unfold not in an organized and mathematically arranged capacity but in an organically longing and destined vein. Hence, according to Spengler, these browns are not just visually perceived but human vision itself; they are not geometry but the soul of space and landscape. In this capacity, one can make sense of Spengler's statement that 'in Rembrandt, objects dissolve into mere coloured impressions, and forms lose their specific humanness and become collocations of strokes and patches that count as elements of a passionate depth-rhythm.'[114] Whether or not one agrees with Spengler's understanding of the symbolic representational abilities of the colour brown, passages such as these reveal what Huizinga would have found harmful ethical implications of Spengler's historiography: a celebration of the whimsical, undisciplined and fleeting impression, an aesthetic legitimization of the irrational, uncontrolled and myopic impulse.

In his 1921 review of Spengler, Huizinga singled out and opposed Spengler's reading and presentation of Rembrandt.[115] According to Huizinga, Spengler had mistaken Rembrandt's interest in the casualness of human life for a 'baroque' longing for metaphysical images of infinity and fleetingness. In fact, not only had Spengler utterly misunderstood the meaning of Rembrandt's interests, but he had anachronistically over-romanticized their implications. In a narrow and art-historical sense, Huizinga conceded that Rembrandt's art bore baroque tendencies, yet Huizinga found that these had been only incidental and, at that, had belonged to Rembrandt's peripheral and least successful endeavours. Rather, Rembrandt's art had to be understood in terms of the civil society to which Rembrandt both belonged and answered. Rembrandt's interests were not the visualization of timeless planes of infinity or a tragic human condition.[116] Such aesthetic interests belonged to the later

112 Spengler, *Der Untergang des Abendlandes: Umrisse einer Morphologie der Weltgeschichte, Erster Band*, 389.

113 Spengler, *Der Untergang des Abendlandes: Umrisse einer Morphologie der Weltgeschichte, Erster Band*, 346.

114 Spengler, *Der Untergang des Abendlandes: Umrisse einer Morphologie der Weltgeschichte, Erster Band* (1923), 370.

115 VW IV: *Twee worstelaars met den engel* (1921): 459. Huizinga's interest in Rembrandt was far from unique in the Netherlands. The public image of Rembrandt had been constructed and reconstructed from the late nineteenth century onwards. E. H. Kossmann, 'De waardering van Rembrandt in de Nederlandse "Traditie"', *Oud Holland* 106, no. 2 (1992): 81–93.

116 Van der Lem, *Het eeuwige verbeeld in een afgehaald bed: Huizinga en de Nederlandse beschaving*, 276–77.

Romantic age and not to the Dutch seventeenth century. Rembrandt and his peers, Huizinga found, were interested in and curious about everyday objects, not metaphysical ideals. They were not on a semi-religious quest for forms but played with lines and colours – no more, no less. Above all else, these seventeenth-century painters were driven by a non-judgemental curiosity and decency:

> It was a home-grown virtue of sensuousness but not a crude materialism, much more the opposite. The quality tied into a strong sense of reality in a deeper sense, of accepting the world and the things as real; a sense of reality, whether or not philosophically underpinned, recognizing and appreciating each and every thing. In this sensuousness is mirrored a hint of the ethical balance typical of the Dutch form of piety.[117]

The above passage conveys and attributes to Rembrandt and his contemporaries a character of almost childlike fascination with the world, wherein 'fascination' could be understood in opposition to judgement and the contraptions of articulation. Huizinga celebrated the ability of Dutch seventeenth-century painters to lovingly show a state of affairs without explicating and formalizing what it is they saw. In those instances when Rembrandt, in Huizinga's perception, departed from those ideals and virtues, Rembrandt became overly metaphorical and mystifying, and that was when he was at his weakest and least persuasive. In other words, in Huizinga's eyes, Rembrandt's ability and cultural-historical significance was cast in terms of very specific virtues, interpreted against the background of modesty, curiosity, love, piety and balance. At his height, Rembrandt was for Huizinga the prime example of Dutch civilization: 'one understands Rembrandt through the Netherlands and the Netherlands through Rembrandt.'[118] Consequently, when Rembrandt was understood to exemplify other virtues – such as Huizinga deemed to be the case in Rembrandt's *Saskia*, *Painter with His Wife* or *Hendrickje Stoffels* – Huizinga considered him a lesser painter. Huizinga's

117 'Het was een huisbakken deugd, die zindelijkheid, maar het was niet een bot materialisme, veeleer het tegendeel. De eigenschap hangt samen met een sterken werkelijkheidszin, in de diepere beteekenis, dat men de wereld en de dingen als werkelijk aanvaardt, de werkelijkheidszin, die, hetzij philosophisch gebaseerd of niet, de dingen als inderdaad en elk voor zich bestaande erkent en waardeert. Er spiegelt zich in die zindelijkheid iets van een ethisch evenwicht, dat den Nederlandschen vorm van vroomheid typeert.' VW II: *Nederland's beschaving in de zeventiende eeuw* (1941): 465.

118 'Men begrijpt Rembrandt uit Nederland, en Nederland uit Rembrandt.' VW II: *Nederland's beschaving in de zeventiende eeuw* (1941): 493.

Rembrandt was culturally significant in his depictions of civil virtues, the virtues of the Peace Palace.[119]

In sum, Huizinga's opposition to Spengler took shape not only on moral and theoretical levels. Through an analysis of Huizinga's and Spengler's respective cultural-historical appreciation of Rembrandt, the present section has shown how these moral-theoretical critiques also played a part in their respective cultural-historical analyses of particular objects. Spengler's Rembrandt was a painter of tragedy who exposed, through a particular usage of brown shading, 'the metaphysical of loneliness' and 'the boundless forlornness of the soul in the world'. And in this 'radiating darkness', Rembrandt had exhibited the powers of baroque aesthetics.[120] Huizinga's Rembrandt could not have been more different: his Rembrandt was curious, modest, hopeful, happy, playful and in love with the world. His most valuable works had the character of a civil society, and they had barely if anything to do with the violence of baroque aesthetics. In the figure of Rembrandt, Huizinga and Spengler clashed on the levels of ethics, historiography and theories of culture. Huizinga's version of this clash cannot be considered as independent from the moral collapse experienced by his generation between 1914 and 1918 and the convictions it instilled in them.

Conclusion

Huizinga's ideals of internationalism took proper shape in the wake of their loss and erosion – and his critiques of Spengler's *The Decline* followed suit. This chapter set out to explore the following question: what exactly turned Spengler into an author of such considerable danger in Huizinga's eyes? For this purpose, Huizinga's written critiques of Spengler have been investigated through the lens of the ethical culture of the Peace Palace generation and their experiences of moral collapse during the Great War. The main body of this chapter worked to establish a number of points, in the following order:

1. Together with his peers, Huizinga was deeply influence by (1) the practices and ideals of internationalism he had encountered from his student days onwards and (2) the experienced fragility of the international

119 VW II: *Nederland's beschaving in de zeventiende eeuw* (1941): 491.
120 'das metaphysische [der] Einsamkeit', 'die grenzenloze Verlorenheit der Seele im Weltraume', 'Helldunkel'. Spengler, *Der Untergang des Abendlandes: Umrisse einer Morphologie der Weltgeschichte, Erster Band*, 260.

sphere following the outbreak of the Great War and its aftermath. This fragility imbued earlier moral ideals of internationalism with a new sense of duty to defend them. The Peace Palace belonged to the experience of a reinvigorated past ideal whose presence could not be taken for granted and had to be protected. This vulnerability was part of what Gafijczuk calls 'temporal hesitation': the palace opened up and mobilized a past in the present and presented the possibility of its continued existence in the future, which is what Huizinga called 'hope'. The Palace, in this sense, evoked through an experience of loss and uncertainty a sense of duty and hope on Huizinga's side. This invocation, in turn, was mediated both by Dutch colonial heritage and Huizinga's Anabaptist upbringing.

2. Spengler's *The Decline of the West* appeared, by a significant degree of chance, just as the Great War concluded. One of the book's central distinctions concerns the opposition between culture and civilization. To the former belonged the remorseless and unbridled ability to create and experience the organic nature of human community; to the latter belonged rules and laws, discipline and moral duties. It is the destiny of cultures to degenerate their civilized forms, as it was the destiny of Europe to descend into the unvirility of the bureaucratic, democratic, industrial society of commodities. In this sense, precisely the object of Huizinga's hope – the Peace Palace, the Permanent Court of International Justice – seemed to Spengler to be not the continuation or beginning of culture. Rather, it was its civilized end.
3. Huizinga's rejection of Spengler can be broken down into two phases. The first phase concerned the vices of Spengler's historiography, while the second phase addressed Spengler's concept of civilization. Both critiques, however, ultimately boiled down to a moral argument for the virtues of love, curiosity, humility, cooperation and, above all, hope. These virtues were the bedrock of Huizinga's arguments against Spengler, and they drew from the Peace Palace morality spelled out above. However, these virtues were not just part of, say, Huizinga's character. He *experienced* them in and through the loss of a way of life, an internationalist and civil way of life. As such, to Huizinga, Spengler not only criticized Huizinga's ideals of virtuous behaviour but also challenged an entire way of life.
4. Huizinga's virtue-ethical objections to Spengler were reflected by his analyses of cultural-historical particulars, as has been exemplified by a comparison of their analyses of Rembrandt van Rijn's work. Precisely those features of Rembrandt's art that Spengler celebrated were held by Huizinga in a certain disdain. Where Spengler championed the mythical,

> boundless and organic ruptures supposedly in Rembrandt's paintings, Huizinga championed those features he found to exemplify virtues of modesty, curiosity and what could be defined as a stoic resignation from judgment. Huizinga's Rembrandt was careful, kind and attentive – the kind of person which the world after the Great War needed, according to Huizinga.

With these observations in mind, one can return to the question with which this chapter commenced: why did Spengler appear to Huizinga as such a grave threat to humanistic thought, even one of the gravest of threats? The observations above help to approximate the answer. To Huizinga, Spengler symbolized something well beyond his writings, namely a set of moral convictions and a particular way of looking at the world, its future and its historical heritage. Not only did Spengler celebrate and perpetuate those virtues harmful to democratic life, but he warped the entirety of human history into a legitimation of his perspective, and precisely this willingness to subject historical fact to mythical narrative was illustrative of a moral deficit: a lack of curiosity, modesty and love. This moral deficit threatened both the political balance of Europe required for academic exchange and the objectivity and serenity of thought required for humanistic research. Huizinga had experienced loss – the loss of a world retrospectively celebrated for its internationalism – and this experience made the ideals and virtues of internationalism more urgent than ever. Even in his most pessimistic works, Huizinga kept emphasizing their continued importance. On 12 October 1944, whilst detained by German forces, Huizinga prayed:

> Lord, keep this Earth, which can be so beautiful, and this poor mankind, capable of so much good, and this poor country, so dear to us, suitable for the soil and breeding of peace, order and justice. *Kyrie Eleison.*[121]

Huizinga died four months later on 1 Febraury, a few months before Nazi Germany's capitulation. Until the end, Huizinga repeated throughout his writings the experiences of losing, as well as the ensuing hope for, 'peace, order and justice'. One could hardly think of words more antithetical to *The Decline of the West.*

121 'Heer, houd deze aarde, die zoo schoon kan zijn, en dit arme menschdom, dat zooveel van het goede vermag, en dit arme land, dat ons zoo dierbaar is, geschikt tot een bodem en een kweekplaats van vrede, orde en recht. *Kyrie Eleison.*' J. Huizinga, 'Gebeden', in *Mijn weg tot de geschiedenis en gebeden*, ed. A. van der Lem (Nijmegen: Vantilt, 2016), 92.

By 1935, however, another author had been granted the dubious honour of becoming the 'greatest threat to humanistic thought' and the virtues of civil society. Spengler's organic conception of *Kultur* had been usurped by another more ferocious and more intelligent creature: Carl Schmitt's *homo homini lupus*. Huizinga's ensuing engagement with Schmitt's work became a negotiation over the very conditions of negotiation. What does it mean to congregate with another human individual? What does it mean to 'battle' with another human? What does it mean to 'play'? In the 1930s, these questions concerned life and death in more than one sense.

Bibliography

Beiser, F. C. *Weltschmerz: Pessimism in German Philosophy 1860–1900*. Oxford: Oxford University Press, 2016.

Berlage, H. P. *Het Pantheon der Menschheid*. Rotterdam: W. L. & J. Brusse, 1919.

Bikker, J., G. J. M. Weber, M. E. Wieseman, and E. Hinterding. *Late Rembrandt*. London: National Gallery Company, 2014.

Dienstag, J. F. *Pessimism*. Princeton: Princeton University Press, 2006.

Eeden, F. van. *Dromenboek*. Amsterdam: Bert Bakker, 1979.

Gafijczuk, D. 'Dwelling Within: The Inhabited Ruins of History'. *History and Theory* 52, no. 2 (2013): 149–70.

Gordon, P. E., and J. P. McCormick. 'Introduction: Weimar Thought: Continuity and Crisis'. In *Weimar Thought: A Contested Legacy*, edited by P. E. Gordon and J. P. McCormick, 1–11. Princeton/London: Princeton University Press, 2013.

Grotius, H. *The Rights of War and Peace; Including the Law of Nature and of Nations*. Edited by A. C. Campbell. Washington/London: M. W. Dunne, 1901.

Hathaway, O. A., and S. J. Shapiro. *The Internationalists and Their Plan to Outlaw War*. London: Penguin Books, 2018.

Hoffmann, S. L. 'Rückblick auf die Menschenrechte'. *Merkur* 71, no. 812 (2017): 5–20.

Huizinga, J. 'Gebeden'. In *Mijn weg tot de geschiedenis en gebeden*, edited by A. van der Lem, 91–105. Nijmegen: Vantilt, 2016.

Keizer, M. de. *Als een meeuw op de golven: Albert Verwey en zijn tijd*. Amsterdam: Prometheus, 2017.

Kennedy, J. C. *Een beknopte geschiedenis van Nederland*. Amsterdam: Prometheus, 2018.

Kossmann, E. H. 'De waardering van Rembrandt in de Nederlandse "traditie"'. *Oud Holland* 106, no. 2 (1992): 81–93.

Krumm, C. *Johan Huizinga, Deutschland und die Deutschen*. Münster/New York/Munich/Berlin: Waxmann, 2011.

Lem, A. van der. *Johan Huizinga: Leven en werk in beelden en documenten*. Amsterdam: Wereldbibliotheek, 1993.
Lem, A. van der. *Het eeuwige verbeeld in een afgehaald bed: Huizinga en de Nederlandse beschaving*. Amsterdam: Wereldbibliotheek, 1997.
Linmans, W. *De oorlog van morgen: Nederlandse beeldvorming van een volgende oorlog 1918–1940*. Amsterdam: Prometheus, 2021.
Mak, G., J. Bank, G. van Es, P. de Rooy, and R. van Stipriaan. *Verleden van Nederland: Een nieuwe geschiedenis*. Amsterdam/Antwerp: Atlas Contact, 2021.
Mann, T. *Reflections of a Nonpolitical Man*. Edited by Morris W. D. New York: New York Review of Books, 2021.
NSB. *Nationaal Socialistische Beweging in Nederland: Toelichting van het programma*. Utrecht: s.n., 1932.
Otterspeer, W. *Het horzelnest: De Leidse Universiteit in oorlogstijd*. Amsterdam: Prometheus, 2019.
Pos, A. 'Dichterlijk cultuurpessimisme: Spengler in het werk van A. Roland Holst en H. Marsman'. In *De pijn van Prometheus*, edited by R. Aerts and K. van Berkel, 116–39. Groningen: Historische Uitgeverij, 1996.
Pree, C. du. *Johan Huizinga en de bezeten wereld*. Leusden: ISVW Uitgevers, 2016.
Roland Holst, H. *Tusschen twee werelden*. Rotterdam: W. L. & J. Brusse, 1923.
Roland Holst-van der Schalk, H. 'Ter gedachtenis'. In *Het Pantheon der Menschheid*, 22–23. Rotterdam: W. L. & J. Brusse, 1919.
Slaa, R. te, and E. Klijn. *De NSB: Ontstaan en opkomst van de Nationaal-Socialistische Beweging, 1931–1935*. Amsterdam: Boom, 2021.
Spaanstra-Polak, B. *Het symbolisme in de Nederlandse schilderkunst 1890–1900*. Bussum: Uitgeverij Thoth, 2004.
Spengler, O. *Der Untergang des Abendlandes: Umrisse einer Morphologie der Weltgeschichte. Erster Band*, of 2 vols. Munich: C. H. Beck, 1920.
Spengler, O. *Der Untergang des Abendlandes: Umrisse einer Morphologie der Weltgeschichte. Erster Band*, of 2 vols. 2nd ed. Munich: C. H. Beck, 1923.
Spengler, O. *Der Mensch und die Technik*. Munich: C. H. Beck, 1931.
Strupp, C. *Johan Huizinga: Geschichtswissenschaft als Kulturgeschichte*. Göttingen: Vandenhoeck & Ruprecht, 2000.
Vervliet, J. *The Peace Palace Library Centennial: The Collection as a Mirror of the Historical Development of International Law, 1904–2004*. The Hague: Martinus Nijhoff, 2004.
Verwey, A. *Het zwaardjaar*. Amsterdam: W. Versluys, 1916.
Vollenhoven, C. van. *De drie treden van het volkenrecht*. The Hague: Martinus Nijhoff, 1918.
Woud, A. van der. *Een nieuwe wereld: Het ontstaan van het moderne Nederland*. Amsterdam: Prometheus, 2021.

6 The Looming Loss of a Democratic Order

Abstract

This chapter examines how Huizinga experienced the electoral success of the Dutch National Socialist party in 1935 as a loss of democracy and shows how this experience reappeared in his critiques of Carl Schmitt (1888–1985) and Schmitt's conception of *Kampf*.

Keywords: Johan Huizinga; National Socialism; Carl Schmitt; agonism; Homo Ludens

In 1935 the Dutch National Socialist party (Nationaal-Socialistische Beweging, or NSB) achieved a political landslide in the Dutch provincial elections.[1] By securing 8 per cent of the vote in their first-ever election, the NSB symbolized a great challenge to the Dutch democratic system and its culture. Since its founding in 1931, the NSB had called for 'a modernization of the state' by 'removing at the very root' the 'impotent' culture of 'parliamentary democracy', and after the disruptive force of National Socialist politics to budding democratic cultures had become palpable in Germany in 1933, Dutch liberals, social democrats, Christian democrats and socialists alike took the NSB's electoral advancement seriously.[2] 'No one would be surprised

1 Dutch democracy may have been young at the time – male suffrage had passed parliament in 1917, female suffrage followed in 1919 – but its electoral inclinations had proven quite consistent over the years: electoral results differed no more than a few per cent each election, and the votes were distributed among historically established parties. F. Boterman, *Tussen utopie en crisis* (Amsterdam: De Arbeiderspers, 2021), 58–65, 389; R. te Slaa and E. Klijn, *De NSB: ontstaan en opkomst van de Nationaal-Socialistische Beweging, 1931–1935* (Amsterdam: Boom, 2021), 65; R. Hartmans, *Schaduwjaren* (Utrecht: Omniboek, 2018), 66–67; J. C. Kennedy, 'Vooruitgang en crisis, 1870–1949', in *Een beknopte geschiedenis van Nederland* (Amsterdam: Bert Bakker, 2017), 304–5.

2 NSB, *Nationaal Socialistische Beweging in Nederland: Toelichting van het programma* (Utrecht: s.n., 1932), 1. An overview of the effects of the NSDAP's surge to power in Dutch culture and politics has been offered by Boterman, *Tussen utopie en crisis*, 377–415.

Rydin, T., *The Works and Times of Johan Huizinga (1872–1945): Writing History in the Age of Collapse.* Taylor & Francis Group 2023
DOI: 10.5117/9789463724593_CH06

if madness suddenly released itself into rage,'[3] Huizinga wrote in 1935 with reference to the anti-democratic movements of his day. Previous chapters have shown that Huizinga was dismissive of democracy throughout most of the 1910s and '20s. Yet, in the face of its imminent erosion and in the face of its most likely successor, 'democracy' acquired another meaning to Huizinga. In the 1930s, the defence of 'democracy' had increasingly less to do with the 'mechanization of culture' than with the 'conservation of courage and trust [and duty]', the preservation of the 'inevitable democratic colour' of Dutch society.[4] New times called for new positions.[5]

But what role exactly did the experiences of the looming loss of democracy play in Huizinga's perspective and authorship? Regarding this question, recent accounts have stressed Huizinga's role as a 'public intellectual' from the mid-1930s on. Politics had entered the cultural sphere, and as a cultural historian, Huizinga could not but respond – in public.[6] This chapter looks to complement this social perspective with another: the experiences of democracy's imminent loss did not merely push Huizinga into a new social role wherein he rehashed pre-existing convictions for a new and wider audience. These experiences renegotiated his political position, his anthropological convictions and even his conception of 'cultural history'. The most powerful example of how these experiences mediated his academic beliefs can be found in Huizinga's critical engagement with the works of Carl Schmitt (1888–1985), the crown jurist of Hitler's Nationalsozialistische Deutsche Arbeiterpartei. From 1935 onwards, Huizinga took up his pen

3 VW VII: *In de schaduwen van morgen* (1935): 315. See p. 11 for further details on the references to Huizinga's collected works and letters as well as to the Huizinga archives.

4 Regarding Huizinga's discussion of the 'mechanization of culture' in relation to democracy before 1936, see e.g. VW V: *Mensch en menigte* (1918): 290–291, 319–29; VW V: *Amerika levend en denkend* (1927): 459; VW VII: *De taak der cultuurgeschiedenis* (1929): 68. For Huizinga's later, more positive appreciation of democracy and its relation to Dutch history, see e.g. VW VII: *Voorrede tot den zevende druk van 'In de schaduwen van morgen'* (1938): 313; VW II: *Nederland's beschaving in de zeventiende eeuw* (1941): 450. In his book on seventeenth-century Dutch society, Huizinga even went as far as arguing that the Dutch 'democratic structure' was part and parcel of the country's 'hydrographic structure'. VW II: *Nederland's beschaving in de zeventiende eeuw* (1941): 420.

5 Recently an extensive study has been devoted to the exploration of how these political developments affected the identity and tasks Leiden University set itself in the 1930s, in W. Otterspeer, *Het horzelnest: De Leidse Universiteit in oorlogstijd* (Amsterdam: Prometheus, 2019). In this book, Otterspeer argues that the university's democratic post-war ideals had been more or less latently present already among Huizinga and his colleagues in the 1930s; the war had only catalysed the maturation of these convictions.

6 C. du Pree, *Johan Huizinga en de bezeten wereld* (Leusden: ISVW Uitgevers, 2016), 13, 218–19; C. Strupp, *Johan Huizinga. Geschichtswissenschaft als Kulturgeschichte* (Göttingen: Vandenhoeck & Ruprecht, 2000), 257–58.

against Schmitt's version of *homo homini lupus* (man is a wolf to man), and the result was published in 1938: *Homo ludens*. Huizinga's world was populated not by *homo homini lupus* but by a 'playing human'. We 'must remember that a human can *want to* not be a predator', Huizinga wrote whilst NSB paramilitary units marched on Dutch streets.[7]

Authoritative accounts of Huizinga and his academic output commonly mention Schmitt only once or twice, if at all.[8] By examining Huizinga's critique of Schmitt in relation to Huizinga's experiences of the Dutch political and religious climate of 1933–37, this chapter, however, offers an alternative to two common interpretations of *Homo ludens*, arguably Huizinga's best-known book. The first of these understands the book as an extrapolation into an anthropological direction of Huizinga's historical writings from the 1910s and '20s.[9] The second views *Homo ludens* primarily as Huizinga's academic refuge from the political reality of his time.[10] This chapter, instead, aligns itself with those few authors who have understood *Homo ludens* as a response to the challenges faced by Dutch democracy in the 1930s,[11] and it is particularly sympathetic to those who have underlined

7 'Zij zullen zich moeten herinneren, dat de mensch *kan willen*, geen roofdier te zijn.' VW VII: *In de schaduwen van morgen* (1935): 425.

8 A. van der Lem, *Johan Huizinga: Leven en werk in beelden en documenten* (Amsterdam: Wereldbibliotheek, 1993), 235; L. Hanssen, *Huizinga en de troost van de geschiedenis: Verbeelding en rede* (Amsterdam: Uitgeverij Balans, 1996), 74; A. van der Lem, *Het eeuwige verbeeld in een afgehaald bed: Huizinga en de Nederlandse beschaving* (Amsterdam: Wereldbibliotheek, 1997), 192; C. Strupp, *Johan Huizinga. Geschichtswissenschaft als Kulturgeschichte* (Göttingen: Vandenhoeck & Ruprecht, 2000), 261; C. du Pree, *Johan Huizinga en de bezeten wereld* (Leusden: ISVW Uitgevers, 2016), 126. No mention of Schmitt is made in W. E. Krul, *Historicus tegen de tijd: Opstellen over leven en werk van J. Huizinga* (Groningen: Historische Uitgeverij Groningen, 1990); W. Otterspeer, *Orde en trouw: Over Johan Huizinga* (Amsterdam: De Bezige Bij, 2006).

9 Kaegi, *Das historische Werk Johan Huizingas*, 1947; D. H. A. Kolff, 'Huizinga's proefschrift en de stemmingen van Tachtig', *BMGN – Low Countries Historical Review* 104, no. 3 (1989): 380–92; J. Huizinga, *De hand van Huizinga*, ed. W. Otterspeer (Amsterdam: Amsterdam University Press, 2009), 15–20.

10 Th. J. G. Locher, 'Johan Huizinga', *Jaarboek van de Maatschappij der Nederlandse Letterkunde* (1946): 88–109; E. E. G. Vermeulen, 'Huizinga over de wetenschap der geschiedenis: Zijn positie als theoreticus van de geschiedenis tussen Fruin en Romein', *Tijdschrift voor Filosofie* 24, no. 1 (1962): 81–90; R. L. Colie, 'Johan Huizinga and the Task of Cultural History', *The American Historical Review* 69, no. 3 (1964): 607–30; R. Anchor, 'History and Play: Johan Huizinga and His Critics', *History and Theory* 17, no. 1 (1978): 63–93.

11 E. Kussbach, 'Recht und Kultur: Der Rechtsbegriff bei Huizinga', *Archiv für Rechts- und Sozialphilosophie* 54, no. 2 (1968): 179–216; A. van Heerikhuizen, 'Johan Huizinga en "Het lot van het Boek *Homo Ludens*"', in *De regels en het Spel: Opstellen over recht, filosofie, literatuur en geschiedenis aangeboden aan Tom Eijsbouts*, ed. J. H. Reestman et al. (The Hague: T.M.C. Asser Press, 2011), 153–65; J. Edwards, 'Play and Democracy: Huizinga and the Limits of Agonism', *Political Theory* 41, no. 1 (2013): 90–115.

the book's Christian undercurrents. 'Schmitt's cynical reasoning appeared to Huizinga merely as an extreme example of the dangers inherent in any type of argumentation that ignores the existence of values embodied in rules,' Gombrich argued, and as '[*Homo ludens*] grew under his hands[,] it changed from a book about man and play to a meditation about man and God.'[12] To Huizinga in the 1930s, theology and anthropology were not refuges from political reality. They were a response to it.

Gombrich's account of Huizinga's relation to Schmitt, however, barely goes beyond surface-level analyses and gut feelings. A considerably more elaborate account of Huizinga's criticism of Schmitt in *Homo ludens* has been offered by Geertjan de Vugt. In his article *Philia and Neikos*, De Vugt reconstructs a 'hidden dialogue' between Huizinga and Schmitt in the 1930s and argues that Huizinga's 'entire treatise on play [had] been building up towards a critique of Schmitt'.[13] In so doing, De Vugt pays particular attention to Schmitt's concept of *Ernstfall* and Huizinga's technical critique thereof. However, by honing in on theoretical intricacies, De Vugt's paper misses what was at stake for Huizinga on an existential level in *Homo ludens*. Gombrich rightly observed that 'for Huizinga the problem [posed by Schmitt's political theory] became urgent only when it became a moral problem' – that is, his opposition to Schmitt was not conditioned by cerebral arguments but born from a moral opposition, and to understand this opposition, Huizinga's experiences of democracy's looming collapse are indispensable.[14] Thus, this chapter looks to complement De Vugt's account with a moral-experiential perspective on how a waning democracy had made urgent a new extra-political, metaphysical structure of virtues and beliefs.

For these purposes, the present chapter's main body has been divided into four sections. The first examines Huizinga's experiences of democracy's looming disintegration during the 1930s. Against this background, the second section introduces Carl Schmitt's theory of *homo homini lupus* and what it symbolized to Huizinga. The third section proceeds to examine Huizinga's subsequent technical objections to Schmitt voiced in *Homo ludens*. The

12 E. Gombrich, 'Huizinga's *Homo ludens*', in *Johan Huizinga 1872–1972*, ed. by W. R. H. Koops, E. H. Kossmann, and G. van der Plaat (The Hague: Martinus Nijhoff, 1973), 146, 153.

13 G. de Vugt, 'Philia and Neikos: Huizinga's "Auseinandersetzung" with Carl Schmitt', *World Literature Studies* 9, no. 1 (2017): 52. This chapter agrees in attributing a prominent role to Schmitt in *Homo ludens*, albeit to a more modest extent. Schmitt was important to Huizinga, but *Homo ludens* had other antagonists, too (e.g. *homo sapiens*, *homo faber*) and even though this chapter examines the role of Schmitt, it holds that the *Homo ludens* at large should be appreciated in its diversity.

14 Gombrich, 'Huizinga's *Homo ludens*', 141.

fourth section shows how the difference between Huizinga's and Schmitt's respective anthropologies manifested itself not only on the level of their theoretical accounts of culture and meaning but also on the applied level of their historical-empirical accounts, such as those of seventeenth-century Dutch maritime culture. The chapter concludes by tying together the material for its central argument: Huizinga's *Homo ludens* – both the book and the concept – was a response to his experience of democracy withering away. The ripples of these experiences can be found in the anthropological, moral and historical implications of *Homo ludens*. On all three levels, Huizinga was invested in showing how virtues conditioned the human ability to play and, in effect, how virtues upheld human culture: 'we know for certain: if we want to keep culture, we must continue to create it.'[15] Experience had shown that culture required duty.

The autumntide of democracy: Huizinga's experience of the political in the 1930s

In 1936 the left-progressive weekly magazine *De Groene Amsterdammer* added a questionnaire to one of its May issues.[16] Four hotly debated issues had been collected: (1) what is the fabric of the 'social order'? (2) Which form should governmental power have? (3) What kind of attitude ought to dictate Dutch foreign policy? (4) Should the Netherlands maintain its position in the Dutch East Indies and, if yes, in what form? The tremendous scope of each question reveals just how corrosive political uncertainty had become; the questions addressed no marginal themes. In the late 1920s, such questions, particularly the first three, would have seemed as irrelevant as their answers would have seemed obvious. Since the 1870s, Dutch civil society had been steadily organized along the lines of four traditional pillars (Protestants, Catholics, socialists and liberals), and politically, a 'liberal-democratic' culture reigned supreme whilst practising a foreign policy of pacifism, cooperation and trade.[17] Throughout the 1930s, however, these truisms eroded, and few political events exemplified how previous political truisms were becoming questionable as poignantly as the Dutch provincial

15 'Wij weten het ten stelligste: willen wij cultuur behouden, dan moeten wij voortgaan met cultuur te scheppen.' VW VII: *In de schaduwen van morgen* (1935): 327.

16 *De Groene Amsterdammer* 15-05-1936. For a discussion of this questionnaire and its significance, see Boterman, *Tussen utopie en crisis*, 416–18.

17 Kennedy, *Een beknopte geschiedenis van Nederland*, 275–331; Mak et al., *Verleden van Nederland: Een nieuwe geschiedenis*, 389–406, 413–61.

election results of 1935. A few days after the results of this election had been made public, an article in the Catholic newspaper *De Tijd* reflected on the 'infection' called National Socialism:

> Astonishingly, many who until recently called themselves liberals, especially students and older academics, have proven susceptible to this infection. But also all kinds of reactionary elements – legislators, who in their hearts still despise labour law, middle class citizens, who consider department stores, large chain stores and corporations as their archenemy, and so many others, who, in circumstances of emergency, seeing no solutions to their problems, would opt for any kind of change, for whatever the outcome may be, it could not be worse than their current predicament – these people soon joined in ever-growing numbers the ranks of this movement, which has already led to complete revolutions in Italy and Germany.[18]

In 1935, in their first-ever election, the Dutch National Socialist Movement had secured 8 per cent of the votes for the provinces. This election was of substantial importance to national politics, as the provincial representatives got to elect the members of the Dutch Senate, and on nearly all issues raised in the above-mentioned questionnaire, the NSB represented a revolutionary stance. According to the NSB, the traditional 'pillarization' of Dutch society had paralysed the country's politics, and thus it had to be replaced by a culture of national solidarity; democracy could not meet the challenges of the day and had to be transformed into an autocratic system. The movement held that the prowess of the Dutch military had to be expanded and transformed from a defensive into an offensive force. The fact that the NSB had succeeded to secure 8 per cent of the vote with utterly unorthodox, or anti-orthodox, views in an electorate traditionally channelled through a pillarized set of political identities seemed indicative of a new era. This was, at any rate, the conclusion Anton Mussert (1894–1946), former civil engineer and leader of the NSB, excitedly disseminated.

18 'Verwonderlijk bleek het, dat juist zoovelen zich tot voor kort nog noemende liberalen, vooral onder de studenten en oudere intellectueelen, zoo vatbaar bleken voor deze infectie. Maar ook allerlei reactionaire elementen – wetgevers, die in hun hart nog steeds de arbeidswetgeving verfoeien, middenstanders, die in warenhuizen, groote filiaalzaken en coöperaties hun erfvijanden zien, en zoovelen, die in nood, geen uitkomst meer wetende, voor elke verandering zijn, die immers voor hen, meenen ze, in elk geval geen verslechtering meer kan brengen – deze allen sloten zich weldra in steeds grooter aantal bij deze beweging aan, die in Italië en Duitschland reeds tot volslagen omwentelingen heeft geleid.' *De Tijd* 03-09-1935.

Where did these votes come from? The suspicions vented in the *De Tijd* article cited above have been corroborated by recent research.[19] The NSB voter came from across the demographic spectrum: well-to-do Catholics worried about communist secularism, industrialists worried about the loss of private ownership and working-class people who were upset, disillusioned and simply bored by queuing all day for food and work. While queuing, these workers would have had enough time to recognize and ponder the messages stated on the NSB posters in the streets: 'do not let your boy grow up to queue. Vote N.S.B.'[20] The economic catastrophe of 1929 had not spared the Netherlands, and its consequences resonated throughout society. The Dutch economy was an open one: over a third of the gross national product came from international trade, and nearly a half of the gross national product was spent on imports. Between 1929 and 1935, Dutch unemployment rose from 40,000 to 630,000, nearly 25 per cent of the Dutch workforce.[21] The NSB strategists particularly targeted these unemployed men. The queues for food stamps, work and unemployment benefits had become important recruiting grounds for the party's campaigners. They would hand out free soup, offer positions in the party, sing, and make sure their posters were clearly visible from wherever people were standing in the queues.[22]

Nevertheless, given that the voting percentage was more or less the same for the 1933 parliamentary election as it was for the 1935 provincial election and given the more or less stable success of the traditional working-class parties, it seems that not many working-class people ended up opting for the NSB. This knowledge, however, came only (long) after the fact, and, at the time, considerable fears existed that this significantly sized group of unemployed people would behave unpredictably. From the 1935 election onwards, the threat posed by the NSB to the traditionally pillarized political order was taken most seriously.[23]

But what did the NSB represent to its voters? Whilst newspapers and public intellectuals were trying to figure out the NSB voter demographic, the NSB itself continued to organize large public gatherings, athletic events, their own media outlets and ostentatious parades of their paramilitary units. Membership of the party was explicitly meant to include a way of

19 Boterman, *Tussen utopie en crisis*, 402–4; Hartmans, *Schaduwjaren*, 74–79.

20 See this sentiment in Figure 6.1.

21 Hartmans, *Schaduwjaren*, 19. Different yet comparable figures are given in Boterman, *Tussen utopie en crisis*, 260; Mak et al., *Verleden van Nederland: Een nieuwe geschiedenis*, 426.

22 Te Slaa and Klijn, *De NSB: Ontstaan en opkomst van de Nationaal-Socialistische Beweging, 1931–1935*, 641–45; Boterman, *Tussen utopie en crisis*, 401–2; Hartmans, *Schaduwjaren*, 69–71.

23 Boterman, *Tussen utopie en crisis*, 404–7.

life, for only through a bodily education could the longed-for 'rejuvenation of the spirit' needed to combat the 'suicidal' spirit of democracy – which was dominated by the vices of a 'lifeless, impotent, obstinate, indifferent, disbelieving, scattered, narrow-minded and quarrelsome' mind[24] – be attained. The NSB said it had an alternative on offer:

> In opposition [to democracy], national socialism wishes to construct a basis of willpower, pride, duty, and a belief in our own power and right to exist, our national spirit, feeling of solidarity, willingness to cooperate, sacrifice, so that our renewed nation can blossom on it, and is able to offer all her sons and daughters, from all standings and professions, the best possibilities for moral and physical well-being.[25]

Determination, pride, duty, solidarity, nationalism, self-sacrifice – these were the ingredients of the virtuous and physical well-being sought by the NSB. In terms of ideology, this ethical dimension followed the Italian fascist example; its outspokenly anti-democratic tenor was suffused with anti-religious and anti-socialist tones. Only later, towards 1937, did their antisemitism grow more pronounced.[26] From its inception, the NSB engaged in open exchanges with their Italian and German peers and adopted their symbolic codes, dress and custom. One of the most outspoken examples of this exchange was the NSB's 'resilience department' (*Weerbaarheidsafdeeling*, or WA), which had been inspired by the Italian 'Blackshirts' and the German *Sturmabteilung*.[27] All members of the NSB could join this department to help protect events, parades and other kinds of gatherings by the NSB. In the Dutch papers of 1931–35, however, the WA was often reported to have been involved in violent altercations, not infrequently in pre-meditated clashes with communist and social democratic groups.[28] Depending on their rank,

24 NSB, *Nationaal Socialistische Beweging in Nederland: Toelichting van het programma*, 1.

25 'Daartegenover wil het nationaal-socialisme een grondslag samenstellen van wilskracht, fierheid, plichtsgevoel, geloof in eigen kracht en bestaansrecht, nationalen zin, solidariteitsgevoel, gezindheid tot samenwerking, offervaardigheid, opdat op deze grondslag opbloeie een hernieuwde natie, welke hare zonen en dochteren, uit elken stand en elk beroep, voor hun zedelijk en lichamelijk welzijn de beste mogelijkheid kan bieden.' NSB, *Nationaal Socialistische Beweging in Nederland: Toelichting van het programma*, 1.

26 Boterman, *Tussen utopie en crisis*, 411–15; E. Klijn and R. te Slaa, *De NSB: Twee werelden Botsen, 1936–1940* (Amsterdam: Boom, 2021), 276–92; Hartmans, *Schaduwjaren*, 74.

27 Boterman, *Tussen utopie en crisis*, 387–89.

28 Klijn and Te Slaa, *De NSB: Twee werelden botsen, 1936–1940*, 44–51. For later forms of organized violence from the NSB, see Klijn and Te Slaa, 126–39.

Figure 6.1. (A) An NSB poster from 1935 stating: 'Do not let your boy grow up [queuing] at the welfare office.' (B) Men queuing on 2 August 1933 to collect a free tax exemption for bike ownership, for which they were eligible due to economic hardship. (C) A poster protesting against the NSB from 1935. The poster stresses the violent nature of the NSB: 'The ends justify the means!' (Lit.: 'The end sanctifies the means!')

the WA members wore caps and badges of different kinds. However, all were dressed in black uniforms, red shoulder pieces and high black boots.

In 1935, four years after its foundation, the WA was ruled illegal, and though newspapers reported on its former members causing occasional disturbances, the most overt form of NSB violence dimmed for the first few

Figure 6.2. **(A) Cartoon in *Het Volk* (03-02-1935) after the existence of the German concentration camp Oranienburg became known. The text reads: 'A rip in the national socialist curtain'. (B) A cartoon in *De Groene Amsterdammer* (06-03-1936). Hitler is portrayed in the image of Loreley and calls: 'Rapprochement! Peace!' At the cliff's feet lies the shipwreck of the Locarno Treaties. In this agreement, Weimar Germany signed to never go to war.**

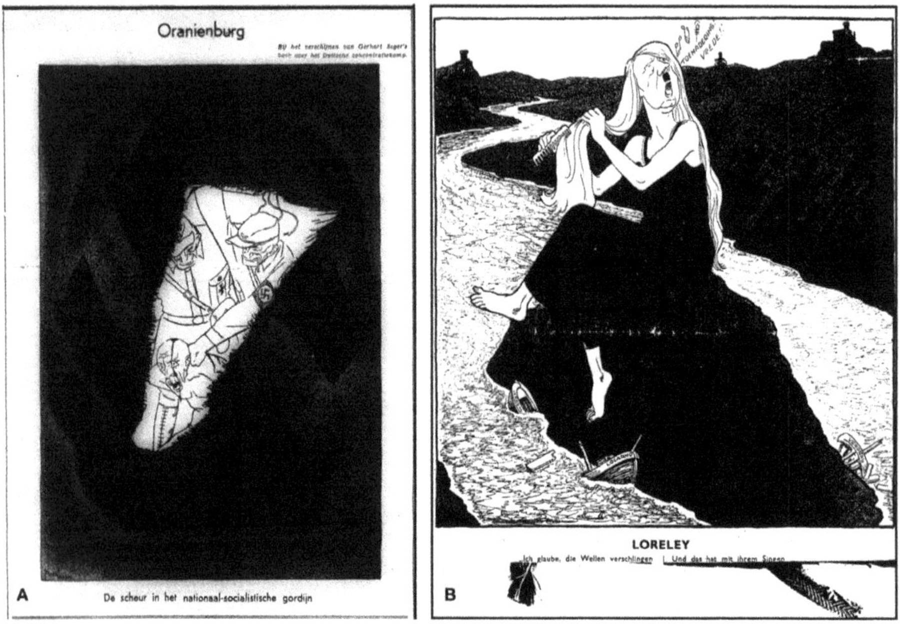

years.[29] Of course, that was not enough to shake the image of an aggressive politics. The ties with the German NSDAP grew stronger, and liberal, Christian and socialist newspapers were keen to remind their readers of the implications. The German concentration camp Oranienburg had become part of Dutch public knowledge in 1934, and as Jewish refugees started fleeing from Germany to the Netherlands in ever-increasing numbers from 1933 onwards, it must have seemed clear to most, regardless of their political identities, that if the NSB came to power, tremendous upheavals would await.[30] After all, with or without the WA, the NSB continued to denounce 'democracy, the cruellest idol to ever rule over men', for 'whoever wishes to

29 Boterman, *Tussen utopie en crisis*, 407; Klijn and Te Slaa, *De NSB: Twee werelden botsen, 1936–1940*, 45.
30 Boterman, *Tussen utopie en crisis*, 246–47.

serve it must sacrifice freedom, justice, character and honour, for money and possession. Our poor people lie in the dust for this idol.'[31]

As has been explored particularly in the fourth chapter of this book, Huizinga had a rather negative attitude towards 'democracy' for the greatest part of his professional life. Until the 1930s, the term represented to him the industrial, bureaucratic world of social engineering and the commodification of culture. 'History may become democratic; she must remain stoic,' Huizinga declared in 1929.[32] Even in a more narrowly defined political sense, Huizinga was suspicious of the Dutch multi-party system, fearing the parties catering to an increasingly specific audience would lose sight of national interests. In both senses, however, cultural and political, Huizinga's relation to the term 'democracy' shifted dramatically during the 1930s. 'Democracy' ceased to refer to the aforementioned socio-cultural and cultural-political developments; against the background of the national-socialist threats to the democratic order, the term became first and foremost the virtue-ethical counterpart to fascist thought, a symbol of anti-dogmatism, of the moderation of political passion and of a willingness to compromise. It was a symbol of those 'character traits' conspicuously absent in the fascist perspective.[33]

Huizinga's changing attitude towards democracy took place against the background of the NSB's surge, but on a more particular and personal level, it tied into at least three events in Huizinga's life throughout 1933. To begin, in April that year, Huizinga's close friend Cornelis van Vollenhoven (1874–1933), discussed in the previous chapter, passed away.[34] To Huizinga, Van Vollenhoven symbolized the immediate counterpart to the political developments in Germany at the time: a character of internationalism and cooperation in the spirit of Erasmus and Grotius.[35] Van Vollenhoven was

31 'De wreedste afgod, die ooit over menschen heeft geregeerd, is de afgod der democratie. Wie dien wil dienen, moet hem brengen als offer voor zijn geld en zijn goed, zijn vrijheid en zijn recht, zijn karakter en zijn eer. Voor dien afgod ligt ons arme volk in het stof gebogen.' *De Tijd* 18-08-1936: 2.

32 VW VII: *De taak der cultuurgeschiedenis* (1929): 65.

33 'Gezag, parlementarisme en democratie konden tot zoover in een drieledige harmonie samengaan. Dat onder dit systeem zekere volksdeelen òf van directen invloed verstoken bleven, òf zich moeten aansluiten bij een groep, welker meening zij niet ten volle konden deelen, was niet, gelijk een ultra-rationale staatsleer meende, louter onbillijk en een nadeel, maar integendeel een zeer wezenlijk voordeel. De beperktheid van doorwerking van iedere groepsmeening of ieder groepsbelang bevorderde het algemeen nationale, niet al te exact geformuleerde karakter van de politiek.' VW VII: *In de schaduwen van morgen* (1935): 307.

34 BW II: Richard Roland Holst–Huizinga (1933): 1007.

35 VW VI: *Mr C. van Vollenhoven* (1933): 496–498; VW VII: *In de schaduwen van morgen* (1935): 380; VW II: *Terugblik op Nederlands groei in de veertig jaren van het regeeringsjubileum* (1938): 553.

a monumental and active figure of what Huizinga would have considered democratic civilization, and his passing made Van Vollenhoven's cause a personal issue for Huizinga.[36] Later that year, Huizinga lost another friend, André Jolles (1874–1946), but this time not to death. Huizinga's friendship with Jolles was probably his longest and most emotionally profound.[37] Jolles had always been the more temperamental, confrontational and judgmental individual, often forcing Huizinga to adopt a more emphatic, modest and relativizing tone.[38] After Jolles became a member of the NSDAP in October 1933, all contact between them ceased on mutually hostile terms.[39] The loss of these two friends, however, took place against the background of an earlier, third event, to which we now turn.

On 29 October 1932, Johan Huizinga was instated as Leiden University's chancellor. One of the many tasks that awaited him was the opening of the International Student Service Conference, an annual conference meant to further the international exchange of ideas and perspectives among students from England, Germany, France and the Netherlands.[40] That year, the German delegation was led not by an academic figure, as was common, but by a thirty-one-year-old civil servant from Germany's Ministry of Propaganda, Johann von Leers (1902–1965). In Leiden, von Leers disseminated a pamphlet he had authored titled *Forderung der Stunde: Juden Raus!* (Demand of the hour: Jews out!) Huizinga learned about the pamphlet and its distribution by the end of 10 April, and the next day, the German delegation was dismissed, and the conference was dissolved.[41] Before dismissing von Leers, Huizinga had a brief conversation with him to confirm that he had been the pamphlet's author, and what would seem to most a minor and insignificant fact about this conversation stuck with Huizinga: he refused to shake von Leers's hand.[42]

36 The importance of Van Vollenhoven and his passing has been discussed in Van der Lem, *Johan Huizinga: Leven en werk in beelden en documenten*, 230–32.

37 They had been close friends and avid correspondents since the late 1890s. Krul, *Historicus tegen de tijd: Opstellen over leven en werk van J. Huizinga*, 16. For an undated picture of Jolles and Huizinga together in their student days, see HA: 135 I: 3.1.

38 The tones of their correspondence have been discussed in W. Thys, *André Jolles (1874–1946) 'Gebildeter Vagant': Briefe und Dokumente* (Amsterdam/Leipzig: Amsterdam University Press/ Leipziger Universitätsverlag GmbH, 2000), 12–14.

39 BW II: R. Roland Holst–Huizinga (1933): 1057.

40 BW II: Huizinga–Dutch Department of Foreign Affairs (1933): 995.

41 The most detailed account of this event to date has been given in Van der Lem, *Johan Huizinga: Leven en werk in beelden en documenten*, 228–30. Another elaborate account has been given in C. du Pree, *Johan Huizinga en de bezeten wereld* (Leusden: ISVW Uitgevers, 2016), 170–74. Huizinga himself also recalled the event in a personal letter to Fritz Saxl (1890–1948), BW III: Huizinga–Saxl (1933): 1170.

42 BW II: Huizinga–Dutch Department of Foreign Affairs (1933): 995.

The decision to dispel von Leers and, in effect, the German delegation led to Huizinga's books being blacklisted in Germany. In a public letter defending his decision, Huizinga stated:

> In order to remain freely true to its calling to safeguard its duty and honour, a University must at certain occasions act so as to defend the spiritual ground on which its stands.[43]

What were the 'spiritual grounds' that needed defending? Huizinga elaborated in *In the Shadows of Tomorrow* (1935):

> Reason, once opponent of and self-proclaimed victor over Belief, must today, in order to elude her own demolition, seek refuge with belief. Only on the unimpaired, immovable basis of a living metaphysical awareness can an absolute concept of truthfulness with its corollaries of absolutely applicable norms of virtuousness and justice remain safe against the continuous currents of instinctive life urges.[44]

Reason needed belief (note the difference between Belief and belief[45]), a susceptibility to the notion of absoluteness. This was not because absolute truth would ever be reached, of course, but because the ideal instils norms of 'virtuousness and justice' strong enough to resist the 'currents of instinctive life urges'. In 1933 Huizinga delivered his inaugural lecture as university chancellor, and on this occasion he argued that the university had grown out of 'play': 'disputation', 'school', 'polemics' and 'problems' come from older practices of competition and feud.[46] After the incident with von Leers, Huizinga added another ingredient to the emergence of universities: a certain religious absolutism, reminding the human soul of its inability to fully grasp the world.

43 'Een Universiteit moet, om haar roeping in vrijheid getrouw te blijven, haar plicht en haar eer naar eigen maatstaven bepalen, en zal zich bij wijlen handelend optreden voorgeschreven vinden, dat enkel dient tot verdediging van den gewijden geestelijken grond waarop zij staat.' BW II: Ten Haeff–Huizinga (1933): 1058. This letter is discussed in Van der Lem, *Het eeuwige afgebeeld in een afgehaald bed*, 231.

44 'De Rede, die eenmaal het Geloof bestreed en meende te hebben verslagen, moet nu, om haar afbraak te ontgaan, toevlucht zoeken bij het geloof. Want het is enkel op de onverzwakte en onwrikbare basis van een levend metaphysisch besef, dat een absoluut waarheidsbegrip, met zijn uitvloeisel van volstrekt geldende normen van zedelijkheid en gerechtigheid, veilig is tegen den wassenden stroom van instinctieven levensdrang.' VW VII: *In de schaduwen van morgen* (1935): 364.

45 The capitalization is Huizinga's own.

46 VW V: *Over de grenzen van spel en ernst in de cultuur* (1933): 12.

Reality – like God, like history – *always* defies the cognitive apparatus of the subject, and only from within this 'metaphysical awareness' can virtues of modesty, curiosity, empathy and a willingness to cooperate and compromise exist.[47] Through the case of von Leers, democratic virtues and academic virtues appeared on the same side of the board. Democracies and universities alike risked a certain cannibalism (the democratic election of anti-democratic parties, the academic freedom of anti-academic voices), and in order to defuse this impending self-detonation, certain virtues were of primary importance:

> The new ascesis will be one not of the world's renunciation and salvation; it will be one of self-command and a tempered appreciation of power and pleasure. The glorification of life will have to be extinguished ever so slightly. One will have to remember how Plato has already described the efficacy of the wise as a preparation for death. A fixed orientation on life's teachings and feelings towards death exalts the right with which one wields life's powers.
>
> The new ascesis will be one of submission – submission to the highest of that which is thinkable. This is not the State, people or class, nor is it one's own personal existence. Praise those, for whom this principle can apply only to Him who spoke: 'I am the way, and the truth, and life itself.'[48]

The passage above comes from Huizinga's *In the Shadow of Tomorrow*, which was published in 1935. To Huizinga in 1935, the most effective way to cool down a democracy heated to the point of combustion, to temper an academic arena on the verge of abolishing open discourse, was the following: a 'mild appreciation of power and pleasure' and the ability to remind ourselves of our finitude. An 'orientation towards death' enables a life of ascetism, the modesty of mind required to deflate any wish to eradicate and silence one's opponent, any wish to force the world to take a particular shape. Between

47 VW VII: *In de schaduwen van morgen* (1935): 364. The importance of this metaphysical interest to Huizinga's later understanding of play has helpfully been pointed out in Hanssen, *Huizinga en de troost van de geschiedenis: Verbeelding en rede*, 170–73.

48 'De nieuwe askese zal een zijn niet van wereldverzaking en om hemelsch heil, wel van zelfbeheersching en getemperde schatting van macht en genot. De verheerlijking van het leven zal men een weinig moeten dooven. Men zal zich moeten herinneren, hoe reeds Plato de werkzaamheid van den wijze beschreef als een bereiding tot den dood. Een vaste oriënteering van levensleer en levensgevoel op den dood verhoogt het recht gebruik van de levenskrachten. De nieuwe askese zal een overgave moeten zijn. Overgave aan dat wat als hoogste te denken valt. Dat kan Staat of volk of klasse evenmin zijn als het eigen persoonlijk bestaan. Gelukkig zij, voor wie dat beginsel slechts den naam kan dragen van Hem die sprak: "Ik ben de weg, en de waarheid, en het leven."' VW VII: *In de schaduwen van morgen* (1935): 425–426.

the lines, this orientation was stoic – *memento mori* – but first and most explicitly, it was Christian. 'I am the way, the truth, and life,' Huizinga cited from the Gospel of John. The university's 'spiritual grounds' were spiritual indeed; they were Christian. In the light of the challenges posed by National Socialism to Dutch democracy and academic life, it had become clear, Huizinga held, that games were not self-sufficient.[49] Games, be they academic, political or of another kind, rest on an eternal structure of virtue to subsist, of which asceticism was an important part.

To many of Huizinga's younger colleagues and differently inclined peers in 1935, Huizinga's appreciation of asceticism as a democratic virtue pointed to no more than his apparent pseudo-aristocratic disengagement with oppression.[50] He was a well-to-do, celebrated professor unwilling to stick his neck out, unable to even accept an invitation from his cousin, the journalist Menno ter Braak, to become a member of the Committee of Vigilance of Anti-National-Socialist Intellectuals, a body responsible for meetings, lectures and media outlets warning against National Socialism.[51] The committee had been founded in 1936, by which time the uncompromising violence of the NSDAP was publicly known. Had Huizinga fallen prey to the 'osmosis between conviction and predisposition' of which he had faulted Erasmus in 1924, the opportunistic turning-into-virtue of character flaws? Ter Braak certainly thought so. Fellow historians such as Pieter Geyl certainly thought so, too.[52] Not much is known of Huizinga's incentives regarding this decision, but outwardly, he seems to have found it the morally correct trajectory. But how? How could a self-proclaimed 'defender' of the university forgo this opportunity? Of course, a variety of more or less speculative explanations exist. An indispensable piece to any explanation, however, lies in Huizinga's stoicism:

> We live in a possessed world. And we know it. No one would be surprised, if madness suddenly released itself into rage, leaving this poor European

49 'Er ligt een pijnlijke tragiek in het feit, dat de triomf van het nationaal-socialisme bereikt is met de middelen der democratie. Want alleen de opeenvolgende en elkander steeds overtreffende stembussuccessen hebben het ten slotte veroorloofd de macht te grijpen.' VW VII: *Geschonden wereld* (1945): 532.

50 P. Geyl, 'Huizinga as Accuser of His Age', *History and Theory* 2, no. 3 (1963): 231–62.

51 'Comité van Waakzaamheid van Anti-Nationaal-Socialistische Intellectueelen'. The most extensive discussions of why and how Huizinga declined involvement in this committee can be found in Du Pree, *Johan Huizinga en de bezeten wereld*, 152, 174–76.

52 See Geyl, 'Huizinga as Accuser of His Age' and the correspondence between Huizinga and Ter Braak in BW III: Huizinga–Ter Braak (1936): 1228. For discussions of this intergenerational opposition, see Du Pree, 42–47, 75–80; Krul, *Historicus tegen de tijd: Opstellen over leven en werk van J. Huizinga*, 264–87.

> humanity behind crippled and confused, while the engines continue to run, and the flags continue to wave, but the spirit has dissipated. [...] All around us, doubts exist over the firmness of the societal system in which we live, an undistinguished fear for the near future, feelings of the wavering and decline of civilization. These are not mere flashes of anguish falling upon us at nighttime from thin air, when the flame of life hesitates. These are carefully weighed expectations, based on observation and judgment. The facts overwhelm us.[53]

Whether or not Huizinga's personal inclinations osmotically diffused into his moral beliefs, his understanding of virtues in the second half of the 1930s was imbued with a profound sense of fatalism. In 1935, after the elections, Huizinga was more or less convinced that horror, however undefined, was imminent. After and even during the Great War, Huizinga held, a sense of optimism had taken hold of the minds of authors, artists and politicians: 'get through *this*, with all powers, and then, when *this* is over, we will improve everything, yes, permanently.'[54] This optimism had given way, at least in Huizinga's mind, to a flooding sense of being overwhelmed by a world (economic, political) acting of its own accord, independently of former ideals of humanity, love and curiosity. The world moved faster, more loudly and more unpredictably, he held. Yet this was not only a negative observation. The world's independence of moral conviction worked two ways. The value of morality and virtue lay not in their earthly manifestation and the effects they had; their value lay in a 'spiritual habitus', which had to be retrieved once again. The ills of the 1930s lay not with people who withdrew themselves from the world, Huizinga found, but with those who so frantically, so violently, so uncompromisingly sought to change it. Paradoxically, to Huizinga the world had to be changed through a stoic acceptance of the world as it was.

In Gafijczuk's terms, the observations above can helpfully be rephrased: Huizinga's virtue-ethical convictions – the virtues he celebrated and, in his

53 'Wij leven in een bezeten wereld. En wij weten het. Het zou voor niemand onverwacht komen, als de waanzin eensklaps uitbrak in een razernij, waaruit deze arme Europeesche menschheid achterbleef in verstomping en verdwazing, de motoren nog draaiende en de vlaggen nog wapperende, maar de geest geweken. [...] Alom de twijfel aan de hechtheid van het maatschappelijk bestel, waarin wij leven, een vage angst voor de naaste toekomst, gevoelens van daling en ondergang der beschaving. Het zijn niet louter benauwingen die ons overvallen in de ijle uren van den nacht, als de levensvlam laag brandt. Het zijn weloverwogen verwachtingen, op waarneming en oordeel gegrond. De feiten overstelpen ons.' VW VII: *In de schaduwen van morgen* (1935): 315.

54 '*Dit* doormaken, met alle krachten, en dan, als *dit* voorbij is, zullen wij alles veel beter maken, ja, blijvend goed!' VW VII: *In de schaduwen van morgen* (1935): 316.

own way, attempted to put into practice – were part of his 'inhabited ruins'.[55] *Because* the political and cultural realities had changed as dramatically as they had, *because* a prior world of certainties and rituals had suddenly expired and become out-dated, *because* of disappearances, the virtues appeared the way they did to Huizinga as character traits that needed to be cultivated independently of the world's trajectory so as to remind the soul of its modest place in the world. This would hopefully inspire the dogmatic preachers of the future to take a less radical and less uncompromising stance. 'We have progressed enough, in our ability to harm this world and our community,' Huizinga wrote in *In the Shadows of Tomorrow*.[56] In the first decades after World War II, Huizinga's 'ruinous' virtues would become ruinous to his reputation.

So far, two important points have been made for the present argument: (1) the struggles both for and against democracy in the Netherlands were understood and phrased in terms of character and virtue and against the background of the parties involved; (2) Huizinga's attitude towards and understanding of 'democracy' changed dramatically. In juxtaposition with the political system and virtues championed by the NSB, democratic practices of compromise and deliberation suddenly did not seem so bad to Huizinga, and not just because the National Socialist alternative seemed atrocious. Because democracy was under pressure, because it was waning, in ruins, the virtues it symbolized achieved a new status in Huizinga's mind: the virtues of democracy now manifested themselves in opposition to seemingly determined fate, that is, as independent of the naturally given world and physical necessity. Democratic virtues had achieved an 'orientation towards death', and thus democracy symbolized more a modest, conservative duty and perseverance than it did a revolutionary force towards progress, as it had symbolized to Huizinga only a decade earlier. History may have had become democratic; it *had* remained stoic. And just when Huizinga had fully committed himself to the democratic cause, he came across the works of Carl Schmitt for the first time.

Schmitt's *Ernstfall*: an agonistic term?

In the autumn of 1932, the name of Carl Schmitt started appearing in Dutch newspapers in relation to the aftermath of the Prussian coup d'état earlier

55 D. Gafijczuk, 'Dwelling Within: The Inhabited Ruins of History', *History and Theory* 52, no. 2 (2013): 151.

56 'Wij zijn genoeg "vooruitgegaan", in het vermogen om deze wereld en onze gemeenschap te bederven.' VW VII: *In de schaduwen van morgen* (1935): 420.

that year. That July the Prussian government proved unable to effectively contain the clashes between the NSDAP's *Sturmabteilung* and communist factions. The ensuing riots had left eighteen people dead, and the German Chancellor Franz von Papen consequently decided to dissolve the Prussian government and place the Freistaat Preußen under direct federal rule. The Prussian authorities protested at the federal intervention, and in an ensuing lawsuit between Prussia and the federal government, the still rather young jurist Carl Schmitt defended the latter's case.[57] On behalf of the Weimar Republic, he argued that the chancellor had a right to dispel certain legal ramifications in an emergency case, in *Ernstfall*.[58] The Dutch national papers followed this 'historical legal battle' with interest,[59] taking the lawsuit to be nothing less than the renegotiation of the 'modern rule of law'.[60] The court ended up ruling against Schmitt. Still, but by other legal means, Prussia remained under federal rule – until January 1933. That month, a new chancellor with legal views similar to Schmitt's took over, and Schmitt rushed to join the new chancellor's ranks to help compose a legal ground for an absolute state. Though Schmitt had acquired a name of his own accord, from that January onwards, he became known mainly as 'Hitler's *Kronjurist*'.[61]

Two years later, in March 1935, Carl Schmitt visited the Netherlands to give a lecture in Utrecht on the legal theory of the 'Führer principle'. Socialist and Christian newspapers rushed to the event and reported on how 'the Nazi professor' had tried to shut down attending journalists with 'Göbbelesque methods'.[62] One reporter commented: 'people may follow Göbbel's methods in Hitlerland, but in the Netherlands, journalists decide themselves how to go about things.'[63] Another reflected on the 'God-like figure' that Hitler had become in Schmitt's legal theory: 'surely, this is a bit much?'[64] In the wake

57 For an chronological overview of these events see D. Dyzenhaus, 'Legal Theory in the Collapse of Weimar : Contemporary Lessons?', *The American Political Science Review* 91, no. 1 (1997): 121–34.

58 Dyzenhaus, 'Legal Theory in the Collapse of Weimar : Contemporary Lessons?', 125–27. For a wider perspective on Schmitt's legal theory at the time and its relation to Schmitt's contemporaries, see J. P. McCormick, 'Legal Theory and the Weimar Crisis of Law and Social Change', in *Weimar Thought: A Contested Legacy*, ed. P. E. Gordon and J. P. McCormick (Princeton/Oxford: Princeton University Press, 2013), 61–69.

59 *De Nederlander* 19-10-1932: 1.

60 *Het Vaderland* 27-10-1932: 2.

61 BW III: Huizinga–Ter Braak (1935): 1148.

62 *Het Volk* 13-03-1935: 7

63 *Het Volk* 13-03-1935: 7

64 *Hepkema's Courant* 15-03-1935: 1

of this lecture, Schmitt's fame, or notoriety (depending on whom one asked), grew in the Netherlands, and it is no coincidence that references to Schmitt start popping up in Huizinga's work from 1935 on, first in a correspondence letter to Menno ter Braak in August 1935 and somewhat later that year in Huizinga's book *In the Shadows of Tomorrow*.[65] In this book, Huizinga addressed Schmitt's *The Concept of the Political* and vented his indignation in particular at Schmitt's now-famous *Freund-Feind-Unterscheidung*, his 'friend-enemy distinction'. This 'purely human foundation of all cultural relations', Huizinga held, 'entailed nothing but a celebration of violence'.[66]

To Huizinga, Schmitt indeed appeared not so much a legal scholar as a theorist of 'cultural relations', however faulty his views may have been. Though Schmitt was, for obvious reasons, most commonly read and interpreted at the time as a jurist and legal-political philosopher, Huizinga identified him rather as a philosopher of culture.[67] To him, Schmitt fitted not only in a tradition of, for example, Machiavelli and Hobbes, political thinkers in favour of an absolutist state, but he also fitted in a line of authors consisting of Ernst Curtius (1814–1896), Jacob Burckhardt (1818–1897) and Friedrich Nietzsche (1844–1900).[68] Like Curtius, Burckhardt and Nietzsche, Schmitt had studied the role of *Spiel und Spielzeuge* in classical antiquity and how competition had helped cultivate virtues of creativity, courage and duty. These Hellenistic pedagogical forms of competition were referred to as 'agonism' (*Agonistik*), and, to Huizinga, Schmitt's usage of the terms 'friend', 'enemy' and 'struggle' were first and foremost part of this historical-anthropological line of inquiry. Thus, to understand what Schmitt symbolized to Huizinga, not only is an understanding of the political conditions of his time discussed above needed, but a sense of the debates had between Curtius, Burckhardt and Nietzsche about agonism is also key.

In 1875 a number of lectures by the German archaeologist and historian Ernst Curtius had been published under the title of *Antiquity and the Present* (*Althertum und Gegenwart*). Most of these lectures had been given

65 BW III: Huizinga–Ter Braak (1935): 1148.

66 BW III: Huizinga–Ter Braak (1935): 1148.

67 Before taking up the position of crown jurist for the NSDAP in September 1933, and also whilst working as a jurist for the Weimar Republic, Schmitt had held a number of academic positions as a professor in law from 1921 onwards. By the time he became *Kronjurist* for the NSDAP, Schmitt had written what would become his most celebrated and infamous works: *Political Theology* (*Politische Theologie*), published in 1922, and *The Concept of the Political* (*Der Begriff des Politischen*), published in 1927.

68 A look at the chapter titles of *Homo ludens* suffices to give a first impression of the background against which Huizinga approached Schmitt, VW V: *Homo ludens* (1938).

earlier in the 1850s and '60s, and one of them was titled 'The Contest' ('Der Wettkampf'), from 1856.[69] In this particular lecture, Curtius dealt with the practices of competition in ancient Greece, and towards its end, he concluded that the ubiquity of games and competitions in Greek music, poetry, sculpting, philosophy, athleticism and war had to be understood in the light of a religious attitude: only by continuously attesting and showcasing their indebtedness to the gods could competitors keep their jousting spirits from succumbing to the wild passions of revenge and rage. By the time Curtius's book came out, Jacob Burckhardt had been teaching a class on ancient Greek culture at the University of Basel for three years.[70] Like Curtius, Burckhardt understood ancient Greek competitiveness as indicating another, second interest beyond that of the game's conclusion. In contrast to Curtius, Burckhardt held that this second interest did not consist of honouring the gods but rather of a pedagogical interest in the cultivation of an honourable self, a self that would be remembered long after one perished, thus defying the cyclical tides of life:

> The true aim of struggle is, however, victory itself, and especially in Olympia this is considered the highest on Earth, in that it guarantees the victor that which is basically the aim of all Greeks, namely to be admired in life and praised in death.[71]

Curtius's antiquity honoured and praised the gods; Burckhardt's antiquity was different, entailing a striving to become gods. Burckhardt's antiquity had no ambition whatsoever to passively rest in an awestruck shadow of immutable principles and larger-than-life beings. The people of Greek antiquity wanted to cast such shadows themselves, Burckhardt contended, and in late nineteenth-century Germany, it seemed Greek antiquity had succeeded in doing just that, especially in the eyes of a young philologist, Friedrich Nietzsche. Nietzsche had only just obtained his doctoral degree at the University of Bonn when he acquired a professorship at the University

69 E. Curtius, 'Der Wettkampf', in *Althertum und Gegenwart* (Berlin: Wilhelm Hertz, 1875), 132–47.

70 L. Gossman, *Basel in the Age of Burckhardt: A Study in Unseasonable Ideas* (London/Chicago: University of Chicago Press, 2000).

71 'Das wahre Ziel des Kampfes aber ist der Sieg an sich, und dieser, namentlich der in Olympia, gilt als das Höchste auf Erden, indem er dem Sieger verbürgt, was im Grunde das Ziel jedes Griechen ist, daß er im Leben angestaunt und im Tode hochgepriesen werden muß.' J. Burckhardt, 'Der koloniale und agonale Mensch', in *Griechische Kulturgeschichte: Der griechische Mensch*, ed. R. Marx (Leipzig: Alfred Kröner Verlag, 1929), 77.

of Basel in 1868 at the unlikely age of 24. In Basel he would soon meet Jacob Burckhardt, and three years into Nietzsche's employment, and only shortly after the publication of his first book *The Birth of the Tragedy*, a short essay by him appeared under the title of 'Homer's Contest' ('Homers Wettkampf'). In this essay, Nietzsche launched a conception of ancient Greek play and competition rather distant from the ideas that Curtius and Burckhardt had been developing:

> The original function [*ursprüngliche Sinn*] of this strange institution [of agon] is not, however, as a safety valve [*der eines Ventils*] but rather as a means of stimulation [*der eines Stimulanzmittels*]: one removes individuals who tower over the others only to reawaken the play of powers [*Wettspiel der Kräfte*] – a thought that is hostile to the 'exclusivity' of genius [*Genius*] in the modern sense, but which presupposes that in a natural order of things, there are always *several* geniuses, who incite each other to reciprocal actions as they keep each other within the limits of measure. That is the crux of the Hellenic idea of contest: it detests autocracy and fears its dangers, it craves protection against the genius – a second genius.[72]

Curtius's *agon* consisted of praise, Burckhardt's *agon* consisted of defying death and Nietzsche's *agon* consisted of a pedagogical aim, 'a means of stimulation' whose aim was not to cultivate individuals strong enough to dominate all others but rather to tease out through rivalry the strongest qualities in all.[73] Unlike in modern times, Nietzsche contended, genius was not a matter of romantic 'exclusivity'. It was, on the contrary, always

72 Translation from F. Nietzsche, 'Homer's Contest', in *Nietzscheana*, ed. C. Davis Acampora (Urbana, IL: North American Nietzsche Society, 1996), 5. The original reads: 'Der ursprüngliche Sinn dieser sonderbaren Einrichtung ist aber nicht der eines Ventils, sondern der eines Stimulanzmittels: man beseitigt den überragenden einzelnen, damit nun wieder das Wettspiel der Kräfte erwache: ein Gedanke, der der 'Exklusivität' des Genius im modernen Sinne feindlich ist, aber voraussetzt, daß in einer natürlichen Ordnung der Dinge es immer mehrere Genies gibt, die sich gegenseitig zur Tat reizen, wie sie sich auch gegenseitig in der Grenze des Maßes halten. Das ist der Kern der hellenischen Wettkampf-Vorstellung: sie verabscheut die Alleinherrschaft und fürchtet ihre Gefahren, sie begehrt, als Schutzmittel gegen das Genie – ein zweites Genie.' F. Nietzsche, 'Homers Wettkampf', in *Friedrich Nietzsche: Werke in drei Bänden III*, ed. K. Schlechta (Munich: Carl Hanser Verlag, 1954), 295.

73 Nietzsche repeated a similar conviction in the late 1880s regarding life itself: 'Die Aufgabe ist nicht, überhaupt über Widerstände Herr zu werden, sondern über solche, an denen man seine ganze Kraft, Geschmeidigkeit und Waffen-Meisterschaft einzusetzen hat – über gleiche Gegner... Gleichheit vor dem Feinde – erste Voraussetzung zu einem rechtschaffnen Duell.' F. Nietzsche, 'Ecce Homo', in *Friedrich Nietzsche: Werke in drei Bänden. II*, ed. K. Schlechta (Munich: Carl Hanser Verlag, 1954), 1078.

plural. Genius comes through play, and play requires several equally worthy contestants. This, Nietzsche held, fed straight into a second difference between Greek antiquity and nineteenth-century European societies: in antiquity another, more productive conception of 'resentment' prevailed. 'Resentment' then drew not from an inclination towards 'hostile fights of annihilation [*feindseligen Vernichtungskampfe*] against one another', the result of jealousy, but rather towards 'the action of contests [*Tat des Wettkampfes*]'.[74] Greatness required other, equally great contestants; worthy opponents were needed to cultivate oneself, and in this capacity, opponents were to be respected. 'Every great Hellene passes on the torch of the contest [*des Wettkampfes*]; every great virtue [*Tugend*] sets afire new greatness [*Größe*],' Nietzsche concluded.[75]

Schmitt started writing about 'agonism' and 'struggle' in the 1920s, albeit from an angle quite alien to the works of Curtius, Burckhardt and Nietzsche: law. Whether Schmitt was familiar with Ernst Curtius's work on the agonistic is hard to say, but he certainly was familiar with Burckhardt's and Nietzsche's works on the concept.[76] Schmitt used the concept of *Kampf* as a cornerstone of his theories of state, society and law. To an extent, one could argue that he distinguished the term only to elevate it to becoming an all-encompassing and existential category. 'Struggle' was the fabric of thought and life, and its central feature consisted of its relation to the possibility of death:

> It does not mean competition [*Konkurrenz*], nor does it mean pure intellectual controversy [*Kampf der Diskussion*] nor symbolic wrestlings [*Ringen*] in which, after all, every human being is somehow always involved, for it is a fact that the entire life of a human being is a struggle [*Kampf*] and every human being symbolically combatant [*Kämpfer*]. The friend, enemy and combat concepts receive their real meaning precisely because they refer to the real possibility of physical killing [*die reale Möglichkeit der physischen Tötung*].[77]

74 Nietzsche, 'Homer's Contest', 3; Nietzsche, 'Homers Wettkampf', 293.

75 Nietzsche, 'Homer's Contest', 4. The original reads: 'Jeder große Hellene gibt die Fackel des Wettkampfes weiter; an jeder großen Tugend entzündet sich eine neue Größe.' Nietzsche, 'Homers Wettkampf', 294.

76 At the outset of his *The Concept of the Political* (*Der Begriff des Politischen*), he lauded Burckhardt's theory of state and society. C. Schmitt, *The Concept of the Political*, trans. G. Schwab (Chicago: The University of Chicago Press, 2007), 23.

77 Schmitt, *The Concept of the Political*, 33. The original reads: 'Es bedeutet nicht Konkurrenz, nicht den "rein geistigen" Kampf der Diskussion, nicht das symbolische "Ringen", das schließlich jeder Mensch irgendwie immer vollführt, weil nun einmal das ganze menschliche Leben ein "Kampf" und jeder Mensch ein "Kämpfer" ist. Die Begriffe Freund, Feind und Kampf erhalten

The defining feature of struggle was the 'real possibility of physical death' – that is, not the event of extermination but the reality of its possibility. 'The political' is born from this struggle and the possibility of its ultimate consequence. 'The political' is an answer to the 'task of distinguishing correctly friend and enemy'.[78] Any human individual is thrown into this task; this antagonism is the origin of human thought and culture. Yet, to Schmitt, this antagonism crystallized nowhere as clearly and forcefully as it did in the 'state', which, when properly conceived, is the embodiment of this distinction on a social level. When a state, through liberal constitutionalism, is limited to a legal norm without exceptions, it withers, becomes de-politicized, loses urgency and will eventually succumb to hostile forces. In the aftermath of the 1932 Prussian coup d'état, Schmitt had the occasion to put this theory into practice.[79] He held that the state, conceived of in this particular political sense, embodies and must embody the possibility of the greatest exception in life: death. Schmitt wrote:

> To the extent that wars today have decreased in number and frequency, they have proportionately increased in ferocity. War is still today the most extreme possibility. One can say that the exceptional case [*Ernstfall*] has an especially decisive meaning [*Bedeutung*] which exposes the core of the matter [*Kern der Dinge*]. For only in real combat [*im wirklichen Kampf*] is the most extreme consequence of the political grouping of friend and enemy [*Freund und Feind*] revealed. From this most extreme possibility, human life derives its specifically political tension.[80]

ihren realen Sinn dadurch, daß sie insbesondere auf die reale Möglichkeit der physischen Tötung Bezug haben und behalten.' C. Schmitt, *Der Begriff des Politischen* (Munich/Leipzig: Duncker & Humblot, 1932), 20.

78 'Aufgabe, Freund und Feind richtig zu unterscheiden.' Schmitt, *Der Begriff des Politischen*, 25. The translation by Georg Schwab leaves out the element of 'task': the political does not reside in the battle itself, which possesses its own technical, psychological and military laws, but in the mode of behaviour which is determined by this possibility, by clearly evaluating the concrete situation and thereby being able to distinguish correctly the real friend and the real enemy. Schmitt, *The Concept of the Political*, 37.

79 T. B. Strong, 'Foreword: Dimensions of the New Debate around Carl Schmitt', in *The Concept of the Political* (Chicago: The University of Chicago Press, 2007), ix–x.

80 Schmitt, *The Concept of the Political*, 35. The original reads: 'Wenn die Kriege heute nicht mehr so zahlreich und alltäglich sind wie früher, so haben sie doch in gleichem oder vielleicht noch stärkerem Maße an überwältigender totaler Wucht zugenommen, wie sie an zahlenmäßiger Häufigkeit und Alltäglichkeit abgenommen haben. Auch heute noch ist der Kriegsfall der "Ernstfall". Man kann sagen, daß hier, wie auch sonst, gerade der Ausnahmefall eine besonders entscheidende und den Kern der Dinge enthüllende Bedeutung hat. Denn erst im wirklichen Kampf zeigt sich die äußerste Konsequenz der politischen Gruppierung von Freund und Feind.

The 'specifically political tension', the defining feature of 'politics', lies in or is revealed by the possibility of real combat, and though wars may have become less frequent, their intensity had increased in reverse proportion, and their possibility had not become any less urgent or real: 'war is still today the exceptional case', the *Ernstfall*. This possibility of annihilation was both an exception and universal. It was an ever-looming possibility in whose shadow every social and cultural relation was transformed. Huizinga commented in *Homo ludens*:

> Schmitt will conceive of his enemy not even as a fellow contestant or opponent. The enemy, according to him, is an adversary only in the most literal sense of the word, that is, someone who needs to be removed. [...] Not war but peace is the '*Ernstfall*'. Only by overcoming this deplorable friend-enemy distinction can humanity lay claim to the full recognition of her dignity.[81]

Huizinga disagreed with Schmitt over the supposed exceptional status of the struggle to death, but not in legal or political terms. Huizinga understood Schmitt's political theory against the background of the cultural-historical debates over *agon* and the human creativity it fostered. He understood Schmitt's political theory as an anthropological claim pertaining to 'humanity' and what it means to be human. Schmitt's world – as well as Huizinga's understanding of Schmitt's world – was one where the possibility of a struggle to death always loomed, and this possibility, in turn, demanded a retardation of social disintegration in order to prepare for an imminent battle with adversary forces. Political forces could not be bound to legal constraints in this amoral world of struggle. Huizinga's world, on the other hand, was one where a struggle to death was a given, not a mere *Ernstfall*. Rather, it was the possibility of peace that loomed, however dimly, and to Huizinga, it was because of this peaceful *Ernstfall* that communities and their antagonists had to keep composure and defuse any passionate flame of violence. Huizinga thus read Schmitt not only as a legal scholar; to him,

Von dieser extremsten Möglichkeit her gewinnt das Leben der Menschen seine spezifisch politische Spannung.' Schmitt, *Der Begriff des Politischen*, 23.

81 'Zelfs als mededinger of tegenspeler wil Schmitt den vijand niet beschouwd zien. Hij is volgens hem enkel tegenstander in den meest letterlijken zin des woords, dus degeen, die uit den weg moet worden geruimd. [...] Niet de oorlog is de 'Ernstfall', maar de vrede. Want eerst door deze jammerlijke vriend-vijand-verhouding te overwinnen verwerft de menschheid aanspraak op volledige erkenning van haar waardigheid.' VW V: *Homo ludens* (1938): 242.

Schmitt fitted into a tradition of authors who discussed the agonistic games whereby humans congregated and diverged, united and killed, danced, sung, played and murdered.

Huizinga strung together and discussed the above-mentioned authors in relation to the concept of 'agonism' with varying degrees of approval. Schmitt was, at any rate, met by the fiercest opposition. Interestingly, Huizinga was curiously ahead of Schmitt when he summed up Schmitt's position in 1935 in terms of *homo homini lupus*, a concept Schmitt only started using in 1950. In the 1920s, Huizinga completed a biography of Erasmus (appearing in 1924), and because Erasmus extensively discussed the proverb *homo homini lupus* in his *Adagia*, Huizinga had had ample time and opportunity to think systematically and historically about this anthropological category.[82] Later, too, in 1931, Huizinga discussed the concept in relation to the Russian author Lev Shestov (1866–1938).[83] The fact that Huizinga ascribed the term to Schmitt's theory so early on is telling of Huizinga's cultural-anthropological angle on Schmitt. To Huizinga, Schmitt addressed not just legal issues of state and politics but also issues of humanity and the dynamics of its cultural trajectory.

To conclude, Huizinga's opposition to Schmitt's theory of culture was, of course, not a merely academic one. In Huizinga's eyes, and the eyes of many others, Schmitt's theory served as the theoretical defence for the most pressing and alarming threats to the Dutch and European political order at the time. The previous section discussed how and why democracy had become a cultural term to Huizinga in the 1930s, and Schmitt was the academic face of democracy's looming loss. Still, Huizinga's academic apprehension about Schmitt must be read seriously; it was by no means a mere academic smokescreen for a moral-political rejection. The present section has shown that Huizinga's opposition to Schmitt was drafted from a cultural-theoretical vernacular, and this vernacular is important to an understanding of Huizinga's cultural-historical output in the late 1930s. Huizinga mobilized and engaged anew with the history and anthropology of culture to barricade the gates against the national-socialist forces that had come knocking, and after the 1935 elections discussed in the previous section, Huizinga started seeking new and other ways to erect such obstructions. By no later than 1935, Huizinga settled on a book project for this purpose, and well before

82 In *Der Nomos der Erde* (1950). For the introduction of his version of the term, see C. Schmitt, *The Nomos of the Earth*, ed. G. L. Ulmen (New York: Telos Press, 2006), 95–96.
83 BW II: Huizinga–Suys (1932): 885.

he completed it, in 1936 at the latest, he had decided on what its title would be: *Homo ludens*, the playing human.[84]

Homo homini lupus versus *homo ludens*

To be sure, Huizinga had more than one antagonist in mind when he wrote *Homo ludens*. The final manuscript starts by stating two: *homo sapiens* and *homo faber*.[85] Yet beyond its most explicit targets, the book had other ones in its scope, too. Though Huizinga mentioned Schmitt explicitly only twice in *Homo ludens*, his additional discussions of 'the friend-enemy principle' (mentioned twice) and the 'agonistic' (mentioned thirteen times) addressed what many of his readers would have recognized to be either Schmitt or a theory related to him. As the two previous sections have argued, Schmitt symbolized at once a political orientation and a cultural-historical perspective on human agonism. With this in mind, we take a critical look at De Vugt's perspective on *Homo ludens*, which was mentioned at this chapter's outset. Since De Vugt's account is the most extensive and articulate account on this topic, its main claim is worth citing in full:

> On the very eve of World War II, a hidden dialogue emerged on the ludic nature of modern warfare. Throughout the 1930s and especially in his *Homo Ludens*, a canonical work [on] the philosophy of play, Huizinga formulated a vehement critique on Carl Schmitt's thought. The Dutch historian interpreted Schmitt's friend-enemy distinction in absolute terms: the enemy is he who must be exterminated. In such a view, he concluded, modern warfare could no longer be understood in terms of play, because it implies the disappearance of precisely that what keeps play going: the rule-bound taking of turns by 'opponents'. With the imminent war in mind, he observed that highly cultivated nation states were withdrawing from

84 BW III: Huizinga–Geyl (1935): 1115; BW III: Johan Huizinga–Jakob Huizinga (1936): 1221.

85 To Huizinga, the former referred to both the behaviouristic accounts of human culture that had become common from the 1920s onwards and to the purely evolutionary accounts of human cognitive features. *Homo faber*, in turn, tied into socialist conceptions of humanity's self-realization through work as well as their counterpart: the capitalist strategies of *new managerialism*. During his travels through the United States in 1926, Huizinga had grown increasingly frustrated with what he considered to be the manifestations of these two anthropological categories, and both, in their own way, were symptomatic of the more general process of culture's 'mechanization', as has been discussed in most of the previous chapters. However, alongside *homo sapiens* and *homo faber*, Huizinga reserved ample space in *Homo ludens* to take stance against another anthropological category, too. VW V: *Homo ludens* (1938): 26.

the conventions of the law of nations. Since the opponent was turned into an enemy, and no longer under the restrictions imposed by international conventions, modern combat lost its ludic character and 'humanity its dignity'. And, he immediately added, employing a term that is not without significance in the intellectual debates of the 1930s, times of peace have become the true exception, the real *Ernstfall*.[86]

De Vugt's 'Huizinga' wrote with anachronistic foresight. His Huizinga was already writing on the 'very eve of World War II' in 1935, anticipating an 'imminent war'. Huizinga indeed anticipated war in his writings from 1935 on, but this anticipation should be contained in proper contextual bounds.[87] The anticipation consisted not of knowledge of what would come – World War II – but of events, memories, temporalities and fears belonging to 1935. Huizinga started working on *Homo ludens* in 1935, and when the book appeared in 1938, it bore the marks of its past, not of 1939. This is, at any rate, the perspective which the present section looks to present and consider.[88] For this purpose, this section compares the definition of play presented in *Homo ludens* with the only other publication by Huizinga on play: his previously mentioned 1933 rectorial address as Leiden University's chancellor.[89] Through this comparison, the present section argues that *Homo*

86 De Vugt, 'Philia and Neikos: Huizinga's "Auseinandersetzung" with Carl Schmitt', 48.
87 A brief but good methodological discussion concerning how to historically study the (Dutch) fears and images in the 1930s of an impending war has been offered in W. Linmans, *De oorlog van morgen: Nederlandse beeldvorming van een volgende oorlog 1918–1940* (Amsterdam: Prometheus, 2021), 17–20. The ills of anachronism do not require further discussion. Yet in this case, Hannah Arendt's words seem particularly apt: 'Just as in our personal lives our worst fears and best hopes will never adequately prepare us for what actually happens,' and 'each event [reveals] new possibilities which together transcend [...] the significance of all origins.' H. Arendt, 'Understanding and Politics', in *Essays in Understanding: 1930–1954* (New York: Schocken, 2005), 320.
88 Huizinga started working on the project that would become *Homo ludens* by no later than January 1935 – that is, after the NSDAP had taken power and just before the NSB had its first electoral success in the Netherlands. BW III: Huizinga–Geyl (1935): 1115.
89 To be sure, Huizinga had been interested in 'play' ever since his doctoral dissertation from 1897. The term figures frequently in *Autumntide of the Middle Ages* (1919) and re-appeared in several of his historiographic works in the 1920s as well as in his biography of Erasmus. Judging from the material in the Huizinga Archive, it seems likely that Huizinga's interest in play received another boost upon reading the historian of religion William Brede Kristensen's work on 'play and mystery'; see HA 74: 1.6 (1926). Kristensen taught in Leiden, and he and Huizinga were close. The effects of their relation on Huizinga's work seem like a potentially rich topic for future research. In his essay *Das historische Werk Johan Huizingas*, published in 1947, Werner Kaegi (1901–1979) wrote that 'der *Homo Ludens* ist beinah in jedem Kapitel des Herbstes [Autumntide of the Middle Ages] schon gegenwärtig. Er gehört indessen mehr dem Gebiet der Geschichtstheorie

ludens bore the mark not of Huizinga's foresight of what would come but of his contemporary experiences of a loss of democracy, as well as of Carl Schmitt's symbolic role in its erosion. This angle links Huizinga's opposition to Schmitt to Huizinga's more intuitive experiences of loss as well as his sense of a looming uncertainty, rather than to the technical intricacies on which De Vugt's account chooses to focus.

Huizinga's rectorial lecture from 1933 was titled *On the Borders Between Play and Seriousness in Culture*.[90] The published lecture offers a window into how Huizinga understood 'play' only a year before the National Socialist threat to Dutch democracy properly materialized. The lecture was given roughly a month before the NSDAP came to power in Germany, two months before the von Leers case and before the NSB's electoral success in the Netherlands – and it shows.[91] A comparison of this lecture with *Homo ludens* allows for an interesting window on what the political events between 1933 and 1938 might have meant to Huizinga's understanding of the concept of 'play'. Most accounts of *Homo ludens* omit critical discussion of Huizinga's 1933 lecture, and if they do include it, the lecture is often interpreted as a mere announcement of the later book.[92] In the lecture, however, Huizinga assembled a four-pronged definition of 'play' quite distinct from the theory presented in *Homo ludens*. Play, Huizinga held in 1933, was:

1. Aggregation and disentanglement.[93] Play facilitates the aggregation of people and announces their dispersion upon its conclusion. 'Play' forms an enclosed world, wherein 'the players move according to their own, coercive rules.'[94]

an als demjenigen der Geschichtschreibung.' Ernst Gombrich (1909–2001) had an altogether different judgement of the relation between *Herfsttij* and *Homo ludens*. Gombrich had studied *Autumntide* closely and had appreciated it. However, while Kaegi found that *Autumntide* had already been pregnant with *Homo Ludens*, Gombrich, upon attending Huizinga's lecture on 'play' at the Warburg Institute in London in 1937, wrote about having met 'a different Huizinga'. Kaegi, *Das historische Werk Johan Huizingas*, 24. Gombrich, 'Huizinga's *Homo ludens*', 134.

90 One of the central reasons for Huizinga's interest in play had to do with a more general interest among his peers at the time: as was discussed in the previous chapter, the postulation of a universal theory of culture and humankind stood in immediate opposition to the ever-increasing subdivision of academic disciplines, with which Huizinga strongly disagreed. See e.g. Otterspeer, *Het horzelnest: De Leidse Universiteit in oorlogstijd*, 31–35; Van der Lem, *Johan Huizinga: Leven en werk in beelden en documenten*, 251–54.

91 In 1934 Huizinga delivered this lecture twice in German, first in Zurich and then in Vienna. HA 83 II: 1 (1934).

92 One example is to be found in Strupp, *Johan Huizinga. Geschichtswissenschaft als Kulturgeschichte*, 184.

93 'binding en ontknooping'. VW V: *Over de grenzen van spel en ernst in de cultuur* (1933): 5.

94 VW V: *Over de grenzen van spel en ernst in de cultuur* (1933): 5.

2. Display.[95] Huizinga made a point of denouncing the idea that 'play' should *imitate*. Rather, 'play' shows; 'a play forms that which had seemed formless,' and thus makes it 'imaginable'.[96] This ability to create and form a world was directly likened by Huizinga to 'a holy act'.[97]
3. Struggle or battle.[98] 'Play' is a struggle among contestants for a 'cosmological consequence'.[99] In art, science, technology and other cultural domains, social interactions do not properly escape this struggle towards an inherently valuable yet unidentified *good*.
4. Style.[100] In play, one finds 'rhythm, repetition, cadence, chorus, closed form, composition and harmony'; they are 'constituents of style'.[101] 'Style' here seems to encompass the forms of play. Interestingly, Huizinga equates aesthetic 'style' with moral 'order and faithfulness'.[102]

Aggregation, display, struggle and style – these were the four characteristics of 'play', Huizinga held in 1933. Consider next the definition of play launched in the first chapter of *Homo ludens* five years later:

1. 'A voluntary act'.[103] Here, Huizinga argued, 'freedom' had to be understood only in 'its loose sense, which leaves the question of determinism untouched'.[104] Yet, 'play' transcends 'the biological'; it is an escape from the biological and practical 'process', which Huizinga described as a 'slowing down' or 'pausing.'[105] Play is 'free time'.
2. A resignation, or 'stepping out'.[106] Through its only loosely defined 'freedom', Huizinga established its second feature: its reclusion or resignation – literally, its 'stepping out'. Play is a stepping out of 'the process of nature';[107] it is a temporality wherein one acts independently from the needs of life.[108]

95 'vertooning'. VW V: *Over de grenzen van spel en ernst in de cultuur* (1933): 5.
96 'verbeelding'. VW V: *Over de grenzen van spel en ernst in de cultuur* (1933): 5.
97 'een heilige handeling'. VW V: *Over de grenzen van spel en ernst in de cultuur* (1933): 5.
98 'strijd'. VW V: *Over de grenzen van spel en ernst in de cultuur* (1933): 6.
99 'kosmisch gevolg'. VW V: *Over de grenzen van spel en ernst in de cultuur* (1933): 6.
100 'stijl'. VW V: *Over de grenzen van spel en ernst in de cultuur* (1933): 7.
101 VW V: *Over de grenzen van spel en ernst in de cultuur* (1933): 7.
102 'orde en trouw'. VW V: *Over de grenzen van spel en ernst in de cultuur* (1933): 7. One of Otterspeer's books on Huizinga was named after this depiction of 'style', see Otterspeer, *Orde en trouw: Over Johan Huizinga*.
103 'een vrije handeling', 'natuurproces'. VW V: *Homo ludens* (1938): 35.
104 VW V: *Homo ludens* (1938): 35.
105 'verpoozing'. VW V: *Homo ludens* (1938): 36.
106 'een uittreden'. VW V: *Homo ludens* (1938): 35, 41.
107 'natuurprocess'. VW V: *Homo ludens* (1938): 35.
108 'levensbehoeften'. VW V: *Homo ludens* (1938): 37.

3. Entanglement and disentanglement.[109] The ability to pause a life of necessities is described in terms of an ability to return to the game: 'repeatability is one of the most essential qualities of play.'[110] By virtue of this repetition, play can pull a part of the world away from its everyday employment; the arena, the stage, the board game, the screen, the playground, the altar: they are scenes of repetition. Repetition ties into both the temporal and the spatial isolation essential to play, and Huizinga describes this contracting ability of repetition in terms of 'entanglements'.
4. Order.[111] The free acts of play in secluded times and spaces are not lawless: 'it creates order, it *is* order.'[112] The act of playing allows the free creation of a world according to ideas of beauty. 'It realizes in the imperfect world and confused life a temporary and limited perfection. The order imposed on the play is absolute.'[113] After having described this last property of play, Huizinga listed play's two main 'functions': 'a showing of something' and a 'struggle for something'.[114]

For present purposes, a comparison between Huizinga's understanding of 'play' in 1933 and 1938 allows for three interesting observations. To begin, one feature of 'play' remained more or less the same: what Huizinga in 1933 called 'aggregation and disentanglement' he in 1938 called, in a similar capacity, 'ravelling and unravelling'. Secondly, two differences can be distinguished. What Huizinga called 'style' in 1933 became a more conservative-sounding appreciation of 'order' in 1938. In 1933 'play' had appeared essentially as a participation in and 'display' of 'struggle'. In 1938, on the other hand, play was characterized by an ability to freely 'resign'. 'Display' and 'struggle' had ceased to belong to play's essential properties. The third observation is that, by 1938, Huizinga had added to his understanding of the functions of 'play': play had to reveal 'something', and this 'something', whatever that may be, had the quality of something 'holy'. It is 'pregnant with a cosmic insight'.[115] Huizinga had increased play's stock in a conservative, pacifist and spiritual frame of mind.

109 'knooping en ontknooping', 'natuurproces'. VW V: *Homo ludens* (1938): 37, 35.
110 VW V: *Homo ludens* (1938): 37.
111 'orde'. VW V: *Homo ludens* (1938): 37–38. For a discussion of the element of 'order' in Huizinga's concept of play in *Homo ludens*, see Otterspeer, *Orde en trouw: Over Johan Huizinga*, 226.
112 VW V: *Homo ludens* (1938): 38, italics added.
113 VW V: *Homo ludens* (1938): 38.
114 VW V: *Homo ludens* (1938): 41.
115 VW V: *Homo ludens* (1938): 54.

Play's transition from 'struggle' to 'resignation' is particularly revealing of how Huizinga's theory of play related to his times. By the mid-1930s, the elevation of 'struggle' to an essential part of human culture had lost its appeal. Confused images of a racial 'struggle' were consuming culture: 'the race thesis is always hostile [*vijandig*], always anti-; for a theory that presents itself as scientific this is a bad omen. Its position is anti-Asian, anti-African, anti-proletarian, anti-Semitic.'[116] The vernacular of 'struggle' between friend and enemy had been hijacked by the National Socialist forces threatening the norms of Huizinga's generation. 'Struggle' became not quite anti-cultural, but it was no longer naturally aligned with culture's cause. And after 1935 Huizinga's 'play' was still competitive. It was essentially agonistic, but it was not *hostile*. It prepared the player for an extra-particular, religious perspective of the world and cultivated a sensitivity to 'the holy sphere' and 'the heights of beauty'.[117] After 1935 Huizinga's play tied straight into a religion of 'resignation', and this could not have been further removed from Schmitt's understanding of culture's inescapable *Kampf* in human thought and politics, a struggle beyond principle and order. Still, Huizinga recognized, the ideas of 'essential struggle' in cultural agonism had come to reign supreme:

> [Modern war] rests on the 'friend-enemy' principle. All truly political relations between peoples and states are dominated by it, or so we are told. The *other* group is *either* your friend *or* your enemy. Enemy means not *inimicus*, that is, the personally hated, let alone the evil [other], but only *hostis*, that is, the stranger standing in the way of one's own group. Schmitt refuses to see the enemy even as a competitor or opponent. According to him, the enemy is the opponent only in the most literal sense of the word: that is, the one who needs to be exterminated.[118]

116 'De ras-these is altijd vijandig, altijd anti-; voor een leer, die zich voor wetenschap uitgeeft, een slecht teeken. Het standpunt is anti-aziatisch, anti-afrikaansch, anti-proletarisch, anti-semitisch.' VW VII: *In de schaduwen van morgen* (1935): 352. 'Contemporary race theory', Huizinga argued, had subjected each and every human to a supposed inter-racial struggle 'utterly undetectable to anthropology'. VW VII: *In de schaduwen van morgen* (1935): 350.
117 VW V: *Homo ludens* (1938): 36.
118 '[Hedendaagse oorlog] berust op het "vriend-vijand"-beginsel. Alle werkelijk politieke betrekkingen tusschen volken en staten worden door dat beginsel beheerscht, aldus luidt het. De *andere* groep is steeds òf uw vriend òf uw vijand. Vijand beteekent niet *inimicus*, d.w.z. persoonlijke gehate, laat staan booze, maar enkel *hostis*, d.w.z. de vreemde, die uw eigen groep in den weg staat of in den weg treedt. Zelfs als mededinger of tegenspeler wil Schmitt den vijand niet beschouwd zien. Hij is volgens hem enkel tegenstander in den meest letterlijken zin des woords, dus degeen, die uit den weg moet worden geruimd.' VW V: *Homo ludens* (1938): 242.

Modern theories of politics and the practice of war rested on a new understanding of the opponent not as *inimicus* but as *hostis*, and it was against the background of this observation that Huizinga introduced Schmitt in *Homo ludens*. Schmitt represented a world where 'anger has been left without bounds, and has come to demand all the fruits of human reason,' and as Huizinga wrote these lines, he might have had his reflections in *Autumntide* (1919) on medieval warfare in mind, when the ideals of chivalry and honour 'as a rule impeded rather than enhanced the conduct of war, imposed as it was to sacrifice strategy to life's beauty'.[119] In these medieval and earlier contexts, 'agonistic spheres' had rested on virtues of respect and temperance: respect for both the opponent, on whose existence the game depended, and for the rules, whose abidance offered glory to victory. That had changed, Huizinga held, and Schmitt was the archetypical example of this transition's outcome.[120] The NSDAP realized what Schmitt had already defended in court in 1932 on behalf of the federal state: a state legally permitted to break its own laws and agreements. Earlier, in 1935, Huizinga wrote:

119 'Het werkte toch in den regel op de krijgsvoering meer belemmerend dan bevorderend, daar het de eischen der strategie opofferde aan die der levensschoonheid.' VW III: *Herfsttij der Middeleeuwen* (1919): 119.

120 Soon Huizinga found support and solace in the writings of a kindred mind. In his *The Revolt of the Masses* (*La rebelión de las masas*) from 1927, José Ortega y Gasset (1883–1955) wrote about modern military violence:

> For this form of violence is none other than reason exasperated. Force was, in fact, the *ultima ratio*. Rather stupidly it has been the custom to take ironically this expression which clearly indicates the previous submission of force to methods of reason. Civilisation is nothing else than the attempt to reduce force to being the *ultima ratio*. We are now beginning to realise this with startling clearness, because "direct action" consists in inverting the order and proclaiming violence as *prima ratio*, or strictly as *unica ratio*. It is the norm which proposes the annulment of all norms, which suppresses all intermediate processes between our purpose and its execution. It is the *Magna Carta* of barbarism.

The traditional conception of the relation between war and politics had been that of war as *ultima ratio*, the ultimate exponent of political dialogue, the last resort. The modern conception of this relation, Ortega y Gasset held, revolved rather around the conception of war as *prima ratio*, the condition of that form of exchange called society and politics. In this modern sense, politics was either already a form of war, or otherwise, and at the very least, the continuous possibility of war, and consequently morality and politics were completely usurped by the notion of war rather than war being a form of political exchange. Like Ortega y Gasset, Huizinga found that the logic of war as *prima ratio* was part and parcel of the new phenomenon of absolutist states, the kind of state promoted by Schmitt. Huizinga met Ortega y Gasset for the first time in July 1934, and from then on, he followed Ortega y Gasset's output. For above block citation, see J. Ortega y Gasset, *The Revolt of the Masses* (London: George Allen & Unwin LTD, 1932), 81–82.

> But the state – an outraged voice of protest sounds, and not only from the voice of modern despotism – the state cannot be criminal. The state cannot be considered subjected to moral norms of human society. Each attempt to call it before moral judgment bounces off of the independence of the state. It stands beyond all morality. – But also above it?[121]

The rhetorical question at the end of this passage was meant to elicit a negative answer. No: a state and its politics do not stand above morality. In fact, a state and its politics *cannot* stand above morality, for the moment politics is deemed autonomous, the moment the struggle for domination becomes *ultima ratio*, the opponent (whoever that may be, whether another political party, a class or an ethnic group) is transformed from *inimicus* to *hostis*. In the former capacity, the opponent enjoys a respect inherent to the ability to play a game, such as a negotiation or boxing match. In the latter capacity, the opponent is a mere hindrance left for annihilation, an obstacle. And why would one debate or engage with such a side in a restricted fashion? The game of politics and international relations had to rest on virtues – respect, curiosity and a willingness to compromise – in order to subsist. Huizinga's experiences of democractic erosion had taught him about games, and culture. Huizinga's experiences of democracy, not his long-winded theoretical engagement with Schmitt's theory, formed his apprehension about Schmitt and the politics he represented.

In sum, two observations made in this section stand out for the present argument: (1) a comparison of Huizinga's views on play from 1933 and 1935–38 reveal a transition from 'struggle' to 'resignation'; (2) Huizinga's critiques of Schmitt in *Homo ludens* (1938) raise precisely this issue against Schmitt: Schmitt's conception of agonism and contest amounted not to the pinnacle but rather to the demise of play. Without an independent set of certain virtues, of which resignation was one, play degenerates into a form of hate and anger that can no longer be called 'play', Huizinga held. Other such virtues were modesty, patience and curiosity. Huizinga's heightened appreciation of virtues in play tied straight into his defence in the 1930s of 'democracy', 'academic freedom' and the culture of internationalism in which he had grown up. After the NSB achieved its momentous results in the 1935

121 'Maar de Staat, aldus klinkt nu een verontwaardigd protest, niet enkel van de zijde van het moderne despotisme, de Staat kan niet misdadig zijn. De Staat kan niet onderworpen worden geacht aan de zedelijke normen der menschelijke samenleving. Elke poging, hem voor de uitspraak van het zedelijk oordeel te roepen, stuit af op de zelfstandigheid van den Staat. Hij staat *buiten* alle moraal. – Ook *boven* alle moraal?' VW VII: *In de schaduwen van morgen* (1935): 379.

election, it had become apparent once again that democracy left alone – that is, left without a certain culture of virtues – could end up consuming itself from the inside. Huizinga's critique of Schmitt, like Huizinga's theory of play, cannot be seen independently of Huizinga's experience of democracy's possible and premature death. Huizinga had not been an avid proponent of democracy at first, but when its *inimicus* took stage, democracy became a game worth defending.

Land and sea: two perspectives on a river delta

Huizinga's critiques of Schmitt's theory of agonism and culture resurfaced in several corners of Huizinga's later writings. In this last section, one such example is discussed to show how Huizinga's post-1935 political consciousness resounded in the history he wrote. For this purpose, the present section examines Huizinga's and Schmitt's respective understandings of the relation between land and sea in seventeenth-century Dutch maritime culture. An examinination of how their analytical differences reappeared in their interpretation of a more particular historical-empirical question – the border between sea and land – demonstrates how Huizinga's experience in the 1930s of an eroding democracy affected not only his thoughts on culture and ethics in relation to Schmitt's theory of agonism but also his perception of historical objects. This observation is, in turn, crucial not only to this chapter but also to this project at large: Huizinga's history conceded and responded to experiences of loss. Consider first the painting below by Simon de Vlieger (1601–1653).

In this painting, a calm sea reflects the activities of voluptuous clouds in a wide and windy sky. On the left foreground, six docked ships await departure; on the right, three larger vessels head to open waters. The larger vessels have been equipped with square rigs: a mast construction consisting of mobile beams perpendicular to the main mast, allowing the sails to be turned somewhat freely. Such square rigs had only recently been developed at the time of this painting's production. The ships on the left have been equipped with older kinds of masts and sails, limiting the vessels' movement more stringently to the direction of backwinds. This was, at any rate, the consensus among the maritime historians of the 1930s and '40s. Nowadays, scholars agree that the square rigs had been around for much longer.[122]

122 I. C. Campbell, 'The Lateen Sail in World History', *Journal of World History* 6, no. 1 (1995): 1–23.

Figure 6.3. *Calm Water* (*Kalm Water*), painted 1640–50 by Simon de Vlieger (1601–1653) and currently part of the Boijmans Van Beuningen collection in Rotterdam. The location of the site painted is unknown, but it is known that De Vlieger spent most of his time in Rotterdam and Amsterdam whilst producing this painting.

Either way, in his *Land and Sea* (*Land und Meer*), published in 1942, Carl Schmitt wrote, according to the consensus of his time, the following about the implementation of the square rig by Dutch shipbuilders:

> From the West Frisian city of Hoorn in North Holland a new type of ship appeared in 1595, a boat with square rigs [*Rahsegeln*] that used tailwinds very differently from old sails. [...] Here lies the true turning point in the history of the relation between land and sea. That which could be achieved with the materials of ship and rigs, was achieved.[123]

The square rigs and the new maritime mobility they enabled amounted to no less than a new relation between sand and water, and between rocks and

123 'Ausgehend von der westfriesischen Stadt Hoorn in Nordholland tritt um 1595 ein neuer Schiffstyp auf, ein Boot mit Rahsegeln, das nicht wie das alte Segel einfach mit rückwärtigem Winde ganz anders auszunützen wußte als das herkömmliche Segel. [...] Hier liegt der eigentliche Wendepunkt in der Geschichte des Verhältnisses von Land und Meer. Was sich mit dem Material, aus dem Schiff und Takelung damals bestanden, überhaupt schaffen ließ, war damit erreicht.' C. Schmitt, *Land und Meer/Land en Zee*, ed. Henry van Sanderburg (Groningen: Blauwe Tijger, 2017), 50–51.

waves, Schmitt argued. With the introduction of square rigs, sailors and captains had the ability to travel with a greater independence from winds and currents. Now they could follow stars, examine routes, explore new naval warfare tactics and – most importantly, according to Schmitt – trace the movements of the leviathans of the sea: whales. 'Without whales, fishermen would have kept to the coast. Whales coaxed them onto the oceans and emancipated them away from the coastline,' Schmitt held.[124] By making the trajectory of travel independent of oceanographic properties through the invention of square rigs, the Dutch whalers were 'the first heroes' to push through what Schmitt called 'a planetary revolution of "space"': they founded a world wherein 'the people and the things can stand and move in a space.'[125] The consequences, Schmitt held, were cataclysmic. This 'revolution of "space"' brought about a new phase in history's agonism. The invention of square rigs turned 'world history' into 'a struggle of naval powers against land powers' in a military but also ontological sense. The Dutch square rigs painted by De Vlieger embodied an agonistic perspective on the world.[126]

Just a year before Schmitt's *Land and Sea* was published, Huizinga completed his *Dutch Civilization in the Seventeenth Century* (*Nederland's Beschaving in de Zeventiende Eeuw*). In this book, Huizinga commented on scenes such as the one depicted in De Vlieger's *Calm Water.* Huizinga owned at least one painting by De Vlieger, and on at least two occasions in this book, Huizinga commented on De Vlieger's oeuvre at large.[127] Like Schmitt, Huizinga interpreted waters such as those depicted in *Calm Water* as a seventeenth-century renegotiation of the border between land and sea, and like Schmitt, Huizinga considered this renegotiation fundamental to a revolution in seventeenth-century Dutch culture, if not Western European culture at large. However, rather than focussing on the struggle between land and sea, Huizinga focussed on the cooperative exchange between land and river, and this decision was not only historically informed but invested in other considerations as well:

124 'Ohne den Walfisch hätten sich die Fischer immer nur an die Küste gehalten. Der Wal hat sie auf die Ozeane gelockt und von der Küste emanzipiert.' Schmitt, *Land und Meer/Land en Zee*, 46–47.

125 'eine planetarische Raumrevolution'. Schmitt, *Land und Meer/Land en Zee*, 76–77. 'die Menschen und Dinge stehen und bewegen sich jetzt in ein Raum.' Schmitt, 96–97.

126 'die Weltgeschichte ist eine Geschichte des Kampfes von Seemächten gegen Landmächte und von Landmächten gegen Seemächte.' Schmitt, *Land und Meer/Land en Zee*, 22–23.

127 In *Dutch Civilization in the Seventeenth Century*, Huizinga refers to him as 'de Vliegher'. VW II: *Nederland's beschaving in de zeventiende eeuw* (1941): 439, 489.

> The hydrographic structure of the land effectuated to a certain extent a democratic structure of its people. A water land such as this one cannot do without the autonomy of one's own circles, and it is destiny's favour that with each renewal or loss of office positions, such as alderman and bailiff, the significant Dutch dike-reeve was preserved. The common farmer or fisherman could travel here similarly to the grand lord, in his own little boat, with always a detour at hand to circumvent toll or obstructions.[128]

The 'hydrographic structure of the land' had fed into the 'democratic structure' of its inhabitants. This 'water land', with its private owners, sellers, buyers and logistics, required a 'local government' (*heemraad*) to ensure that 'the petty farmer or fisherman' could travel over river and swamp just as easily as the 'lords'.[129] From this inland structure of local communities radiated 'a certain democratic influence on Dutch society as a whole', including on its 'fleet'.[130] This influence, Huizinga held, consisted of a shared culture of consensus and equality among negotiating parties, inspired to an extent by the practice of trade at this scale: 'whether or not one had studied said nothing about whether or not one was "civilized"'; a considerable group of 'traders actively took part in the process of culture', and many of them had not had a formal education.[131] The consequence of this equality in status and commercial possibilities was an expansion of certain 'social spaces' that had already existed in some medieval communities among the military and tradesmen. These spaces were defined by 'courtesy and honour', and through the increasing accessibility of trade, they had become 'the tone of civilization' throughout the Netherlands. Both the celebrated lord and the

128 'Deze hydrografische structuur van het land had tot zekere hoogte een democratische structuur der bevolking tot gevolg. Een waterland als dit kan niet zonder zelfbestuur in eigen kring, en het is een gunst van het lot, dat bij alle vernieuwing of verlies van onze oude ambtstitels, als schepenen, drost enz., juist de veelbeteekenende van heemraden gebleven is. De kleinste boer of visscher kon hier reizen als elders de groote heer, in zijn eigen kleine schuit, met altijd wel een omweg om een tol of een versperring te ontgaan.' VW II: *Nederland's beschaving in de zeventiende eeuw* (1941): 420.

129 VW II: *Nederland's beschaving in de zeventiende eeuw* (1941): 420.

130 'Een zekere democratiseerende invloed op de Nederlandsche samenleving als geheel is van deze gemeenschap van alle lagen der bevolking op de vloot ongetwijfeld uitgegaan.' VW II: *Nederland's beschaving in de zeventiende eeuw* (1941): 439.

131 'Gestudeerd of niet gestudeerd beteekende geenszins een tegenstelling beschaafd of onbeschaafd. Buiten de geleerde beroepen nam nog een breede laag van den handeldrijvenden of neringdoenden middenstand aan het cultuurproces actief deel.' VW II: *Nederland's beschaving in de zeventiende eeuw* (1941): 445.

'common man' were addressed as 'sir', for 'as poets they were equal among each other like Greek and Romans.'[132]

In *Dutch Civilization in the Seventeenth Century*, Huizinga reserved a certain term for these 'social spaces', their virtues and the hydrographic structure that accompanied them. Together, these formed a conservative 'play form' (*spelvorm*).[133] Huizinga had used this and related terms in *Autumntide* to describe late medieval Burgundian and French culture, but this time around, the term had another connotation. As has been explored above, 'play' had become an antagonist to those agonistic theories of absolutist politics and violence. Huizinga's discussion and celebration of the historical seventeenth-century 'play forms' of Dutch society served as a historical investigation as much as a moral lesson: to be cultured and civilized required a moral attitude of equality, justice and respect, and nothing, truly nothing, would have seemed more repugnant to this conception of human flourishing than the degradation of people to *Untermenschen*. In 1941 the Dutch seventeenth century revealed a 'democratic' way of life. Interestingly, when Huizinga delivered a lecture series on seventeenth-century Dutch culture in Cologne in 1933, no mention of a Netherlandish democratic culture was made.[134] Threats to democracy had enabled a new cultural-historical dimension along which to reconstruct the past.

In sum, two perspectives on seventeenth-century Dutch maritime culture have been discussed. Whilst Schmitt emphasized the technical developments that led to a reorganization of space causing a collision of entire societies, Huizinga's angle emphasized how the developments of the time rested on geographically conditioned customs and systems of virtues. To be sure, in themselves, and considered from internalist and logical perspectives, these images are not necessarily mutually exclusive, but the differences in emphases reveal a fundamental moral divergence: the same moral difference previously discussed in terms of Huizinga's virtuous 'play' and Schmitt's 'friend-enemy' principle. Schmitt considered

132 'De toon der beschaving bleef ook bij de patriciërs burgerlijk. Voor dit behoud van een zeker sociaal amalgama was het van groot voordeel, dat het gansche beschavingsstreven zijn alpha en omega vond in een naarstig klassicisme. Het klassicisme nu wist niet van die minieme standsonderscheidjes tusschen aanzienlijk, deftig en eerzaam, zooals zij hier golden. Het had zijn groote allure krachtens Romeinsche en Grieksche traditie, en veroorloofde de nabootsing daarvan evengoed aan den kleinen man als aan den voornamen heer. Als dichters was men allen onder elkaar Grieken en Romeinen.' VW II: *Nederland's beschaving in de zeventiende eeuw* (1941): 445–446.

133 VW II: *Nederland's beschaving in de zeventiende eeuw* (1941): 445.

134 J. Huizinga, *Holländische Kultur. Das siebzehnten Jahrhundert* (Jena: Eugen Diederichs Verlag, 1933). These lectures have not been included in Huizinga's collected works.

the achievements of the seventeenth-century Netherlands through the lens of clash: between sea and land, between one society and the next. Huizinga, on the other hand, considered the same achievements in terms of the practices and virtues dearest to him: modesty, cooperation, consensus and respect. Huizinga and Schmitt engaged in an agonistic exchange on several levels. The experience of democracy withering away had inspired in Huizinga not only a new understanding of human playfulness but also a new and 'democratic' understanding of Dutch history.

Conclusion

Huizinga's appreciation and understanding of 'democracy' changed dramatically during 1930s. Until the late 1920s and early '30s, Huizinga understood 'democracy' in direct relation to the commercialization of politics and the commodification of culture. As National Socialist threats to democracy swelled, however, 'democracy' gained a different symbolic content. In fact, the experience of its looming loss tied into a broader transition in Huizinga's vocabulary: words such as 'culture', 'history' and 'humanity', too, underwent a transition as this experience grew ever more pronounced, and nowhere in Huizinga's later oeuvre did these transitions intersect and interact as explicitly as they did in his *Homo ludens*.

Homo ludens is a complicated book. It addresses several debates, draws from numerous academic fields, includes a tremendously wide variety of empirical materials, uses arguments of different kinds (philological, anthropological and historical) and often appears encyclopaedic rather than directed. Still, the book had one central claim, to which Huizinga could return time and again: culture is *in* play. This chapter has argued that to appreciate the dimensions of this claim and the significance it had to the author, Huizinga's relation to Schmitt is of great help. In his critique of Schmitt, Huizinga mobilized two antagonisms: his aversion to National Socialist politics and the threat it posed to Dutch democracy, as well as his aversion to a more particular, academic theory of agonism associated with it through Schmitt. In the 1930s, politics had become culture, and by defending a cooperative theory of play and culture, Huizinga addressed at once a political reality and an academic school. Unlike what has been argued by De Vugt, this chapter thus holds that *Homo ludens* was a response not to Huizinga's supposed foresight of war and what it could bring but to his very own contemporary times. The arguments to this end can be summarized as follows:

1. During the 1930s, convictions that had long been political and cultural truisms withered away under the pressure of mass employment and new means of political communication. Inspired, and later facilitated, by their German counterparts, the Dutch National Socialist Movement enjoyed a remarkable election result in 1935. Their election programme called for a moral re-education of the Dutch, championed autocracy over democracy and wished to abolish the four traditional identity pillars (Catholicism, Protestantism, socialism, liberalism), under whose banner most social activities had been 'pillarized'. Huizinga had not been sympathetic towards democracy in the 1910s and '20s, but when democracy found itself opposed to the trends of his times, his appreciation of it underwent a transformation. This transformation partly consisted of certain qualities becoming synonymous with a 'democratic' attitude against the background of the National Socialist threat: virtues of patience, modesty, curiosity and a willingness to compromise. However, this transformation was equally affected by the very fact that democracy seemed to be eroding. By expressing a sympathy for democracy, Huizinga could mobilize his stoic-Christian inclinations: to support democracy, this ineffective, compromising game, was to reserve a playing field disengaged from the whimsical fashions and unrest of his times. Democracy, precisely because it was experienced as being under the pressure of erosion, had become a stoic practice to Huizinga.
2. Carl Schmitt (1888–1985) was already the NSDAP's *Kronjurist* by the time Huizinga learned about his work, most likely in 1935. To Huizinga, however, Schmitt was as much a theorist of culture as he was a legal philosopher. Huizinga read Schmitt's work on agonism against the background of Burckhardt and Nietzsche rather than of Machiavelli and Hobbes. Thus, Schmitt's friend-enemy thesis and his understanding of the enemy's physical destruction as both the *Ernstfall* and condition for politics and culture were not only a reflection on, say, constitutionalism and international law; they were also a reflection on the virtue-ethical education of cultures. Schmitt had subjected all domains of human activity to his principle, and against the background of a self-detonating democratic system and his Anabaptist upbringing, Huizinga had more than one reason to reject this submission. To begin, Schmitt's theory amounted to an understanding of humanity as consisting of *homo homini lupus*.
3. For democracy to persist, Huizinga held, a virtue-ethical system had to exist outside of and in support of political practice, and it was cultural history's duty to remind its researchers and readers of the extra-political

domains of life: ethics, religion and 'culture'. Without an external point of reference, political concerns become total, pre-emptively deflating all possibility to execute the most human behaviour: play. Practices of 'play', Huizinga held, helped develop those virtues that democracy needed most: a respect for one's opponent, the rules and the ideals served. Huizinga's theory of play in *Homo ludens* not only opposed Schmitt but also served as an anthropological defence for the democratic world that, however young, was no longer promised. The case of von Leers was Huizinga's first personal encounter with this new political reality, and it is more than telling that his theory of play changed after this event. Von Leers revealed democracy's stoic virtues precisely by defying them. Only during its *looming demise* did democracy appear worth preserving.

4. The difference between Huizinga's *homo ludens* and Schmitt's *homo homini lupus* stretched beyond a moral and anthropological dispute into academic debates of a different kind, too. Huizinga's experience of 'democracy's' erosion mobilized not only moral convictions and anthropological theories but affected the cultural history he wrote. This was exemplified in this chapter by juxtaposing Schmitt's and Huizinga's understanding of seventeenth-century Dutch maritime culture. Here the once morally and politically informed considerations of their respective theories of agonism and play spilled over into analyses of a less straightforwardly political kind. In De Vlieger's painting *Calm Water* (1640–50), Huizinga and Schmitt each found a very different history being depicted. Whilst Schmitt saw a world of war and conquering projected onto the seventeenth-century Dutch square rig sails, Huizinga saw in them a world of co-operation and equality. The German occupation under which the Netherlands had been since May 1940 was repugnant to the Dutch national-historical identity, and only in democracy's *autumntide* did this appear to be the case to Huizinga.

Bibliography

Anchor, R. 'History and Play: Johan Huizinga and His Critics'. *History and Theory* 17, no. 1 (1978): 63–93.

Arendt, H. 'Understanding and Politics'. In *Essays in Understanding: 1930–1954*, 307–27. New York: Schocken, 2005.

Boterman, F. *Tussen utopie en crisis: Nederland in het interbellum 1918–1940*. Amsterdam: De Arbeiderspers, 2021.

Burckhardt, J. 'Der Koloniale und Agonale Mensch'. In *Griechische Kulturgeschichte: Der griechische Mensch*, edited by R. Marx, 46–108. Leipzig: Alfred Kröner Verlag, 1929.

Campbell, I. C. 'The Lateen Sail in World History'. *Journal of World History* 6, no. 1 (1995): 1–23.

Colie, R. L. 'Johan Huizinga and the Task of Cultural History'. *The American Historical Review* 69, no. 3 (1964): 607–30.

Curtius, E. 'Der Wettkampf'. In *Althertum und Gegenwart*, 132–47. Berlin: Wilhelm Hertz, 1875.

Dyzenhaus, D. 'Legal Theory in the Collapse of Weimar : Contemporary Lessons?' *The American Political Science Review* 91, no. 1 (1997): 121–34.

Edwards, J. 'Play and Democracy: Huizinga and the Limits of Agonism'. *Political Theory* 41, no. 1 (2013): 90–115.

Gafijczuk, D. 'Dwelling Within: The Inhabited Ruins of History'. *History and Theory* 52, no. 2 (2013): 149–70.

Geyl, P. 'Huizinga as Accuser of His Age'. *History and Theory* 2, no. 3 (1963): 231–62.

Gombrich, E. 'Huizinga's *Homo ludens*'. In *Johan Huizinga 1872–1972*, edited by W. R. H. Koops, E. H. Kossmann, and G. van der Plaat, 133–54. The Hague: Martinus Nijhoff, 1973.

Gossman, L. *Basel in the Age of Burckhardt: A Study in Unseasonable Ideas*. London/Chicago: University of Chicago Press, 2000.

Hanssen, L. *Huizinga en de troost van de geschiedenis: Verbeelding en rede*. Amsterdam: Uitgeverij Balans, 1996.

Hartmans, R. *Schaduwjaren*. Utrecht: Omniboek, 2018.

Heerikhuizen, A. van. 'Johan Huizinga en "Het lot van het boek *Homo Ludens*"'. In *De regels en het spel: Opstellen over recht, filosofie, literatuur en geschiedenis, aangeboden aan Tom Eijsbouts*, edited by J. H. Reestman, A. Schrauwen, M. van Montfrans, and J. Jans, 153–65. The Hague: T.M.C. Asser Press, 2011.

Huizinga, J. *Holländische Kultur. Das siebzehnten Jahrhundert*. Jena: Eugen Diederichs Verlag, 1933.

Huizinga, J. *De hand van Huizinga*. Edited by W. Otterspeer. Amsterdam: Amsterdam, 2009.

Kaegi, W. *Das historische Werk Johan Huizingas*. Leiden: Leiden University Press, 1947.

Kennedy, J. C. *Een beknopte geschiedenis van Nederland*. Amsterdam: Prometheus, 2018.

Klijn, E., and R. te Slaa. *De NSB: Twee werelden botsen, 1936–1940*. Amsterdam: Boom, 2021.

Kolff, D. H. A. 'Huizinga's proefschrift en de stemmingen van Tachtig'. *BMGN – Low Countries Historical Review* 104, no. 3 (1989): 380–92.

Krul, W. E. *Historicus tegen de tijd: Opstellen over het leven en werk van J. Huizinga*. Groningen: Historische Uitgeverij Groningen, 1990.

Kussbach, E. 'Recht und Kultur: Der Rechtsbegriff bei Huizinga'. *Archiv für Rechts- und Sozialsphilosophie* 54, no. 2 (1968): 179–216.

Lem, A. van der. *Johan Huizinga: Leven en werk in beelden en documenten*. Amsterdam: Wereldbibliotheek, 1993.

Lem, A. van der. *Het eeuwige verbeeld in een afgehaald bed: Huizinga en de Nederlandse beschaving*. Amsterdam: Wereldbibliotheek, 1997.

Linmans, W. *De oorlog van morgen: Nederlandse beeldvorming van een volgende oorlog 1918–1940*. Amsterdam: Prometheus, 2021.

Locher, Th. J. G. 'Johan Huizinga'. *Jaarboek van de Maatschappij der Nederlandse Letterkunde* (1946): 88–109.

Mak, G., J. Bank, G. van Es, P. de Rooy, and R. van Stipriaan. *Verleden van Nederland: Een nieuwe geschiedenis*. Amsterdam/Antwerp: Atlas Contact, 2021.

McCormick, J. P. "Legal Theory and the Weimar Crisis of Law and Social Change." In *Weimar Thought: A Contested Legacy*, edited by P. E. Gordon and J. P. McCormick, 55–72. Princeton/Oxford: Princeton University Press, 2013.

Nietzsche, F. 'Ecce Homo'. In *Friedrich Nietzsche: Werke in drei Bänden. II*, edited by K. Schlechta, 1065–159. Munich: Carl Hanser Verlag, 1954.

Nietzsche, F. 'Homers Wettkampf'. In *Friedrich Nietzsche: Werke in drei Bänden. III*, edited by K. Schlechta, 291–99. Munich: Carl Hanser Verlag, 1954.

Nietzsche, F. 'Homer's Contest'. In *Nietzscheana* 5, edited by C. Davis Acampora. Urbana, IL: North American Nietzsche Society 1996.

NSB. *Nationaal Socialistische Beweging in Nederland: Toelichting van het programma*. Utrecht: s.n., 1932.

Ortega y Gasset, J. *The Revolt of the Masses*. London: George Allen & Unwin LTD, 1932.

Otterspeer, W. *Orde en trouw: Over Johan Huizinga*. Amsterdam: De Bezige Bij, 2006.

Otterspeer, W. *Het horzelnest: De Leidse Universiteit in oorlogstijd*. Amsterdam: Prometheus, 2019.

Pree, C. du. *Johan Huizinga en de bezeten wereld*. Leusden: ISVW Uitgevers, 2016.

Schmitt, C. *Der Begriff des Politischen*. Munich/Leipzig: Verlagen von Duncker & Humblot, 1932.

Schmitt, C. *The Nomos of the Earth*. Edited by G. L. Ulmen. New York: Telos Press, 2006.

Schmitt, C. *The Concept of the Political*. Edited by G. Schwab. Chicago: The University of Chicago Press, 2007.

Schmitt, C. *Land und Meer/Land en Zee*. Edited by Henry van Sanderburg. Groningen: Blauwe Tijger, 2017.

Slaa, R. te, and E. Klijn. *De NSB: Ontstaan en opkomst van de Nationaal-Socialistische Beweging, 1931–1935*. Amsterdam: Boom, 2021.

Strong, T. B. 'Foreword: Dimensions of the New Debate around Carl Schmitt'. In *The Concept of the Political*, 3–16. Chicago: The University of Chicago Press, 2007.

Strupp, C. *Johan Huizinga. Geschichtswissenschaft als Kulturgeschichte*. Göttingen: Vandenhoeck & Ruprecht, 2000.

Thys, W. *André Jolles (1874–1946) 'Gebildeter Vagant': Briefe und Dokumente*. Amsterdam/Leipzig: Amsterdam University Press/Leipziger Universitätsverlag GmbH, 2000.

Vermeulen, E. E. G. 'Huizinga over de wetenschap der geschiedenis: Zijn positie als theoreticus van de geschiedenis tussen Fruin en Romein'. *Tijdschrift Voor Filosofie* 24, no. 1 (1962): 81–90.

Vugt, G. de. 'Philia and Neikos: Huizinga's "Auseinandersetzung" with Carl Schmitt'. *World Literature Studies* 9, no. 1 (2017): 48–59.

Conclusion: In the Image of Loss

Abstract
This concluding chapter presents the book's most central contention: by examining the various ways in which 'experiences of loss' informed Huizinga's work, this research shows not only that Huizinga's histories responded to the challenges of his own times. Through an illustration of how experiences migrated into his historical output, the work also shows that Huizinga's historical investigations became a way of living through sorrow and uncertainty. Huizinga's historical output responded to the experiences by narrating them in a historically meaningful way. As such, his histories were not only passive recipients of experiences but also transformed his present and became a way of life in times of loss.

Keywords: Johan Huizinga; interwar culture; cultural history; historical experience; history as a way of life

How do change, uncertainty, sorrow and experiences of loss tie into the way 'the past' appears to us? How do the rapid transformations of cities, ways of travel, communication and the organization of labour inform the shape and content of 'history'? And what do these ties tell us about the location of 'the past' and the roles its investigation may have in our lives? These broader questions lay in silence behind those two more specific ones to which this book has been devoted first and foremost.

1. What role did experiences of loss play in the cultural-historical works of Johan Huizinga (1872–1945)?
2. What image of Huizinga the author can one extract accordingly?

The previous chapters have, each in turn, insisted on answering the first question. They have identified, traced and weighed Huizinga's 'experiences of loss' and their respective parts in his academic output. Each chapter has shown how Huizinga's lived experiences of loss migrated time and again from his personal writings to his historical works. This concluding chapter

Rydin, T., *The Works and Times of Johan Huizinga (1872–1945): Writing History in the Age of Collapse.* Taylor & Francis Group 2023
DOI: 10.5117/9789463724593_CONC

lists and discusses these experiences and their resonances again, but only briefly. The main purpose of this chapter is to address the second question by inducing from these observations an image of Huizinga the author. Thus, this chapter can finally return explicitly to the two 'images' of Huizinga that this book aimed to critically assess: that of Carla du Pree and that of Willem Otterspeer. Only now, when the experiences of loss discussed in previous chapters can be lined up and compared, can their 'images' be challenged on their own terms, not by launching particular empirical incongruences at them but by offering an alternative portrait. By executing this comparison, this chapter also serves this book's purpose of showcasing the potential of an 'experiential' method when studying European authors in the period of 1900 to 1940 as well as other times of rapid change.

Experiences of loss

The preceding five chapters have discussed a number of experiences and their relationship to the historical perspective in Huizinga's work. The first of these chapters examined Huizinga's critique of Burckhardt's *Renaissancebegriff* against the background of Huizinga's eye-witness accounts of Amsterdam's changing cityscapes in 1903–5. Huizinga mourned the historical heritage that was lost and the carelessness with which such sights were erased. Regarding the city's aesthetic transformation, Huizinga sympathized deeply with his friend and author Jan Veth, who warned against 'the tyranny of the present'.[1] In order to remain truly creative, Veth argued, the existence of certain codes, rules and traditions is necessary, and not only from a pedagogical perspective. Without the preservation of certain norms and codes, all that is created will at once be destroyed; creativity must always contain at least an element of re-creation. Huizinga agreed wholeheartedly, and against this background, Burckhardt's *uomo unico* – a courageous yet beastly, divine though daunting iconoclast – seemed not only morally repugnant but also historically and anthropologically unconvincing; this was not and could not have been how the fantastic creativity of the Renaissance had come about, Huizinga held. The experience of a loss of heritage in his own world produced a moral and anthropological predisposition towards a cyclical and traditionalist appreciation of human artistry.

Huizinga's appreciation of and trust in historical reproduction soon conceded a serious blow, however, as the third chapter has shown. When

1 J.Veth, 'Vredes-verwoestingen', *De Gids* 80, no. 1 (1916): 24.

Huizinga opened the envelope sent to him by Hoste in 1918 and pulled out the photographs of what became of Ypres after the Battle of Passchendaele, he felt convinced that 'the catastrophe' was 'irreparable'.[2] At this time, Huizinga was still suffering the heartaches of his wife Mary's premature death in 1914, and in the following years, these experiences of personal and geopolitical catastrophe mediated a new image of the past as something irretrievably lost. Huizinga's methodological writings mirrored this transition. Before the war, Huizinga had objected to Karl Lamprecht's historiographical anti-aestheticism. Then, Huizinga had argued that such aestheticism and imagination was not only permissible but even necessary to retrieve 'objective truth'.[3] After the war, on the other hand, Lamprecht's *Methode* was faulted less for its alleged empirical insensitivity than for its supposed inability to cultivate a virtuous frame of mind. History, Huizinga started arguing during the war, first and foremost had the ability and purpose to inspire a stoic spirit. The loss of his wife and Ypres was mirrored in his ensuing critiques of Lamprecht and in the very value Huizinga attributed to historical research more generally.

Only a couple of months before the Battle of Passchendaele had wreaked havoc over Ypres, the United States announced its military intervention in the Great War. From that moment on, Huizinga started preparing course material on American culture and history, and his first book on the United States grew from this material. To Huizinga, 'America' was synonymous with an impending masculine world of mechanics and management, commercialized sports and arts, popular democracy, instrumental ethics and 'frontier' thought[4] – and now American 'mechanization' was on its way to Europe.[5] Against this background, Huizinga launched his version of American history as a criticism of Tocqueville's proto-sociological understanding of American democracy. Tocqueville, he held, had been too occupied with the social realm to witness the 'revenge' of technology that was being exacted on its human artificer in America.[6] In this critique, Huizinga mobilized his own experiences with managerialism, popular democracy and the industrialization of both urban and rural life. The 'loss' that America symbolized drew from the 'loss' Huizinga experienced in his own time. The mechanical present mediated an image of the past that was both non-mechanical yet destined

2 BW I: Huizinga–Hoste (1918): 205. See p. 11 for further details on the references to Huizinga's collected works and letters as well as to the Huizinga archives.

3 VW VII: *Het aesthetische bestanddeel van geschiedkundige voorstellingen* (1905): 26–27.

4 VW V: *Mensch en menigte* (1918): 292–35.

5 VW V: *Mensch en menigte* (1918): 290.

6 'wraak'. VW V: *Mensch en menigte* (1918): 292.

to become mechanized. The past became a silence that had drowned in the whir of thumbing cylinders. An 'American' future mediated a 'European' history.

The Great War also figured in a third and different kind of front in Huizinga's life and work: the European internationalism of the 1890s and 1900s in which Huizinga had matured, which had been ground to a halt by the war as international travel and communication collapsed. The ensuing sense of disillusionment, the fifth chapter has shown, was in part a generational one. To Huizinga's generation of academics and educated elites, international correspondence and conference networks, travel and cultural exchange had become part of a self-evident fabric of life. The challenges posed to this way of life by the war made urgent and distinct the ideals and virtues that had formerly seemed a given. In this light, the Peace Palace and Hugo de Groot soon symbolized the need for historical authorities, and against this background, Spengler's critique of the values of 'civilitas' and the world citizen was repugnant to Huizinga from the outset.[7] To Huizinga, Spengler's historical determinism symbolized a threat to the norms of an international world of cooperation and exchange. Spengler's book was an assault on intellectual modesty, love and duty. *Zivilisation* was not the death of *Kultur* – on the contrary, it was its historical, anthropological and moral bedrock. However, only after the war and the temporary demise of internationalism did the past appear to Huizinga as such.

In the 1930s, the term 'culture' became politicized in yet another way. Throughout the 1920s, Huizinga had been suspicious and sceptical of all things 'democratic', which he deemed to be part of a wider commercialization and Americanization of Dutch and other European societies. His antipathy evolved into sympathy only when the Dutch national-socialist party achieved a significant result in the 1935 provincial elections. From then on, Huizinga started drafting anew his definitions of culture to accommodate his newfound appreciation of democracy and its historical trajectory. A compelling example of these renegotiations can be found in Huizinga's critique of Carl Schmitt's theory of cultural agonism in *Der Begriff des Politischen* (1927). Schmitt had become the NSDAP's crown jurist in 1933, and after the 1935 election results, Huizinga took up his pen against Schmitt's understanding of the relation between rule, virtue, creativity, 'culture' and *homo homini lupus*. What followed were the writings and reflections that would later partly grow into *Homo ludens* (1938). Compared to his theory of play from 1933, Huizinga's post-1935 understanding laid a new and substantial emphasis on

7 VW VII: *Der Mensch und die Kultur* (1938): 452.

'order' (instead of style) and 'resignation' (instead of struggle).[8] A looming loss of democratic order reorganized Huizinga's moral-historical perspective on culture.

The experiences above are not altogether congruent with one another, neither in their anatomy nor in their effect. The changing cityscapes, for one, strengthened Huizinga's belief in the importance of historical reconstruction, whilst the sights of Ypres in 1918 shattered Huizinga's optimism regarding history's retrievability. Still, all these experiences share at least one feature: in Dariusz Gafijczuk's terminology, they cast the world in the shape of an 'afterlife'.[9] Experiences of loss and shock, often following Huizinga's direct encounter with upsetting scenes, had the potential to transform objects and events of scientific study into 'frontiers' where 'a present has partially collapsed under its own weight.'[10] The rapid transformation of cityscapes in the 1900s, Hoste's photographs of Ypres in 1918, the halt of international academic communication in 1914–18 and again after 1933: Huizinga reported on such events and developments in metaphors, tropes and narratives that reappeared later in his historiographical writing. To be absolutely clear, I here do not mean to imply in any way a direct and exhaustive causal relation between experience and view, but I do wish to emphasize a relation of repeated congruence: the image of the world having 'expired', both, and often first, in the experiential domain and again, often later, in the historical domain.[11] The worlds Huizinga had grown up in were receding, and the histories he wrote followed suit. So what do these catastrophes teach us about the author that Huizinga was? What does loss tell us about the history he wrote?

Writing in the image of loss: a way of life

'History' appeared to Huizinga as experiences of loss in his own time. Huizinga's narration of, say, the late Middle Ages was the narration of decadence and mechanization in his own day. American history became urgent to Huizinga only when its trajectory symbolized Europe's historical demise. Dutch history appeared to Huizinga as democratic only when

8 VW V: *Homo ludens* (1938): 37–41

9 D. Gafijczuk, 'Dwelling Within: The Inhabited Ruins of History', *History and Theory* 52, no. 2 (2013): 167.

10 Gafijczuk, 'Dwelling Within: The Inhabited Ruins of History', 164.

11 Gafijczuk, 'Dwelling Within: The Inhabited Ruins of History', 169.

democracy seemed to be at risk. At such moments, the historical periods under consideration, however long ago they may have been, were ended anew in the present, and in effect, these periods were reconstructed again and in opposition to the loss that seemed to define Huizinga's present. For example, the urban modernization projects of 1890–1910 mediated a new ending to the Middle Ages, and consequently, the early 1900s elicited their own 'playful' medieval times by ending them anew in a period of perceived instrumental calculation.

In Gafijczuk's words, Huizinga's images of history emerged the moment that 'the presence' was no longer associated with 'consecutive forms'; it was in moments of 'rupture' that Huizinga's images of history were 'evoked.'[12] This relation to rupture – this 'logic of fragmentation' – conditioned a 'history' that was dialectically opposed to those concerns associated with 'contemporary life'.[13] The 'experiences of loss' enabled an understanding of 'history' that could, in turn, become a refuge away from sorrow and panic. History, in this sense, could become a means of coming to terms with loss, of narrating collapse by offering a space that presented itself as 'ahistorical'. As such, this 'refuge' was not a negation of life and its trials. On the contrary: because this 'history' opened up a world that was dialectically opposed to contemporary life, 'history' could itself become a way of living. Regarding this point, Gafijczuk cites from Georg Simmel's essay 'The Problem of Historical Time' ('Das Problem der historischen Zeit') from 1916:

> Although the distance between life and history may remain an ultimate epistemological dichotomy, it is not an ultimate metaphysical postulate. This is because, in the final analysis, history is also an expression and an act of life, precisely the same life to which it is originally juxtaposed. The juxtaposition or contrast to life is also a form of life.[14]

Huizinga's history was narrated in terms of loss, commonly drafted in a language congruent with the language in which his experiences of loss in his own lived world were drafted. In effect, Huizinga's historical Rome, medieval Burgundy and seventeenth-century Netherlands were assembled from precisely those features that were being expelled from his contemporary world. These ages were silent, balanced, patient and at peace with the inevitable tragedy that awaits all humans and human artifices. In this sense,

12 Gafijczuk, 'Dwelling Within: The Inhabited Ruins of History', 157, 165.
13 Gafijczuk, 'Dwelling Within: The Inhabited Ruins of History', 151.
14 As cited in Gafijczuk, 'Dwelling Within: The Inhabited Ruins of History', 164.

the historical world was, as described by Simmel above, juxtaposed with his contemporary life, not so as to hide from life but so as to offer a virtuous realm from which the present could be met once again. History, in this sense, was a way of resisting the world's freneticism, a way of not letting oneself be thrown around by the passions of opportunity. Gafijczuk's description of this 'modern history' applies well to my understanding of Huizinga's history:

> This is history as life, one that is able to rebel against any form of presence if necessary, not history as the dialysis machine for time.[15]

Huizinga's history could become a way of 'dealing' with and 'resisting' loss. It could be a means of cultivating virtues that might have been at odds with practicalities and fashion but that, precisely because of its independence, allowed for a 'serene' and 'balanced' mind. History understood in this way was not a linear space through which events unfolded (the 'dialysis machine for time'), but it was a breaking-away from the present. In Simmel's words: the ensuing juxtaposition of history and present became itself 'a form of life'. Huizinga's history – the research, writing and reading – became a way of leading a 'virtuous' life, of surviving tumultuous times.

Drawing from these two features of Huizinga's history – its inception in loss while also being a way of living with loss – I suggest extracting the following image of Huizinga the historian: to Huizinga, 'history' was a 'way of life', by which I, drawing from Pierre Hadot (1922–2010), mean to say that Huizinga's 'history' turned historical investigation into 'a mode of existing-in-the-world, which had to be practiced at each instant, and the goal of which was to transform the whole of the individual's life'.[16] His history became a way of setting a moral example for himself, of retrieving confidence and serenity in ideals, of arresting himself when passions ran wild, of stepping outside of the practical concerns of daily life and of turning to a historically retrieved space of non-functional ideals, rituals and virtues. 'History' was a pedagogical practice of his mind. His works were not just excursions into the ways of life in bygone times; Huizinga's 'histories', both the research and the resulting works, were themselves ways of life amidst experiences of a present risking a fall into an irretrievable past. His works were his lived version of *memento mori*, and in effect, these works became his signposts for how to cultivate virtues of persistence, duty, love and

15 Gafijczuk, 'Dwelling Within: The Inhabited Ruins of History', 170.
16 P. Hadot, *Philosophy as a Way of Life: Spiritual Exercises from Socrates to Foucault*, ed. A. I. Davidson (Oxford: Blackwell, 1995), 265.

curiosity – that is, virtues resisting opportunistic action and fashionable whim, the virtues required for resigned serenity amidst loss and change.

By making this point, I am not suggesting that Huizinga consciously and intentionally assembled his works for the purpose of self-cultivation, though I do not doubt that this occasionally may have been the case later in his life. My point is, rather, that 'history' could become 'a way of life' to Huizinga because his images were mediated through the 'experiences of loss' discussed here. In this sense, his works were more commonly authored not so much by a weighing, self-reflexive mind as they were conditioned by the experiences of loss that shook and reorganized his perspectives in the 1900s and 1910s. Caught between 'experience' and his written output, Huizinga the author only later and gradually developed a reflexive understanding of 'history' in terms of its self-cultivating potential. Huizinga's virtue-ethical understanding of history came only towards the late 1920s and early 1930s, by which time he had come to consider the act of writing history one among many human acts of 'playfulness'. At this point, he had already written what would later prove to be most of his historical works. His historical theory was a retrospective reflection on the historical research he had conducted in the 1900s, 1910s and 1920s. Against this background, the historical works written by Huizinga, including their moral-political content, should not be considered merely the fruits of a psychological being, of character traits and personal inclinations. Huizinga's history had to engage with experiences of loss to realize its self-cultivating potential.

With the image of 'history as a way of life' in place, I now turn to its comparison with those two other images of Huizinga discussed above and in the introduction: those of Carla du Pree and Willem Otterspeer. Their differences notwithstanding, both accounts, I have argued, draft an understanding of Huizinga's histories in terms of his character traits. In the case of Du Pree's account, Huizinga's social role as a public intellectual is understood in terms of his *Zivilcourage*, namely his ability to dutifully carry out his tasks *contrecoeur*. She mobilizes this and other character traits to explain his behaviour:

> The means of public performance did not suit Huizinga's person, convictions and character. He was too much an individualist to this end. [...] But when push came to shove, he had the courage to opt as an individual for the choice deemed ethical by him, in his written works and lectures, but also, when necessary, to set an example.[17]

17 'Het middel van de publieke actie paste niet bij de persoon Huizinga, bij zijn overtuigingen en karakter. Daar was hij te veel individualist voor. [...] Maar als het erop aankwam had hij de

Du Pree's Huizinga is a psychological being navigating a socially conditioned world. At times, these social conditions mediated his beliefs and public person, but most commonly, the social conditions prescribed his instruments and the expectations of him (held by both himself and others) rather than his actual beliefs. Otterspeer's Huizinga, on the other hand, was an even less socially conditioned being:

> Huizinga's anti-modernism cannot be understood without grasping his view of humanity. This view was determined in large measure by the role he accorded love and loyalty, friendship and faithfulness.[18]

Otterspeer's Huizinga had views that resulted primarily from the virtues particularly dear to him. One among many reasons adduced by Otterspeer to this end is Huizinga's self-proclaimed interest in virtues: 'for over forty years,' Huizinga wrote in 1943, 'I have believed that virtues and vices' are 'now as much as 2000 years ago among the most precious means of thought when it comes to theories of the soul and human mores.'[19] Otterspeer takes this and similar passages most seriously: his Huizinga was through-and-through defined by those character traits he deemed virtuous.

Though Du Pree's and Otterspeer's accounts of Huizinga rest on a careful, wide and contextualist reading, their shared interest in character invites a question of primary importance to a properly critical understanding of Huizinga: Why? Why were these virtues important to him? Why did he care so much about love and temperance, duty and justice? Both their accounts, I find, do not push hard enough. Both accounts interpret Huizinga's words on their terms and with great respect. This book has shown that certain convictions, narratives and even virtues became urgent to Huizinga only after experiences of loss. Huizinga's character – as much as any other author's

moed om als invidividu de voor hem ethisch juiste keuze te maken, in geschriften en lezingen, maar ook door indien nodig een daad te stellen.' C. du Pree, *Johan Huizinga en de Bezeten Wereld* (Leusden: ISVW Uitgevers, 2016), 250.

18 'Wie Huizinga als antimodernist wil begrijpen, moet zich verdiepen in het mensbeeld dat hij erop nahield. Dat werd in hoge mate bepaald door de rol die emoties als liefde en loyaliteit, vriendschap en trouw, erin speelden.' W. Otterspeer, *Orde en trouw: Over Johan Huizinga* (Amsterdam: De Bezige Bij, 2006), 227. W. Otterspeer, *Reading Huizinga*, ed. B. Jackson. (Amsterdam: Amsterdam University Press, 2010), 230.

19 'Het is al meer dan veertig jaar mijn overtuiging en ik heb die meer dan eens geuit, dat deze reeksen van deugden en ondeugden, of men er dan van beide zeven telt of acht, een van onze kostbaarste denkmiddelen beteekenen, heden nog evengoed als voor tweeduizend jaar, voor de kennis van alles wat de zielkunde en de zedenleer van den mensch betreft.' VW VII: *Geschonden wereld* (1945): 574. Cited in Otterspeer, *Orde en trouw: Over Johan Huizinga*, 229.

– should not be edified, and an experiential approach has the advantage of stretching out an individual beyond their psychological boundaries into a wider social and cultural life without deflating a person's features with the sharp corners of structures.

Huizinga's celebration of modesty, duty and love and his appreciation of tragedy were extensions of and responses to his experiences of loss – the uncompromising, inevitable and irreparable losses that made up both his personal life and his social world in transition. These inclinations, in turn, took shape in what I have described as his historical 'ways of life': his research became a ritual and moral-pedagogical device in dealing with tragic decline that, in Huizinga's understanding, awaits all human artifice; his histories were monuments for those virtues that depend not on worldly opportunism and calculation but on timeless ideals that are independent of whim, fashion and function. These written works, in turn, became spaces wherein Huizinga could take refuge from overwhelming sensations of loss and shock, wherein he could find stability amidst passions, a bannister when he trembled. From these works, he could muster the inspiration to offer resistance, time and again, to the shockwaves dominating his day and age. In the face of loss, Huizinga's history was 'designed to raise [us] from the temporal and the visible to an apprehension of the eternal and invisible'.[20]

Bibliography

Augustine, St. *City of God*. Edited by D. Knowles. London: Penguin Books, 1972.

Gafijczuk, D. 'Dwelling Within: The Inhabited Ruins of History'. *History and Theory* 52, no. 2 (2013): 149–70.

Hadot, P. *Philosophy as a Way of Life: Spiritual Exercises from Socrates to Foucault*. Edited by A. I. Davidson. Oxford: Blackwell, 1995.

Otterspeer, W. *Orde en trouw: Over Johan Huizinga*. Amsterdam: De Bezige Bij, 2006.

Otterspeer, W. *Reading Huizinga*. Edited by B. Jackson. Amsterdam: Amsterdam University Press, 2010.

Pree, C. Du. *Johan Huizinga en de bezeten wereld*. Leusden: ISVW Uitgevers, 2016.

Veth, J. 'Vredes-verwoestingen'. *De Gids* 80, no. 1 (1916): 507–27.

20 St Augustine, *City of God*, ed. D. Knowles (London: Penguin Books, 1972), X.16.14.

Bibliography

Cited works by Huizinga

From J. Huizinga, Verzamelde werken I–IX, ed. L. Brummel, W. R. Juynboll, and Th. J. G. Locher (Haarlem: H. D. Tjeenk Willink & Zoon N.V., 1949–53):

VW I: *De Vidûshaka in het Indisch toneel* (1897)
VW I: *Iets over de waardeering der Indische letterkunde* (1898)
VW I: *Over studie en waardeering van het Buddhisme* (1903)
VW I: *De opkomst van Haarlem* (1905)
VW I: *Over de oudste geschiedenis van Haarlem* (1907)
VW I: *Mijn weg tot de historie* (1947)
VW II: *De beteekenis van 1813 voor Nederland's geestelijke beschaving* (1913)
VW II: *Grotius' plaats in de geschiedenis van den menschelijken geest* (1925)
VW II: *Hugo de Groot en zijn eeuw* (1925)
VW II: *Terugblik op Nederlands groei in de veertig jaren van het regeeringsjubileum* (1938)
VW II: *Nederland's beschaving in de zeventiende eeuw* (1941)
VW III: *Herfsttij der Middeleeuwen* (1919)
VW IV: *Over historische levensidealen* (1915)
VW IV: *Het probleem der Renaissance* (1920)
VW IV: *Twee worstelaars met den engel* (1921)
VW IV: *Erasmus* (1926)
VW IV: *Patriotisme en nationalisme in de Europeesche geschiedenis tot het einde der negentiende eeuw* (1941)
VW V: *Mensch en menigte in Amerika* (1918)
VW V: *Amerika levend en denkend* (1926)
VW V: *Over de grenzen van spel en ernst in de cultuur* (1933)
VW V: *Homo ludens* (1938)
VW VI: Artikel in NRC, *Jan Veth* (1924)
VW VI: *Jan Veth* (1927)
VW VI: *Leven en werk van Jan Veth* (1927)
VW VI: *Mr C. van Vollenhoven* (1933)
VW VII: *Het aesthethische bestanddeel van geschiedkundige voorstellingen* (1905)
VW VII: *Boekbespreking van K. Lamprecht's* Einführung in das historische Denken (1913)
VW VII: *Taak en termen der beschavingsgeschiedenis* (1926)
VW VII: *De taak der cultuurgeschiedenis* (1929)

VW VII: *Over een definitie van het begrip geschiedenis* (1929)
VW VII: *In de schaduwen van morgen* (1935)
VW VII: *Nederland's geestesmerk* (1935)
VW VII: *De wetenschap der geschiedenis* (1937)
VW VII: *Kort begrip van den aard der geschiedenis als cultuurverschijnsel en als wetenschap* (1937)
VW VII: *Der Mensch und die Kultur* (1938)
VW VII: *Historische grootheid* (1940)
VW VII: *Over vormverandering der geschiedenis* (1941)
VW VII: *Geschonden wereld* (1945)
VW VIII: *Geschiedenis der universiteit gedurende de derde eeuw van haar bestaan, 1814–1914* (1914)
VW VIII: *Toespraak tot de Maatschappij der Nederlandsche Letterkunde te Leiden* (1919)

Cited and not included in the collected works:

Huizinga, J. *Holländische Kultur. Das siebzehnten Jahrhundert.* Jena: Eugen Diederichs Verlag, 1933.
Huizinga, J. 'Gebeden'. In *Mijn weg tot de geschiedenis en gebeden*, edited by A. van der Lem, 91–105. Nijmegen: Vantilt, 2016.

Translations used for reference:

Huizinga, J. *Autumntide of the Middle Ages*. Translated by D. Webb. Edited by G. Small and A. van der Lem. Leiden: Leiden University Press, 2020.

Primary literature

Berlage, H. P. *Het Pantheon der Menschheid*. Rotterdam: W. L. & J. Brusse, 1919.
Berlage, H. P., W. M. Dudok, J. Gratama, A. R. Hulshoff, H. van der Kloot Meiburg, J. F. Staal, and J. Luthmann. 'Het doel van deze uitgave'. In *Moderne bouwkunst in Nederland*, 5–7. Rotterdam: W.L. & J. Brusse's Uitgeversmaatschappij, 1932.
Brugmann, K. *Grundriss der vergleichenden Grammatik der indogermanischen Sprachen*. Strasbourg: Karl J. Trübner, 1886.
Burckhardt, J. 'Der koloniale und agonale Mensch'. In *Griechische Kulturgeschite: Der griechische Mensch*, edited by R. Marx, 46–108. Leipzig: Alfred Kröner Verlag, 1929.
Burckhardt, J. *Die Kultur der Renaissance in Italien*. Vienna: Phaidon-Verlag, 1934.

Burckhardt, J. *The Civilization of the Renaissance in Italy*. Oxford: Phaidon Press, 1981.

Curtius, E. 'Der Wettkampf'. In *Althertum und Gegenwart*, 132–47. Berlin: Wilhelm Hertz, 1875.

Doesburg, T. van, Mondriaan, P. 'Manifesto I of De Stijl (1922)'. In *100 Artists' Manifestos: From the Futurists to the Stuckists*, edited by A. Danchev. London: Penguin Books, 2011.

Dilthey, W. *Der Afbau der Geschichtlichen Welt in den Geisteswissenschaften*. Leipzig/Berlin: B. G. Teubner, 1927.

Eeden, F. van. *Dromenboek*. Amsterdam: Bert Bakker, 1979.

Eeden, F. van. *De kleine Johannes*. Edited by D. Mok. Amsterdam: Abraxas, 2009.

Grotius, H. *The Rights of War and Peace; Including the Law of Nature and of Nations*. Edited by A. C. Campbell. Washington/London: M. W. Dunne, 1901.

Hegel, G. W. F. *Lectures on the History of Philosophy*. Edited by E. S. Haldane and F. H. Simson. London: Kegan Paul, Trench, Trübner & Co., 1895.

Heijmans, B. *De Renaissance in Italië*. Zutphen: W. J. Thiemen & Cie, 1919.

Huizinga, D. 'Over vitalisme en mechanisme'. In *Handelingen van het vierde Nederlandsch Natuur- en Geneeskundig Congres*, 18–36. The Hague: H. L. Smits, 1893.

Huizinga, L. *Herinneringen aan mijn vader*. The Hague: H. P. Leopold's Uitgeversmij, 1963.

Jolles, A. *Einfache Formen*. Berlin: De Gruyter, 2006.

Kloos, W. *Nieuwere literatuur-geschiedenis. Deel 2*, of 5 vols. Amsterdam: L. J. Veen, 1925.

Lamprecht, K. *Die kulturhistorische Methode*. Berlin: R. Gaertners, 1900.

Mann, T. *Betrachtungen eines Unpolitischen*. Frankfurt: Fischer, 1974.

Mann, T. *Reflections of a Nonpolitical Man*. Edited by Morris W. D. New York: New York Review of Books, 2021.

Müller, M. *Lectures on the Science of Language*. London: Longmans, Green & Co., 1864.

Müller, M. *Biographies of Words and The Home of the Aryans*. London: Longmans, Green & Co., 1912.

Nietzsche, F. 'Ecce Homo'. In *Friedrich Nietzsche: Werke in drei Bänden. II*, edited by K. Schlechta, 1065–159. Munich: Carl Hanser Verlag, 1954.

Nietzsche, F. 'Homers Wettkampf'. In *Friedrich Nietzsche: Werke in drei Bänden. III*, edited by K. Schlechta, 291–99. Munich: Carl Hanser Verlag, 1954.

NSB. *Nationaal Socialistische Beweging in Nederland: Toelichting van het programma*. Utrecht: s.n., 1932.

Ortega y Gasset, J. *The Revolt of the Masses*. London: George Allen & Unwin Ltd., 1932.

Osthoff, H., and K. Brugmann. *Morphologische Untersuchungen auf dem Gebiete der indogermanischen Sprachen*. Cambridge: Cambridge University Press, 2013.

Pannekoek, A. 'Het historisch materialisme'. *De Nieuwe Tijd* 24, no. 2 (1919): 15–22.

Ranke, L. *The History of the Popes During the Last Four Centuries*. London: G. Bell and Sons, 1913.

Rickert, H. *Kulturwissenschaft und Naturwissenschaft*. Tübingen: Verlag von J. C. B. Mohr, 1926.

Roland Holst, H. *Tusschen twee werelden*. Rotterdam: W. L. & J. Brusse, 1923.

Roland Holst-van der Schalk, H. 'Ter gedachtenis'. In *Het Pantheon der Menschheid*, 22–23. Rotterdam: W. L. & J. Brusse, 1919.

Schmitt, C. *Der Begriff des Politischen*. Munich/Leipzig: Duncker & Humblot, 1932.

Schmitt, C. *The Nomos of the Earth*. Edited by G. L. Ulmen. New York: Telos Press, 2006.

Schmitt, C. *The Concept of the Political*. Edited by G. Schwab. Chicago: The University of Chicago Press, 2007.

Schmitt, C. *Land und Meer/Land en Zee*. Edited by Henry van Sanderburg. Groningen: Blauwe Tijger, 2017.

Spengler, O. *Der Untergang des Abendlandes: Umrisse einer Morphologie der Weltgeschichte. Erster Band*, of 2 vols. Munich: C. H. Beck, 1920.

Spengler, O. *Der Untergang des Abendlandes: Umrisse einer Morphologie der Weltgeschichte. Erster Band*, of 2 vols. 2nd ed. Munich: C. H. Beck, 1923.

Spengler, O. *Der Mensch und die Technik*. Munich: C. H. Beck, 1931.

Steding, C. *Das Reich und die Krankheit der Europäischen Kultur*. Hamburg: Hanseatische Verlagsanstalt Hamburg, 1942.

Tocqueville, A. de. *Democracy in America*. Edited by G. E. Bevan. London: Penguin Books, 2003.

Vollenhoven, C. van. *De drie treden van het volkenrecht*. The Hague: Martinus Nijhoff, 1918.

Verwey, A. *Het zwaardjaar*. Amsterdam: W. Versluys, 1916.

Veth, J. 'Vredes-verwoestingen'. *De Gids* 80, no. 1 (1916): 507–27.

Zweig, S. *Die Welt von Gestern: Erinnerungen eines Europäers*. Frankfurt: Fischer Taschenbuch Verlag, 2019.

Secondary literature

Aerts, R. 'Op zoek naar een Nederlands fin de siècle'. *De Gids*, no. 156 (1993): 91–101.

Anchor, R. 'History and Play: Johan Huizinga and His Critics'. *History and Theory* 17, no. 1 (1978): 63–93.

Ankersmit, F. R. 'Historiography and Postmodernism'. *History and Theory* 28, no. 2 (1989): 137–53.

Ankersmit, F. R. *Sublime Historical Experience*. Edited by M. Bal and H. de Vries. Stanford: Stanford University Press, 2005.

Arendt, H. 'Understanding and Politics'. In *Essays in Understanding: 1930–1954*, 307–27. New York: Schocken, 2005.

Bambach, C. R. *Heidegger, Dilthey, and the Crisis of Historicism*. Ithaca/London: Cornell University Press, 1995.

Beiser, F. C. *Weltschmerz: Pessimism in German Philosophy 1860–1900*. Oxford: Oxford University Press, 2016.

Bikker, J., G. J. M. Weber, M. E. Wieseman, and E. Hinterding. *Late Rembrandt*. London: National Gallery Company, 2014.

Bosch, M. *Strijd! De vrouwenkiesrechtbeweging in Nederland, 1882–1922*. Hilversum: Verloren, 2019.

Boterman, F. *Tussen utopie en crisis*. Amsterdam: De Arbeiderspers, 2021.

Clark, G. N. 'Preface'. In *Erasmus of Rotterdam*, translated by F: Hopman, vii–x. London: Phaidon, 1952.

Burke, P. *What Is Cultural History?* Cambridge: Polity Press, 2004.

Campbell, I. C. 'The Lateen Sail in World History'. *Journal of World History* 6, no. 1 (1995): 1–23.

Chickering, R. *Karl Lamprecht: A German Academic Life (1856–1915)*. Atlantic Highlands, NJ: Humanities Press, 1993.

Colie, R. L. 'Johan Huizinga and the Task of Cultural History'. *The American Historical Review* 69, no. 3 (1964): 607–30.

Daniel, U. *Kompendium Kulturgeschichte*. Frankfurt: Suhrkamp Verlag, 2001.

Daston, L. 'The History of Science and the History of Knowledge'. *KNOW* 1, no. 1 (2017): 131–54.

Dienstag, J. F. *Pessimism*. Princeton: Princeton University Press, 2006.

Domanska, E. 'Frank Ankersmit: From Narrative to Experience'. *Rethinking History* 13, no. 2 (2009): 175–95.

Dorléac, L. B. 'L'histoire de l'art ou l'éloge de la faiblesse'. In *L'histoire culturelle en France et en Espagne*, edited by B. Pellistrandi and J. F. Sirinelli, 77–86. Madrid: Casa de Velázquez, 2008.

Dorsman, L. J. 'Periodisering als integrale benadering: Nederlandse historici in het Fin-de-Siècle'. *Theoretische geschiedenis* 16, no. 3 (1989): 277–95.

Dumolyn, J., and É. Lecuppre-Desjardin. 'Huizinga's Silence'. In *Rereading Huizinga: Autumn of the Middle Ages, a Century Later*, edited by P. Arnade, M. Howell, and A. van der Lem, 65–84. Amsterdam: Amsterdam University Press, 2019.

Dunk, H. W. von der. 'Huizinga als kultuurpessimist'. *Groniek*, no. 26 (1973): 150–54.

Dupré, S., and G. Somsen. 'The History of Knowledge and the Future of Knowledge Societies'. *Berichte zur Wissenschaftsgeschichte* 42, no. 2–3 (2019): 186–99.

Dyzenhaus, D. 'Legal Theory in the Collapse of Weimar: Contemporary Lessons?' *The American Political Science Review* 91, no. 1 (1997): 121–34.

Edwards, J. 'Play and Democracy: Huizinga and the Limits of Agonism'. *Political Theory* 41, no. 1 (2013): 90–115.

Ellwood, D. W. 'How the American Century Started'. In *The Shock of America: Europe and the Challenge of the Century*, 22–71. Oxford: Oxford University Press, 2012.

Febvre, L. 'Un moment avec Huizinga'. *Annales. Histoire, sciences sociales* 6, no. 4 (1951): 493–96.

Ferguson, W. K. *The Renaissance in Historical Thought*. New York/Chicago/Dallas/Atlanta/San Francisco: The Riverside Press Cambridge, 1948.

Gafijczuk, D. 'Dwelling Within: The Inhabited Uirns of History'. *History and Theory* 52, no. 2 (2013): 149–70.

Geyl, P. 'Huizinga als aanklager van zijn tijd'. *Mededelingen der Koninklijke Nederlandse Akademie van Wetenschappen, Afdeling Letterkunde* 24, no. 4 (1961): 137–77.

Geyl, P. 'Huizinga as Accuser of His Age'. *History and Theory* 2, no. 3 (1963): 231–62.

Gombrich, E. *In Search of Cultural History*. Oxford: Oxford University Press, 1969.

Gombrich, E. 'Huizinga's *Homo ludens*'. In *Johan Huizinga 1872–1972*, edited by W. R. H. Koops, E. H. Kossmann, and G. van der Plaat, 133–54. The Hague: Martinus Nijhoff, 1973.

Gordon, P. E., and J. P. McCormick. 'Introduction: Weimar Thought: Continuity and Crisis'. In *Weimar Thought: A Contested Legacy*, edited by P. E. Gordon and J. P. McCormick, 1–11. Princeton/London: Princeton University Press, 2013.

Gossman, L. *Basel in the Age of Burckhardt: A Study in Unseasonable Ideas*. London/Chicago: University of Chicago Press, 2000.

Grethlein, J. 'Experientiality and "Narrative Reference", with Thanks to Thucydides'. *History and Theory* 49, no. 3 (2010): 315–35.

Guggisberg, H. R. 'Burckhardt und Huizinga – Zwei Historiker in der Krise ihrer Zeit'. In *Johan Huizinga 1872–1972*, edited by W. R. H. Koops, E. H. Kossmann, and G. van der Plaat, 155–76. The Hague: Martinus Nijhoff, 1973.

Hanssen, L. *Huizinga en de troost van de geschiedenis: Verbeelding en rede*. Amsterdam: Uitgeverij Balans, 1996.

Hartmans, R. *Schaduwjaren*. Utrecht: Omniboek, 2018.

Hathaway, O. A., and S. J. Shapiro. *The Internationalists and Their Plan to Outlaw War*. London: Penguin Books, 2018.

Heerikhuizen, A. van. 'Johan Huizinga en "Het lot van het boek *Homo Ludens*"'. In *de regels en het spel: Opstellen over recht, filosofie, literatuur en geschiedenis, aangeboden aan Tom Eijsbouts*, edited by J. H. Reestman, A. Schrauwen, M. van Montfrans, and J. Jans, 153–65. The Hague: T.M.C. Asser Press, 2011.

Hockney, D., and C. M. Falco. 'Optical Insights into Renaissance Art'. *Optics and Photonics News* 11, no. 7 (2000): 52–59.

Hoffmann, S. L. 'Koselleck, Arendt, and the Anthropology of Historical Experience'. *History and Theory* 49, no. 2 (2010): 212–36.

Hoffmann, S. L. 'Koselleck in America'. *New German Critique* 44, no. 3 (2017): 167–87.

Hoffmann, S. L. 'Rückblick auf die Menschenrechte'. *Merkur* 71, no. 812 (2017): 5–20.

Hollander, J. den, H. Paul, and R. Peters. 'Introduction: The Metaphor of Historical Distance'. *History and Theory* 50, no. 4 (2011): 1–10.

Hugenholtz, F. W. N. 'The Fame of a Masterwork'. In *Johan Huizinga 1872–1972*, edited by W. R. H. Koops, E. H. Kossmann, and G. van der Plaat, 91–103. The Hague: Martinus Nijhoff, 1972.

Huistra, P. *Bouwmeesters, zedenmeesters: Geschiedbeoefening in Nederland tussen 1830 en 1870*. Nijmegen: Uitgeverij Vantilt, 2019.

Hübinger, G. 'Konzepte und Typen der Kulturgeschichte'. In *Geschichtsdiskurs Band 4: Krisenbewußtsein, Katastrophenerfahrungen und Innovationen 1880–1945*, edited by W. Küttler, J. Rüsen, and E. Schulin. Frankfurt: Fischer Taschenbuch Verlag, 1997.

Ireland, C. *The Subaltern Appeal to Experience: Self-Identity, Late Modernity, and the Politics of Immediacy*. Montreal/London/Ithaca: McGill-Queen's University Press, 2004.

Ivry, B. 'The Forward-Looking Anachronist: Johan Huizinga and the Autumntide of the Middle Ages'. *Salmagundi*, no. 212/213 (2021): 165–193.

Jay, M. *Songs of Experience*. Berkeley/Los Angeles: University of California Press, 2005.

Jay, M. 'Historical Explanation and the Event: Reflections on the Limits of Contextualization'. *New Literary History* 42, no. 4 (2011): 557–71.

Kaegi, W. *Das historische Werk Johan Huizingas*. Leiden: Leiden University Press, 1947.

Kaminsky, H. 'From Lateness to Waning to Crisis: The Burden of the Later Middle Ages'. *Journal of Early Modern History* 4, no. 1 (2000): 85–125.

Kantorowicz, E. 'Review of Johan Huizinga, "Geschichte und Kultur"'. *American Historical Review* 60, no. 4 (1955): 853–55.

Keedus, L. *The Crisis of German Historicism*. Cambridge: Cambridge University Press, 2015.

Keizer, M. de. *Als een meeuw op de golven: Albert Verwey en zijn tijd*. Amsterdam: Prometheus, 2017.

Kelley, D. R. *Fortunes of History: Historical Inquiry from Herder to Huizinga*. New Haven/London: Yale University Press, 2003.

Kennedy, James C. 'Vooruitgang En crisis, 1870–1949'. In *Een beknopte geschiedenis van Nederland*. Amsterdam: Bert Bakker, 2017.

Kolff, D. H. A. 'Huizinga's proefschrift en de stemmingen van Tachtig'. *BMGN – Low Countries Historical Review* 104, no. 3 (1989): 380–92.

Koops, W. R. H., E. H. Kossmann, and A. van der Lem, eds. 'Johan Huizinga 1872–1972'. The Hague: Martinus Nijhoff, 1973.

Kooy, P. 'Groningen en de industrialisatie van Nederland'. *Groniek*, no. 64 (1979): 13–22.

Koselleck, R. '"Space of Experience" and "Horizon of Expectation": Two Historical Categories'. In *Futures Past: On the Semantics of Historical Time*, edited by K. Tribe, 255–75. New York: Columbia University Press, 2004.

Kossmann, E. H. 'Postscript'. In *Johan Huizinga 1872–1972*, edited by W. R. H. Koops, E. H. Kossmann, and G. van der Plaat, 223–34. The Hague: Martinus Nijhoff, 1973.

Kossmann, E. H. 'Romeins "Breukvlak" en de Nederlandse geschiedenis'. *BMGN – Low Countries Historical Review* 106, no. 4 (1991): 652–58.

Kossmann, E. H. 'De waardering van Rembrandt in de Nederlandse "traditie"'. *Oud Holland* 106, no. 2 (1992): 81–93.

Krul, W. E. 'Ter Braak contra Huizinga. Over de grenzen van de ironie'. *Groniek*, no. 100 (1988): 61–79.

Krul, W. E. 'Het leven der woorden. Taalkunde en geschiedenis in Huizinga's vroegste wetenschappelijke werk'. *BMGN – Low Countries Historical Review* 104, no. 3 (1989): 365.

Krul, W. E. *Historicus tegen de tijd: Opstellen over het leven en werk van J. Huizinga*. Groningen: Historische Uitgeverij Groningen, 1990.

Krul, W. E. 'Huizinga in zijn brieven: Beeld en zelfbeeld'. *Tijdschrift voor Geschiedenis* 106, no. 1 (1993): 23–37.

Krul, W. E. 'In the Mirror of Van Eyck: Johan Huizinga's "Autumn of the Middle Ages"'. *Journal of Medieval and Early Modern Studies* 27, no. 3 (1997): 353–84.

Krul, W. E. 'Het *fin-de-siècle* als vertoning'. *BMGN – Low Countries Historical Review* 117, no. 4 (2002): 519–25.

Krumm, C. *Johan Huizinga, Deutschland und die Deutschen*. Münster/New York/Munich/Berlin: Waxmann, 2011.

Kussbach, E. 'Recht und Kultur: Der Rechtsbegriff bei Huizinga'. *Archiv für Rechts- und Sozialsphilosophie* 54, no. 2 (1968): 179–216.

Lammers, A. *Uncle Sam en Jan Salie: Hoe Nederland Amerika ontdekte*. Amsterdam: Uitgeverij Balans, 1989.

Lem, A. van der. *Johan Huizinga: Leven in werk in beelden en documenten*. Amsterdam: Wereldbibliotheek, 1993.

Lem, A. van der. *Het eeuwige verbeeld in een afgehaald bed: Huizinga en de Nederlandse beschaving*. Amsterdam: Wereldbibliotheek, 1997.

Lem, A. van der. 'The Making of "The Autumn of the Middle Ages" II: The Eagle and His Pigeonholes: How Huizinga Organized His Sources'. In *Rereading Huizinga: Autumn of the Middle Ages, a Century Later*, edited by P. Arnade, M. Howell, and A. van der Lem, 211–26. Amsterdam: Amsterdam University Press, 2019.

Lem, A. van der. *Verbreek nooit een oude rooilijn! Huizinga over stadhuis en stadsschoon in Leiden*. Leiden: Antiquariaat Klikspaan, 2021.

Lente, D. van. 'Huizinga's Children: Play and Technology in Twentieth Century Dutch Cultural Criticism (from the 1930s to the 1960s)'. *Icon*, no. 19 (2013): 52–74.

Linmans, W. *De oorlog van morgen: Nederlandse beeldvorming van een volgende oorlog 1918–1940*. Amsterdam: Prometheus, 2021.

Locher, Th. J. G. 'Johan Huizinga'. *Jaarboek van de Maatschappij der Nederlandse Letterkunde* (1946): 88–109.

Mah, H. 'The Predicament of Experience'. *Modern Intellectual History* 5, no. 1 (2008): 97–119.

Mak, G., J. Bank, G. van Es, P. de Rooy, and R. van Stipriaan. *Verleden van Nederland: Een nieuwe geschiedenis*. Amsterdam/Antwerp: Atlas Contact, 2021.

McCormick, J. P. 'Legal Theory and the Weimar Crisis of Law and Social Change'. In *Weimar Thought: A Contested Legacy*, edited by P. E. Gordon and J. P. McCormick, 55–72. Princeton/Oxford: Princeton University Press, 2013.

McGarry, J. 'Autumntide of the Middle Ages'. *The Hopkins Review* 15, no. 1 (2022): 127–38.

Meertens, P. J. 'Nederlandse filologie'. In *Geestelijk Nederland 1920–1940 II*, edited by K. F. Proost and J. Romein, 1–22. Amsterdam/Antwerp: Kosmos, 1948.

Moeyes, P. *Buiten schot: Nederland tijdens de Eerste Wereldoorlog*. Amsterdam/Antwerp: Arbeiderspers, 2001.

Murray, A. 'Love and Death: How Huizinga Came to Write His Masterpiece'. *Times Literary Supplement*, no. 6161 (2021): 11–13.

Noordegraaf, J. '"On Light and Sound": Johan Huizinga and Nineteenth-Century Linguistics'. In *The Dutch Pendulum. Linguistics in the Netherlands 1740–1900*, 130–58. Münster: Nodus Publikationen, 1996.

Östling, J., D. Larsson Heidenblad, E. Sandmo, A. Nilsson Hammar, and K. Nordberg. 'The History of Knowledge and the Circulation of Knowledge: An Introduction'. In *Circulation of Knowledge: Explorations in the History of Knowledge*, edited by J. Östling, E. Sandmo, D. Larsson Heidenblad, A. Nilsson Hammar, and K. Nordberg, 9–33. Lund: Nordic Academic Press, 2018.

Otterspeer, W. *Orde en trouw: Over Johan Huizinga*. Amsterdam: De Bezige Bij, 2006.

Otterspeer, W. *Reading Huizinga*. Edited by B. Jackson. Amsterdam: Amsterdam University Press, 2010.

Otterspeer, W. *Het horzelnest: De Leidse Universiteit in oorlogstijd*. Amsterdam: Prometheus, 2019.

Otterspeer, W. 'Huizinga and Tocqueville'. Paper presented at conference *Huizinga heute. Hundert Jahre* Herbst des Mittelalters, Vienna, 15–17 May 2019.

Otterspeer, W. *De kleine Huizinga*. Amsterdam/Antwerp: Atlas Contact, 2019.

Paul, H. J. 'A Collapse of Trust: Reconceptualizing the Crisis of Historicism'. *Journal of the Philosophy of History* 2, no. 1 (2008): 63–82.

Paul, H. J. 'Performing History: How Historical Scholarship Is Shaped by Epistemic Virtues'. *History and Theory*, no. 50 (2011): 1–19.

Paul, H. J. 'The Scholarly Self: Ideals of Intellectual Virtue in Nineteenth-Century Leiden'. In *The Making of the Humanities Volume II: From Early Modern to Modern Disciplines*, edited by R. Bod, J. Maat, and T. Weststeijn, 397–412. Amsterdam: Amsterdam University Press, 2012.

Paul, H. J. 'Weber, Wöhler, and Waitz: Virtue Language in Late Nineteenth-Century Physics, Chemistry, and History'. In *Epistemic Virtues in the Sciences and the Humanities*, edited by H. Paul and J. van Dongen, 397–412. Amsterdam: Amsterdam University Press, 2017.

Polybius. *The Histories*. Edited by W. R. Paton. Cambridge, MA/London: Harvard University Press, 1922.

Pos, A. 'Dichterlijk cultuurpessimisme: Spengler in het werk van A. Roland Holst en H. Marsman'. In *De pijn van Prometheus*, edited by R. Aerts and K. van Berkel, 116–39. Groningen: Historische Uitgeverij, 1996.

Pree, C. du. *Johan Huizinga en de bezeten wereld*. Leusden: ISVW Uitgevers, 2016.

Raedts, P. *De ontdekking van de Middeleeuwen: Geschiedenis van een illusie*. Amsterdam: Wereldbibliotheek, 2014.

Romein, J. *Geschiedenis*. Edited by K. F. Proost and J. Romein. *Geestelijk Nederland 1920–1940 II*. Amsterdam: Kosmos, 1948.

Romein, J. 'De Geschiedschrijving in Nederland tijdens het interbellum'. In *Carillon der tijden: Studies en toespraken op cultuurhistorisch Terrein*. Amsterdam: Querido, 1953.

Romein, J. *Op het breukvlak van twee eeuwen: De westerse rereld Rond 1900*. Amsterdam: Querido, 1976.

Rooy, P. de. 'Een hevig gewarrel. Humanitair idealisme en socialisme in Nederland rond de Eeuwwisseling'. *BMGN – Low Countries Historical Review* 106, no. 4 (1991): 625.

Rossum, V. van. 'De stad gebouwd: De oude binnenstad vernieuwd'. In *Amsterdam in de tweede Gouden Eeuw*, 36–58. Bussum: Thoth, 2000.

Ruehl, M. A. *The Italian Renaissance in the German Historical Imagination, 1860–1930*. Cambdrige: Cambridge University Press, 2015.

Rydin, T. 'Huizinga's "Heimwee": Responding to Burckhardt's "Die Kultur Der Renaissance in Italien" in Times of Loss'. *History of European Ideas* (2020): 732–47.

Rüsen, J. *Historische Vernunft. Grunzüge einer Historik I: Die Grundlagen der Geschichtswissenschaft*. Göttingen: Vandenhoeck & Ruprecht, 1983.

Santing, C. 'Het liefkozen van schone vormen. Johan Huizinga en het vrouwenvraagstuk'. *De Gids* 168, no. 2 (2005): 121–30.

Sas, N. C. F. van. 'Fin-de-siècle als nieuw begin. Nationalisme in Nederland rond 1900'. *BMGN – Low Countries Historical Review* 106, no. 4 (1991): 595–609.

Schnädelbach, H. *Philosophy in Germany 1831–1933*. Edited by E. Matthews. Cambridge: Cambridge University Press, 2009.

Schulin, E. 'Weltkriegserfahrung und Historikerreaktion'. Frankfurt: Fischer Taschenbuch Verlag, 1997.

Scott, J. W. 'The Evidence of Experience'. *Critical Inquiry* 17, no. 4 (1991): 773–97.

Shaw, D. G. 'Huizinga's Timeliness'. *History and Theory* 37, no. 2 (1998): 245–58.

Sicking, J. M. J. 'Woede en verontwaardiging: De Eerste Wereldoorlog in de Nederlandse literatuur'. *Armada* 14, no. 52 (2008): 44–52.

Small, G. 'The Making of "The Autumn of the Middle Ages" I: Narrative Sources and Their Treatment in Huizinga's "Herfsttij"'. In *Rereading Huizinga: Autumn of the Middle Ages, a Century Later*, edited by P. Arnade, M. Howell, and A. van der Lem, 169–210. Amsterdam: Amsterdam University Press, 2019.

Small, G. 'Epilogue'. In *Autumntide of the Middle Ages*, 538–79. Leiden: Leiden University Press, 2020.

Spaanstra-Polak, B. 'De atmosfeer van het fin-de-siècle'. In *Het symbolisme in de Nederlandse schilderkunst 1890–1900*. Bussum: Uitgeverij Thoth, 2004.

Spaanstra-Polak, B. *Het symbolisme in de Nederlandse schilderkunst 1890–1900*. Bussum: Uitgeverij Thoth, 2004.

Strong, T. B. 'Foreword: Dimensions of the New Debate around Carl Schmitt'. In *The Concept of the Political*, 3–16. Chicago: The University of Chicago Press, 2007.

Strupp, C. *Johan Huizinga. Geschichtswissenschaft als Kulturgeschichte*. Göttingen: Vandenhoeck & Ruprecht, 2000.

Slaa, R. te, and E. Klijn. *De NSB: Ontstaan en opkomst van de Nationaal-Socialistische Beweging, 1931–1935*. Amsterdam: Boom, 2021.

Torpey, J. 'Huizinga on America'. *Journal of Classical Sociology* 15, no. 3 (2015): 293–301.

Thys, W. 'Huizinga en de Beweging van Negentig'. In *Johan Huizinga 1872–1972*, edited by W. R. H. Koops, E. H. Kossmann, and G. van der Plaat, 29–52. The Hague: Martinus Nijhoff, 1973.

Thys, W. *André Jolles (1874–1946) "Gebildeter Vagant": Briefe und Dokumente*. Amsterdam/Leipzig: Amsterdam University Press; Leipziger Universitätsverlag GmbH, 2000.

Tollebeek, J. 'Wetenschap en waardering: Nederlandse historici in verzet tegen het positivisme (1890–1910)'. *Groniek*, no. 124 (1994): 69–96.

Tollebeek, J. *De toga van Fruin. Denken over geschiedenis in Nederland sinds 1860*. Amsterdam: Wereldbibliotheek, 1996.

Tollebeek, J. '"Renaissance" and "Fossilization": Michelet, Burckhardt, and Huizinga'. *Renaissance Studies* 15, no. 3 (2001): 354–66.

Tollebeek, J. *Een nieuwe wereld: Het ontstaan van het moderne Nederland*. Amsterdam: Prometheus, 2021.

Vanwesenbeeck, B. 'Huizinga, Theorist of Lateness?' In *Rereading Huizinga: Autumn of the Middle Ages, a Century Later*, edited by P. Arnade, M. Howell, and A. van der Lem, 245–58. Amsterdam: Amsterdam University Press, 2019.

Velde, H. te. 'Fin de siècle in de Nederlandse politiek'. *Leidschrift* 1, no. 14 (1998): 45–55.

Vermeulen, E. E. G. 'Huizinga over de wetenschap der geschiedenis: Zijn positie als theoreticus van de geschiedenis tussen Fruin en Romein'. *Tijdschrift voor Filosofie* 24, no. 1 (1962): 81–90.

Veen, A. van. *Amsterdam 1900: Foto's van Olie, Breitner, Eilers en tijdgenoten*. Bussum: Thoth, 2016.

Verburgt, L. 'The History of Knowledge and the Future History of Ignorance'. *KNOW* 4, no. 1 (2020): 1–24.

Vervliet, J. *The Peace Palace Library Centennial: The Collection as a Mirror of the Historical Development of International Law, 1904–2004*. The Hague: Martinus Nijhoff, 2004.

Voerman, G. 'Anton Pannekoek: A "Principled Theorist"'. In *Anton Pannekoek: Ways of Viewing Science and Society*, 51–74, 2019.

Vree, F. van. 'De Stad van het Betere Leven. Cultuur en Samenleving in Nederland Rond 1900'. *BMGN – Low Countries Historical Review* 106, no. 4 (1991): 641–51.

Vries, J. de, ed. *Nederland 1913: Een reconstructie van het culturele leven*. Haarlem: Meulenhoff/Landshoff, 1989.

Vugt, G. de. 'De droomtekeningen van Johan Huizinga'. *De Witte Raaf*. 2021.

Vugt, G. de. 'Philia and Neikos: Huizinga's "Auseinandersetzung" with Carl Schmitt'. *World Literature Studies* 9, no. 1 (2017): 48–59.

Wagenaar, M. 'De stad ontworpen: Stadsontwerp tussen wens en werkelijkheid'. In *Amsterdam in de tweede Gouden Eeuw*, edited by M. Bakker, R. Kistemaker, H. van Nierop, W. Vroom, and P. Witteman, 9–35. Bussum: Thoth, 2000.

White, H. 'Historical Pluralism'. *Critical Inquiry* 12, no. 3 (1986): 480–93.

White, H. *Metahistory*. Baltimore: John Hopkins University Press, 2014.

Wittkau, A. *Historismus: zur Geschichte des Begriffs und des Problems*. Göttingen: Vandenhoeck & Ruprecht, 1994.

Woud, A. van der. *Het landschap en de mensen: Nederland 1850–1940*. Amsterdam: Prometheus, 2020.

Zammito, J. H. 'Reading "Experience": The Debate in Intellectual History among Scott, Toews, and LaCapra'. In *Reclaiming Identity: Realist Theory and the Predicament of Postmodernism*, edited by P. M. L. Moya and M. R. Hames-García, 279–311. Berkeley/Los Angeles/London: University of California Press, 2000.

Cited newspapers

Algemeen Handelsblad 31-01-1848
Algemeen Handelsblad 14-09-1857
Algemeen Handelsblad 28-01-1918
Algemeen Handelsblad 19-02-1918
Algemeen Handelsblad 26-11-1918
Algemeen Handelsblad 04-04-1912
Algemeen Handelsblad 29-09-1913
Algemeen Handelsblad 01-02-1921
Arnhemsche Courant 13-09-1905
Arnhemsche Courant 16-09-1905
Arnhemsche Courant 06-04-1912
Arnhemsche Courant 01-12-1914
De Courant 21-11-1906
De Groene Amsterdammer 15-05-1936
De Maasbode 21-07-1923
De Nederlander 19-10-1932
De Standaard 05-10-1920
De Tijd 28-08-1914
De Tijd 16-12-1914
De Tijd 03-09-1935
De Tijd 18-08-1936
Eindhovensch Dagblad 11-09-1914
Hepkema's Courant 15-03-1935
Het Nieuws van den Dag voor Nederlandsch-Indië 15-07-1912
Het Vaderland 27-10-1932
Het Vaderland 07-03-1935
Het Volk 16-10-1914
Het Volk 13-03-1935
Hoornsche Courant 08-10-1914
Nieuws van de Dag 22-06-1903
Rotterdamsche Courant 26-02-1848

Index of Names

For Product Safety Concerns and Information please contact our EU
representative GPSR@taylorandfrancis.com
Taylor & Francis Verlag GmbH, Kaufingerstraße 24, 80331 München, Germany

www.ingramcontent.com/pod-product-compliance
Ingram Content Group UK Ltd.
Pitfield, Milton Keynes, MK11 3LW, UK
UKHW022027061126
477531UK00002B/74
9789463724593